Springer Proceedings in Mathematics & Statistics

Volume 178

Springer Proceedings in Mathematics & Statistics

This book series features volumes composed of selected contributions from workshops and conferences in all areas of current research in mathematics and statistics, including operation research and optimization. In addition to an overall evaluation of the interest, scientific quality, and timeliness of each proposal at the hands of the publisher, individual contributions are all refereed to the high quality standards of leading journals in the field. Thus, this series provides the research community with well-edited, authoritative reports on developments in the most exciting areas of mathematical and statistical research today.

More information about this series at http://www.springer.com/series/10533

Sergei Silvestrov · Milica Rančić
Editors

Engineering Mathematics I

Electromagnetics, Fluid Mechanics, Material Physics and Financial Engineering

Springer

Editors
Sergei Silvestrov
Division of Applied Mathematics
The School of Education, Culture and Communication
Mälardalen University
Västerås
Sweden

Milica Rančić
Division of Applied Mathematics
The School of Education, Culture and Communication
Mälardalen University
Västerås
Sweden

ISSN 2194-1009 ISSN 2194-1017 (electronic)
Springer Proceedings in Mathematics & Statistics
ISBN 978-3-319-82496-3 ISBN 978-3-319-42082-0 (eBook)
DOI 10.1007/978-3-319-42082-0

Mathematics Subject Classification (2010): 00A69, 60-XX, 65-XX, 74-XX, 78-XX, 78A25, 78A50, 91Gxx, 00B15, 00B10

Printed on acid-free paper

This Springer imprint is published by Springer Nature
The registered company is Springer International Publishing AG
The registered company address is: Gewerbestrasse 11, 6330 Cham, Switzerland

Preface

This book highlights the latest advances in engineering mathematics with a main focus on the mathematical models, structures, concepts, problems and computational methods and algorithms most relevant for applications in modern technologies and engineering. In particular, it features mathematical methods and models of applied analysis, probability theory, differential equations, tensor analysis and computational modelling used in applications to important problems concerning electromagnetics, antenna technologies, fluid dynamics, material and continuum physics and financial engineering.

The individual chapters cover both theory and applications, and include a wealth of figures, schemes, algorithms, tables and results of data analysis and simulation. Presenting new methods and results, reviews of cutting-edge research, and open problems for future research, they equip readers to develop new mathematical methods and concepts of their own, and to further compare and analyse the methods and results discussed.

Chapters 1–10 are concerned with applied mathematics methods and models applied in electrical engineering, electromagnetism and antenna technologies. Chapter 1 by Dragan Poljak is concerned with applications of integro-differential equations and numerical analysis methods to the analysis of grounding systems important in the design of lightning protection systems. The analysis of horizontal grounding electrodes has been carried out using the antenna theory approach in the frequency and time domain respectively. The formulation is based on the corresponding space-frequency and space-time Pocklington integro-differential equations. The integro-differential relationships are numerically handled via the Galerkin–Bubnov scheme of the Indirect Boundary Element Method. Frequency domain and time domain analysis is illustrated by computational examples. Chapter 2 by Silvestar Šesnić and Dragan Poljak deals with the use of analytical methods for solving various integro-differential equations in electromagnetic compatibility, with the emphasis on the frequency and time domain solutions of the thin-wire configurations buried in a lossy ground. Solutions in the frequency domain are carried out via certain mathematical manipulations with the current function appearing in corresponding integral equations. On the other hand, analytical solutions in the time domain are undertaken using the Laplace transform and Cauchy residue theorem. Obtained analytical results are compared to those calculated using the numerical solution of the frequency domain Pocklington equation, where applicable. Also, an overview of analytical solutions to the Grad-Shafranov equation for tokamak plasma is provided. In Chap. 3 by Milica Rančić, Radoslav Jankoski, Sergei Silvestrov and Slavoljub Aleksić, a new simple approximation that can be used for modelling one type of Sommerfeld integrals typically occurring in the expressions that describe sources buried in the lossy ground, is proposed. The proposed approximation has a form of a weighted exponential function with an additional complex constant term. The derivation procedure for this approximation is explained in detail, and the validation is supplied by applying it to the analysis of a bare conductor fed in the centre and immersed in the lossy ground at arbitrary depth. In Chap. 4 by Radoslav Jankoski, Milica Rančić, Vesna Arnautovski-Toseva and Sergei Silvestrov, high frequency analysis of a horizontal dipole antenna buried in lossy ground is performed. The soil is treated as a homogenous half-space of known electrical parameters. The authors compare the range of applicability of two forms of transmission line models, a hybrid circuit method, and a point-matching method in this context. Chapter 5 by Pushpanjali G. Metri pertains to an experimental implementation and evaluation of geometrically designed antennas. A novel design for an equilateral triangular microstrip antenna is proposed and tested. The antenna is designed, fabricated and tested for single and multiband operation. A theory for such antennas based on the experimental results is also considered. Chapter 6 by Nenad Cvetković, Miodrag Stojanović, Dejan Jovanović, Aleksa Ristić, Dragan Vučković and Dejan Krstić provides a brief review of the derivation of two groups of approximate closed form expressions for the electrical scalar potential Green's functions that originates from the current of the point ground electrode in the presence of a spherical ground inhomogeneity, proposes approximate solutions and considers known exact solutions involving infinite series sums. The exact solution is reorganized in order to facilitate comparison to the closed form solutions, and to estimate the error introduced by the approximate solutions, and error estimation is performed comparing the results for the electrical scalar potential obtained applying the approximate expressions and the accurate calculations. This is illustrated by a number of numerical experiments. In Chap. 7 by Mario Cvetković and Dragan Poljak, the electromagnetic thermal dosimetry model for the human brain exposed to electromagnetic radiation is developed. The electromagnetic model based on the surface integral equation formulation is derived using the equivalence theorem for the case of a lossy homogeneous dielectric body. The thermal dosimetry model of the brain is based on the form of Pennes' equation of heat transfer in biological tissue. The numerical solution of the electromagnetic model is carried out using the Method of Moments, while the bioheat equation is solved using the finite element method. The electromagnetic thermal model developed here has been applied in internal dosimetry of the human brain to assess the absorbed electromagnetic energy and consequent temperature rise. In Chap. 8 by Mirjana Perić, Saša Ilić and Slavoljub Aleksić, multilayered shielded structures are analysed using the hybrid boundary element method. The approach is based on the equivalent electrodes method, on the point-matching method for the potential of the perfect electric conductor electrodes and for the normal component of electric field at the boundary surface between any two dielectric layers. In order to verify the obtained results, they have been compared with the finite element method and results that have already been reported in the literature. In Chap. 9 by Vesna Javor, new engineering modified transmission line models of lightning strokes are presented. The computational results for lightning electromagnetic field at various distances from lightning discharges are in good agreement with experimental results that are usually employed for validating electromagnetic, engineering and distributed-circuit models. Electromagnetic theory relations, thin-wire antenna approximation of a lightning channel without tortuosity and branching, as well as the assumption of a perfectly conducting ground, are used for electric and magnetic field computation. An analytically extended function, suitable for approximating channel-base currents in these models, is also considered. Chapter 10 by Karl Lundengård, Milica Rančić, Vesna Javor and Sergei Silvestrov explores the properties of the multi-peaked analytically extended function for approximation of lightning discharge currents. According to experimental results for lightning discharge currents, they are classified into waveshapes representing the first positive, first and subsequent negative strokes, and long-strokes. A class of analytically extended functions is presented and used for the modelling of lightning currents. The basic properties of this function with a finite number of peaks are examined. A general framework for estimating the parameters of the analytically extended function using the Marquardt least-squares method for a waveform with an arbitrary (finite) number of peaks as well as for the given charge transfer and specific energy is described and used to find parameters for some common single-peak waveforms.

In turn, Chaps. 11–15 address the mathematical modelling and optimisation of technological processes with applications of partial differential equations, ordinary differential equations, numerical analysis, perturbation methods and special functions in fluid mechanics models that are important in engineering applications and technologies. Chapter 11 by Jüri Olt, Olga Liivapuu, Viacheslav Maksarov, Alexander Liyvapuu and Tanel Tärgla, is devoted to the mathematical modelling of the process system which paves the way for research on the selection and optimisation of machining conditions. The subject of this chapter is the method of dynamic process approximation method, which makes it possible to analyse the behaviour of the machining process system in the process of chip formation at a sufficient level of accuracy. In Chap. 12 by Prashant G. Metri, Veena M. Bablad, Pushpanjali G. Metri, M. Subhas Abel and Sergei Silvestrov, a mathematical analysis is carried out to describe mixed convection heat transfer in magnetohydrodynamic non-Darcian flow due to an exponential stretching sheet embedded in a porous medium in the presence of a non-uniform heat source/sink. Approximate analytical similarity solutions of the highly nonlinear momentum and energy equations are obtained. The governing system of partial differential equations is first transformed into a system of nonlinear ordinary differential equations using similarity transformation. The transformed equations are nonlinear coupled differential equations and are solved very efficiently by employing a fifth order Runge–Kutta–Fehlberg method with shooting technique for various values of the governing parameters. The numerical solutions are obtained by considering an exponential dependent stretching velocity and prescribed boundary temperature on the flow directional coordinate. The computed results are compared with the previously published work on various special cases of the problem and are in good agreement with the earlier studies. The effects of various physical parameters, such as the Prandtl number, the Grashof number, the Hartmann number, porous parameter, inertia coefficient and internal heat generation on flow and heat transfer characteristics are presented graphically to reveal a number of interesting aspects of the physical parameter. Chapter 13 by Prashant G. Metri, M. Subhas Abel and Sergei Silvestrov presents an analysis of the boundary layer flow and heat transfer over a stretching sheet due to nanofluids with the effects of the magnetic field, Brownian motion, thermophoresis, viscous dissipation and convective boundary conditions. The transport equations used in the analysis take into account the effect of Brownian motion and thermophoresis parameters. The highly nonlinear partial differential equations governing flow and heat transport are simplified using similarity transformation, and the ordinary differential equations obtained are solved numerically using the Runge–Kutta–Fehlberg and Newton–Raphson schemes based on the shooting method. The solutions for velocity temperature and nanoparticle concentration depend on parameters such as Brownian motion, thermophoresis parameter, magnetic field and viscous dissipation, which have a significant influence on controlling of the dynamics. In Chap. 14 by Jawali C. Umavathi, Kuppalapalle Vajravelu, Prashant G. Metri and Sergei Silvestrov, the linear stability of Maxwell fluid-nanofluid flow in a saturated porous layer is examined theoretically when the walls of the porous layers are subjected to time-periodic temperature modulations. A modified Darcy-Maxwell model is used to describe the fluid motion, and the nanofluid model used includes the effects of the Brownian motion. The thermal conductivity and viscosity are considered to be dependent on the nanoparticle volume fraction. A perturbation method that is based on a small amplitude of an applied temperature field is used to compute the critical value of the Rayleigh number and the wave number. The stability of the system, characterized by a critical Rayleigh number, is calculated as a function of the relaxation parameter, the concentration Rayleigh number, the porosity parameter, the Lewis number, the heat capacity ratio, the Vadász number, the viscosity parameter, the conductivity variation parameter, and the frequency of modulation. Three types of temperature modulations are considered, and the effects of all three types are found to destabilize the system as compared to the unmodulated system. Chapter 15 by J. Pratap Kumar, Jawali C. Umavathi, Prashant G. Metri and Sergei Silvestrov is devoted to a study of magneto-hydrodynamic flow in a vertical double passage channel taking into account the presence of the first order chemical reaction. The governing equations are solved by using a regular perturbation technique valid for small values of the Brinkman number and a differential transform method valid for all values of the Brinkman number. The results are obtained for velocity, temperature and concentration. The effects of various dimensionless parameters such as the thermal Grashof number, mass Grashof number, Brinkman number, first order chemical reaction parameter, and Hartman number on the flow variables are discussed and presented graphically for open and short circuits. The validity of solutions obtained by the differential transform method and regular perturbation method are in good agreement for small values of the Brinkman number. Further, the effects of governing parameters on the volumetric flow rate, species concentration, total heat rate, skin friction and Nusselt number are also observed and tabulated.

Chapters 16–18 are concerned with mathematical methods of stochastic processes, probability theory, differential geometry, tensor analysis, representation theory, differential equations, algebra and computational mathematics for applications in materials science and financial engineering. In Chap. 16 by Anatoliy Malyarenko and Martin Ostoja-Starzewski, a random field model of the 21-dimensional elasticity tensor is considered, and representation theory is used to obtain the spectral expansion of the model in terms of stochastic integrals with respect to random measures. The motivation for treating this tensor as a random field is that nearly all the materials encountered in nature as well those produced by man, except for the purest crystals, possess some degree of disorder or inhomogeneity. At the same time, elasticity is the starting point for any solid mechanics model. Chapter 17 by Anatoliy Malyarenko, Jan Röman and Oskar Schyberg is devoted to mathematical models for catastrophe bonds which are an important instrument in the fields of finance, insurance and reinsurance, where the natural risk index is described by the Merton jump-diffusion while the risk-free interest rate is governed by the Hull–White stochastic differential equation. The sensitivities of the bond price with respect to the initial condition, volatility of the diffusion component, and jump amplitude are calculated using the Malliavin calculus approach. Lastly, in Chap. 18 by Betuel Canhanga, Anatoliy Malyarenko, Jean-Paul Murara and Sergei Silvestrov, stochastic volatilities models for pricing European options are considered as a response to the weakness of the constant volatility models, which have not succeeded in capturing the effects of volatility smiles and skews. A model with two-factor stochastic volatilities where the correlation between the underlying asset price and the volatilities varies randomly is considered, and the first order asymptotic expansion methods are used to determine the price of European options.

The book consists of carefully selected and refereed contributed chapters covering research developed as a result of a focused international seminar series on mathematics and applied mathematics, as well as three focused international research workshops on engineering mathematics organised by the Research Environment in Mathematics and Applied Mathematics at Mälardalen University from autumn 2014 to autumn 2015: the International Workshop on Engineering Mathematics for Electromagnetics and Health Technology; the International Workshop on Engineering Mathematics, Algebra, Analysis and Electromagnetics; and the 1st Swedish-Estonian International Workshop on Engineering Mathematics, Algebra, Analysis and Applications.

This book project has been realised thanks to the strategic support offered by Mälardalen University for the research and research education in Mathematics, which is conducted by the research environment Mathematics and Applied Mathematics (MAM), in the established research area of Educational Sciences and Mathematics at the School of Education, Culture and Communication at Mälardalen

University. We also wish to extend our thanks to the EU Erasmus Mundus projects FUSION, EUROWEB and IDEAS, the Swedish International Development Cooperation Agency (Sida) and International Science Programme in Mathematical Sciences, Swedish Mathematical Society, Linda Peetre Memorial Foundation, as well as other national and international funding organisations and the research and education environments and institutions of the individual researchers and research teams who contributed to this book.

We hope that this book will serve as a source of inspiration for a broad spectrum of researchers and research students in the field of applied mathematics, as well as in the specific areas of applications of mathematics considered here.

Västerås, Sweden
July 2016

Sergei Silvestrov
Milica Rančić

Contents

Contributors

M. Subhas Abel Department of Mathematics, Gulbarga University, Gulbarga, Karnataka, India

Slavoljub Aleksić Faculty of Electronic Engineering, University of Niš, Niš, Serbia

Vesna Arnautovski-Toseva Ss. Cyril and Methodius University, FEIT, Skopje, Macedonia

Veena M. Bablad Department of Mathematics, PDA College of Engineering, Gulbarga, Karnataka, India

Betuel Canhanga Faculty of Sciences, Department of Mathematics and Computer Sciences, Eduardo Mondlane University, Maputo, Mozambique; Division of Applied Mathematics, School of Education, Culture and Communication, Mälardalen University, Västerås, Sweden

Mario Cvetković Department of Power Engineering, University of Split, FESB, Split, Croatia

Nenad Cvetković Faculty of Electronic Engineering, University of Niš, Niš, Serbia

Saša Ilić Faculty of Electronic Engineering, University of Niš, Niš, Serbia

Radoslav Jankoski FEIT, Ss. Cyril and Methodius University, Skopje, Macedonia

Vesna Javor Faculty of Electronic Engineering, Department of Power Engineering, University of Niš, Niš, Serbia

Dejan Jovanović Faculty of Electronic Engineering, University of Niš, Niš, Serbia

Dejan Krstić Faculty of Occupational Safety, University of Niš, Niš, Serbia

Olga Liivapuu Institute of Technology, Estonian University of Life Sciences, Tartu, Estonia

Alexander Liyvapuu Institute of Technology, Estonian University of Life Sciences, Tartu, Estonia

Karl Lundengård Division of Applied Mathematics, School of Education, Culture and Communication, Mälardalen University, Västerås, Sweden

Viacheslav Maksarov Saint-Petersburg Mining University, St. Petersburg, Russia

Anatoliy Malyarenko Division of Applied Mathematics School of Education Culture and Communication, Mälardalen University, Västerås, Sweden

Prashant G. Metri Division of Applied Mathematics, School of Education, Culture and Communication, Mälardalen University, Västerås, Sweden

Pushpanjali G. Metri Department of Physics, Sangameshwar College, Solapur, Maharashtra, India

Jean-Paul Murara Division of Applied Mathematics, School of Education, Culture and Communication, Mälardalen University, Västerås, Sweden; Department of Applied Mathematics, School of Sciences, College of Science and Technology, University of Rwanda, Kigali, Rwanda

Jüri Olt Institute of Technology, Estonian University of Life Sciences, Tartu, Estonia

Martin Ostoja-Starzewski University of Illinois at Urbana-Champaign, Champaign, USA

Mirjana Perić Faculty of Electronic Engineering, University of Niš, Niš, Serbia

Dragan Poljak Department of Electronics, University of Split, FESB, Split, Croatia

J. Pratap Kumar Department of Mathematics Gulbarga University Gulbarga, Karnataka, India

Milica Rančić Division of Applied Mathematics, School of Education, Culture and Communication, Mälardalen University, Västerås, Sweden

Aleksa Ristić Faculty of Electronic Engineering, University of Niš, Niš, Serbia

Jan Röman Swedbank, Stockholm, Sweden

Oskar Schyberg Division of Applied Mathematics, School of Education, Culture and Communication, Mälardalen University, Västerås, Sweden

Silvestar Šesnić Department of Power Engineering, University of Split, FESB, Split, Croatia

Sergei Silvestrov Division of Applied Mathematics, School of Education, Culture and Communication, Mälardalen University, Västerås, Sweden

Miodrag Stojanović Faculty of Electronic Engineering, University of Niš, Niš, Serbia

Tanel Tärgla Institute of Technology, Estonian University of Life Sciences, Tartu, Estonia

Jawali C. Umavathi Department of Mathematics Gulbarga University Gulbarga, Karnataka, India; Department of Engineering, University of Sannio, Benevento, Italy

Kuppalapalle Vajravelu Department of Mathematics, University of Central Florida, Orlando, FL, USA; Department of Mechanical, Material and Aerospace Engineering, University of Central Florida, Orlando, USA

Dragan Vučković Faculty of Electronic Engineering, University of Niš, Niš, Serbia

Chapter 1
Frequency Domain and Time Domain Response of the Horizontal Grounding Electrode Using the Antenna Theory Approach

Dragan Poljak

Abstract The analysis of horizontal grounding electrode has been carried out using the antenna theory (AT) approach in the frequency and time domain, respectively. The formulation is based on the corresponding space-frequency and space-time Pocklington integro-differential equations. The integro-differential relationships are numerically handled via the Galerkin–Bubnov scheme of the Indirect Boundary Element Method (GB-IBEM). Some illustrative computational examples related to frequency domain (FD) and time domain (TD) analysis are given in the paper.

Keywords Transient response · Grounding systems · Frequency domain analysis · Time domain analysis · Pocklington integro-differential equation · Numerical solution

1.1 Introduction

Analysis of grounding systems is rather important issue in the design of lightning protection systems (LPS). Particularly important application is related to LPS for environmentally attractive wind turbines. In general, analysis of grounding systems can be carried out by using the transmission line (TL) model [1, 5, 6] or the full wave model, also referred to as the antenna theory (AT) model (AM) [3, 4, 11]. The latter is considered to be the rigorous one, while the principal advantage of TL approach is simplicity [14]. Both TL and AT models can be formulated in either frequency domain (FD) or time domain (TD) [9].

This paper reviews FD-AT and TD-AT approach, respectively, for the study of horizontal grounding electrode being an important component in many realistic

D. Poljak (✉)
Department of Electronics, University of Split, FESB, R. Boskovica 32, 21000 Split, Croatia
e-mail: dpoljak@fesb.hr

S. Silvestrov and M. Rančić (eds.), *Engineering Mathematics I*, Springer Proceedings in Mathematics & Statistics 178, DOI 10.1007/978-3-319-42082-0_1

grounding systems of complex shape. The key-parameter in the study of horizontal grounding electrode is the equivalent current distribution along the electrode. Once the current distribution along the electrode is determined, other parameters of interest, such as voltage distribution or transient impedance, can be calculated. Within the AT approach the effect of an earth-air interface is taken into account via the corresponding reflection coefficient thus avoiding the rigorous approach based on the Sommerfeld integrals. The space-frequency and space-time integro-differential expressions arising from the AT model are numerically treated by means of the Galerkin–Bubnov scheme of the Boundary Element Method (GB-IBEM) [9]. Some illustrative FD and TD numerical results for the current distribution and subsequently the scattered voltage along the electrode are obtained.

1.2 Frequency Domain Analysis

The configuration of interest, shown in Fig. 1.1, is the horizontal grounding electrode of length L and radius a, buried in a lossy medium at depth d and energized by an equivalent current generator I_g.

The corresponding integral relationships for the current and voltage induced along the electrode can be derived by enforcing the continuity conditions for the tangential components of the electric field along the electrode surface.

Total tangential electric field at the buried conductor surface given by a sum of the excitation field $\mathbf{E}^{exc}$ and scattered field $\mathbf{E}^{sct}$ is equal to the product of the current along the electrode $I(x)$ and surface internal impedance $Z_s(x)$ per unit length of the conductor [13]

$$\mathbf{e}_x \cdot \left(\mathbf{E}^{exc} + \mathbf{E}^{sct}\right) = Z_s(x) I(x), \tag{1.1}$$

where the surface internal impedance $Z_s(x)$ is given by [12, 14]

$$Z_s(x) = \frac{Z_{cw}}{2\pi a} \frac{I_0(\gamma_{wa})}{I_1(\gamma_{wa})}. \tag{1.2}$$

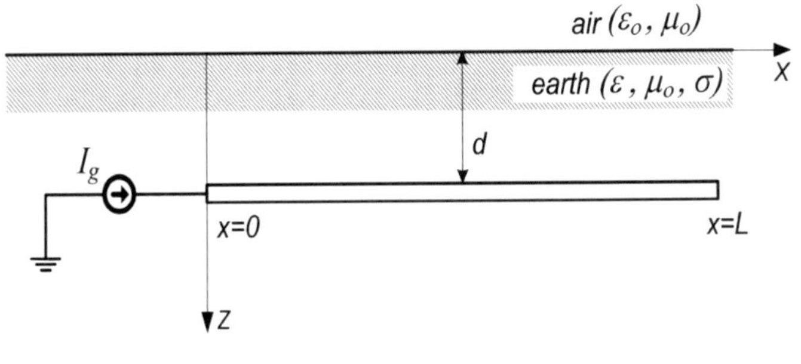

Fig. 1.1 Horizontal grounding wire excited by a current generator I_g

Note that $I_0(\gamma_w)$ and $I_1(\gamma_w)$ are modified Bessel functions of the zero and first order respectively, while Z_{cw} and γ_w are given by [9, 12–14]:

$$Z_{cw} = \sqrt{\frac{j\omega\mu_w}{\sigma_w + j\omega\varepsilon_w}}, \tag{1.3}$$

$$\gamma_w = \sqrt{j\omega\mu(\sigma_w + j\omega\varepsilon_w)}. \tag{1.4}$$

For the case of good conductors the surface impedance $Z_s(x)$ can be neglected. The scattered electric field can be expressed in terms of the vector potential **A** and the scalar potential φ, and according to the thin wire approximation [8, 9, 12, 13] only the axial component of the scattered field exists, i.e. it follows

$$E_x^{sct} = -j\omega A_x - \frac{\partial\varphi}{\partial x}, \tag{1.5}$$

where the vector and scalar potential are given by:

$$A_x = \frac{\mu}{4\pi}\int_0^L I(x')g(x,x')dx', \tag{1.6}$$

$$\varphi(x) = \frac{1}{4\pi\,\varepsilon_{eff}}\int_0^L q(x')\,g(x,x')\,dx', \tag{1.7}$$

while $q(x)$ denotes the charge distribution along the electrode, $I(x')$ is the induced current along the electrode.

The complex permittivity of the lossy ground ε_{eff} is

$$\varepsilon_{eff} = \varepsilon_r\varepsilon_0 - j\frac{\sigma}{\omega}, \tag{1.8}$$

where ε_{rg} and σ denotes the corresponding permittivity and conductivity, respectively.

The Green function $g(x, x')$ is given by

$$g(x,x') = g_0(x,x') - \Gamma_{ref}\,g_i(x,x'), \tag{1.9}$$

where $g_0(x, x')$ is the lossy medium Green function

$$g_0(x,x') = \frac{e^{-\gamma R_1}}{R_1}, \tag{1.10}$$

and $g_i(x, x')$, due to the image electrode in the air, is

$$g_i(x, x') = \frac{e^{-\gamma R_2}}{R_2}. \tag{1.11}$$

The propagation constant of the lower medium is defined as:

$$\gamma = \sqrt{j\omega\mu\sigma - \omega^2\mu\varepsilon}, \tag{1.12}$$

and R_1 and R_2 are given by:

$$R_1 = \sqrt{(x - x')^2 + a^2},\ R_2 = \sqrt{(x - x')^2 + 4d^2}. \tag{1.13}$$

The effect of a ground-air interface is taken into account in terms of the reflection coefficient (RC) [11]:

$$\Gamma_{ref} = \frac{\frac{1}{\underline{n}}\cos\theta - \sqrt{\frac{1}{\underline{n}} - \sin^2\theta}}{\frac{1}{\underline{n}}\cos\theta + \sqrt{\frac{1}{\underline{n}} - \sin^2\theta}};\quad \theta = arctg\frac{\left|x - x'\right|}{2d};\quad \underline{n} = \frac{\varepsilon_{eff}}{\varepsilon_0}. \tag{1.14}$$

The principal advantage of RC approach versus rigorous Sommerfeld integral approach is a simplicity of the formulation and appreciably less computational cost within the numerical solution of related integral expression [8, 9, 12, 13].

Combining the continuity equation. [9]

$$q = -\frac{1}{j\omega}\frac{dI}{dx} \tag{1.15}$$

with (1.7) yields:

$$\varphi(x) = -\frac{1}{j4\pi\omega\varepsilon_{eff}}\int_0^L \frac{\partial I(x')}{\partial x'} g(x, x')\, dx'. \tag{1.16}$$

Furthermore, inserting (1.6) and (1.16) into (1.5) gives an integral relationship for the scattered field

$$E_x^{sct} = -j\omega\frac{\mu}{4\pi}\int_0^L I(x')\, g(x, x')\, dx' + \frac{1}{j4\pi\omega\varepsilon_{eff}}\frac{\partial}{\partial x}\int_0^L \frac{\partial I(x')}{\partial x'} g(x, x')\, dx'. \tag{1.17}$$

Finally, as for the case of grounding electrodes the excitation function is given in the form of a current source the tangential field at the electrode surface does not exist, i.e. it can be written [3]

$$E_x^{exc} = 0. \tag{1.18}$$

Combining Eqs. (1.1), (1.17) and (1.18) leads to the homogeneous Pocklington integro-differential equation for the electrode current

$$j\omega\frac{\mu}{4\pi}\int_0^L I\left(x'\right)g\left(x,x'\right)dx' - \frac{1}{j4\pi\omega\varepsilon_{eff}}\frac{\partial}{\partial x}\int_0^L \frac{\partial I\left(x'\right)}{\partial x'}g\left(x,x'\right)dx' + Z_s(x)I(x) = 0. \quad (1.19)$$

Knowing the current distribution along the electrode the scattered voltage can be determined by computing the line integral of a scattered vertical field component from the remote soil to the electrode surface:

$$V^{sct}(x) = -\int_{\infty}^{d} E_z^{sct}(x,z)dz. \quad (1.20)$$

The vertical field component is expressed by the scalar potential gradient

$$E_z^{sct} = -\frac{\partial\varphi}{\partial z}, \quad (1.21)$$

and the scattered voltage along the electrode can be written

$$V^{sct}(x) = \int_{-\infty}^{d}\frac{\partial\varphi}{\partial z}dz = \frac{d}{dz}\int_{-\infty}^{d}\varphi(x,z)dz. \quad (1.22)$$

Integrating the scattered field from the infinite soil to the electrode surface and assuming the scalar potential in the remote soil to be zero [13] from (1.16) and (1.22) it follows

$$V^{sct}(x) = -\frac{1}{j4\pi\omega\varepsilon_{eff}}\int_0^L\frac{\partial I(x')}{\partial x'}g(x,x')dx'. \quad (1.23)$$

The grounding electrode is energized by an equivalent ideal current generator with one terminal connected to the grounding electrode and the other one grounded at infinity, as depicted in Fig. 1.1.

The current generator is included into the integro-differential equation formulation in terms of the following boundary conditions [11]:

$$I(0) = I_g, \quad I(L) = 0, \quad (1.24)$$

where I_g stands for the impressed unit current generator.

1.2.1 Numerical Solution

The current $I^e(x)$ along the wire segment can expressed, as follows

$$I^e(x') = \{f\}^T\{I\}. \quad (1.25)$$

Applying the weighted residual approach, performing certain mathematical manipulations and eventually assembling the contributions from all segments the integro-differential equation (1.19) is transferred into following matrix equation [9]

$$\sum_{j=1}^{M} [Z]_{ji}\{I\}_i = 0, \qquad \text{and} \quad j = 1, 2, \ldots, M, \tag{1.26}$$

where M is the total number of segments and $[Z]_{ji}$ is the mutual impedance matrix representing the interaction of the i-th source with the j-th observation segment, respectively:

$$[Z]_{ji} = -\frac{1}{4j\pi\omega\varepsilon_{eff}} \left(\int_{\Delta l_j} \{D\}_j \int_{\Delta l_i} \{D'\}_i^T g(x, x') dx' dx + \right.$$

$$\left. + k^2 \int_{\Delta l_j} \{f\}_j \int_{\Delta l_i} \{f\}_i^T g(x, x') dx' dx \right) + \int_{\Delta l_j} Z_L(x) \{f\}_j \{f\}_i^T dx. \tag{1.27}$$

Matrices $\{f\}$ and $\{f'\}$ contain the shape functions, while $\{D\}$ and $\{D'\}$ contain their derivatives, and Δl_i and Δl_j are the widths of i-th and j-th boundary elements.

A linear approximation over a boundary element is used in this work:

$$f_i = \frac{x_{i+1} - x'}{\Delta x} \quad f_{i+1} = \frac{x' - x_i}{\Delta x}, \tag{1.28}$$

as this choice was proved to be optimal one in modeling various wire structures [9].

The excitation function in the form of the current generator I_g is taken into account through the forced boundary condition at the first node of the solution vector, i.e.:

$$I_1 = I_g; \; I_g = 1e^{j0}. \tag{1.29}$$

Once the current distribution is obtained the scattered voltage (1.23) can be readily evaluated using the boundary element formalism.

As the current distribution derivative on the segment is simply given by

$$\frac{\partial I(x')}{\partial x'} = \frac{I_{i+1} - I_i}{\Delta x}, \tag{1.30}$$

the scattered voltage can be computed from the following formula:

$$V^{sct}(x) = -\frac{1}{j4\pi\omega\varepsilon_{eff}} \sum_{i=1}^{M} \frac{I_{i+1} - I_i}{\Delta x} \int_{x_i}^{x_{i+1}} g(x, x') dx'. \tag{1.31}$$

The integral on the right hand side of (1.31) is solved via the standard Gaussian quadrature.

1.2.2 Computational Examples

Figure 1.2 shows the frequency response at the center of the electrode with: $L = 20\,\text{m}$, $d = 1\,\text{m}$, $a = 5\,\text{mm}$ and $I_g = 1\,\text{A}$. The ground conductivity is $\sigma = 0.01\,\text{S/m}$ while the permittivity is $\varepsilon_r = 10$. The results computed via the GB-IBEM are compared to the results obtained via NEC using Sommerfeld integral approach and the Modified Transmission Line Model (MTLM) [13]. The results obtained via different approaches agree satisfactorily for the given set of parameters.

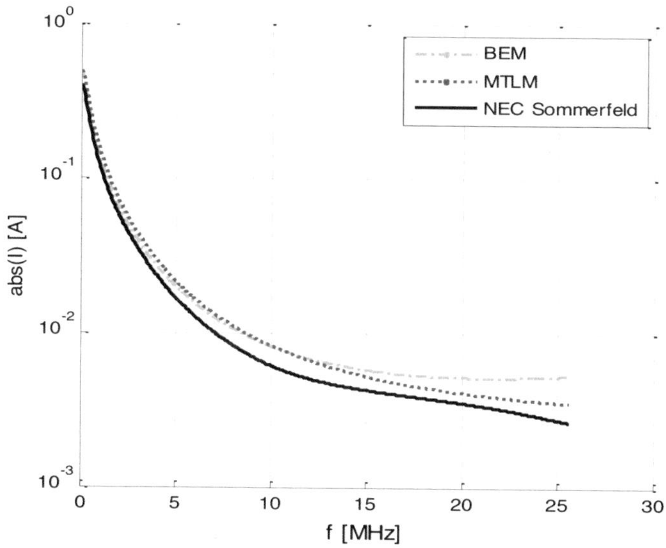

Fig. 1.2 Current induced at the center of the grounding electrode versus frequency ($L = 20\,\text{m}$, $d = 1\,\text{m}$, $a = 5\,\text{mm}$, $\sigma = 0.01\,\text{S/m}$, $\varepsilon_r = 10$)

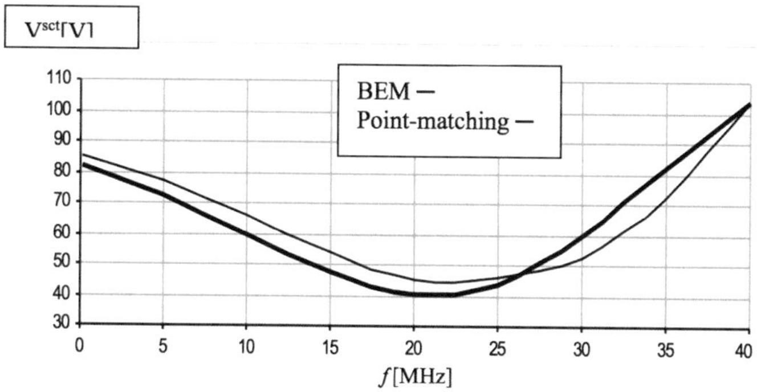

Fig. 1.3 Voltage spectrum at the grounding electrode driving point ($L = 10\,\text{m}$, $d = 1\,\text{m}$, $a = 5\,\text{mm}$, $\sigma = 0.01\,\text{S/m}$, $\varepsilon_r = 10$)

Figure 1.3 shows the voltage spectrum at the injection point of the horizontal grounding electrode with: $L = 10\,\text{m}$, $d = 1\,\text{m}$, $a = 5\,\text{mm}$ and $I_g = 1\,\text{A}$. The ground conductivity is $\sigma = 0.01\,\text{S/m}$, while the permittivity is $\varepsilon_r = 10$.

The agreement between the results obtained via GB-IBEM with linear approximation is in a good agreement with the results calculated via the point matching technique.

1.3 Time Domain Analysis

The geometry of interest is shown in Fig. 1.1 and the time domain counterpart of (1.1) is given by:

$$\mathbf{e}_x \cdot \left[\mathbf{E}^{exc}(x,t) + \mathbf{E}^{sct}(x,t)\right] = \int_0^t z_s(x, t-\tau) I(x,\tau) d\tau, \tag{1.32}$$

where $z_s(x,t)$ is the time domain counterpart of the surface impedance Z_s (1.2).

The axial component of the scattered field is given by

$$E_x^{sct}(x,t) = -\frac{\partial A_x(x,t)}{\partial t} - \frac{\partial \varphi(x,t)}{\partial x}, \tag{1.33}$$

where $A_x(x,t)$ and $\varphi(x,t)$ are time domain counterparts of the vector potential (1.6) and scalar potential (1.7).

Utilizing the time domain counterpart of the continuity Eq. (1.15) and taking into account that the electric field excitation along the electrode does not exist (1.18) the transient current induced along the electrode is governed by the homogeneous space-time Pocklington integro-differential equation

$$\left[-v^2\frac{\partial^2}{\partial x^2} + \frac{\partial^2}{\partial t^2} + \frac{\sigma}{\varepsilon}\frac{\partial}{\partial t}\right] \cdot \left[\frac{\mu}{4\pi}\int_0^L I(x', t - R/v)\frac{e^{-\frac{t}{\tau_g}\frac{R}{v}}}{R}dx' - \right.$$
$$\left. - \int_{-\infty}^{t}\int_0^L \Gamma_{ref}(\theta,\tau)\frac{I(x', t - R^*/v - \tau)}{4\pi R^*}\frac{e^{-\frac{t}{\tau_g}\frac{R^*}{v}}}{R^*}dx'd\tau\right] = 0, \tag{1.34}$$

where the reflection coefficient is given by [10]:

$$\Gamma_{ref}(t) = -\left[\frac{\tau_1}{\tau_2}\delta(t) + \frac{1}{\tau_2}\left(1 - \frac{\tau_1}{\tau_2}\right)e^{-t/\tau_2}\right], \tag{1.35}$$

while τ_1 and τ_2 are the time constants of a lossy medium [10]:

$$\tau_1 = \frac{\varepsilon_r - 1}{\sigma}\varepsilon_0, \quad \tau_2 = \frac{\varepsilon_r + 1}{\sigma}\varepsilon_0. \tag{1.36}$$

Note that the current source is included into the integral equation scheme through the boundary condition:

$$I(0,t) = I_g, \tag{1.37}$$

which is inserted subsequently in the global matrix system [9].

1.3.1 BEM Procedure for Pocklington Equation

The implementation of GB-IBEM to the solution of the Pocklington equation suffers from numerical instabilities. The origin of these instabilities is the existence of space-time differential operator [7, 9].

For the sake of simplicity, this paper deals with the case of an infinite lossy medium.

The space-time dependent current along the electrode can be expressed, as follows:

$$I(x', t - R/v) = \sum_{i=1}^{N} I(t - R/v) f_i(x'). \tag{1.38}$$

Applying the weighted residual approach and performing space-discretization yields

$$\begin{aligned}\sum_{i=1}^{N} I_i(t - \tau_{ij}) \Big[& \tfrac{\mu}{4\pi} \int_{\Delta l_j} \int_{\Delta l_i} \tfrac{\partial f_j(x)}{\partial x} \tfrac{\partial f_i(x')}{\partial x'} \tfrac{e^{-\frac{\sigma}{2\varepsilon v} R}}{R} dx' dx + \\ & + \tfrac{1}{v^2} \tfrac{\partial^2}{\partial t^2} \int_{\Delta l_j} \int_{\Delta l_i} f_j(x) f_i(x') \tfrac{e^{-\frac{\sigma}{2\varepsilon v} R}}{R} dx' dx + \\ & + \tfrac{\sigma}{\varepsilon} \tfrac{\partial}{\partial t} \int_{\Delta l_j} \int_{\Delta l_i} f_j(x) f_i(x') \tfrac{e^{-\frac{\sigma}{2\varepsilon v} R}}{R} dx' dx \Big] = 0 \qquad j = 1,\ 2,\ \ldots, N.\end{aligned} \tag{1.39}$$

Performing the discretization in the time domain the following set of time domain differential equations is obtained

$$[M] \frac{\partial^2}{\partial t^2} \{I(t')\} + [C] \frac{\partial}{\partial t} \{I(t')\} + [K] \{I(t')\} = 0, \tag{1.40}$$

where the corresponding space dependent matrices are:

$$M_{ji} = \frac{1}{v^2} \int_{\Delta l_j} \int_{\Delta l_i} \{f\}_j \{f\}_i^T \frac{e^{-\frac{T}{\tau}}}{R} dx' dx, \tag{1.41}$$

$$C_{ji} = \frac{\sigma}{\varepsilon} \int_{\Delta l_j} \int_{\Delta l_i} \{f\}_j \{f\}_i^T \frac{e^{-\frac{T}{\tau}}}{R} dx' dx, \tag{1.42}$$

$$K_{ji} = \frac{\mu}{4\pi} \int_{\Delta l_j} \int_{\Delta l_i} \{D\}_j \{D\}_i^T \frac{e^{-\frac{T}{\tau}}}{R} dx' dx, \tag{1.43}$$

where $\{D\}$ contains the shape functions derivatives and: $\tau = \frac{2\varepsilon}{\sigma}$ and $T = \frac{R}{v}$.

Set of differential equation (1.40) is solved by using the marching-on-in-time procedure [2]

$$\sum_{i=1}^{n} \left[M_{ji} + \beta \Delta t^2 K_{ji}\right] I_i^k = -\sum_{i=1}^{n} \left[-2M_{ji} + \left(\tfrac{1}{2} - 2\beta + \gamma\right) \Delta t^2 K_{ji}\right] I_i^{k-1}, \tag{1.44}$$

where Δt stands for the time increment and the stability of the procedure is achieved by choosing $\gamma = 1/2$ and $\beta = 1/4$ [4].

1.3.2 Numerical Results for Grounding Electrode

Computational example is related to the transient response of the electrode with length $L = 10\,\text{m}$, radius $a = 5\,\text{mm}$, immersed in the lossy ground with $\varepsilon_r = 10$, and $\sigma = 0.001\,\text{S/m}$. The electrode is energized with the double exponential current pulse:

$$i_g(t) = I_0 \cdot (e^{-at} - e^{-bt}),\ t \geq 0 \tag{1.45}$$

with $I_0 = 1.1043\,\text{A}$, $a = 0.07924 \cdot 10^7\,\text{s}^{-1}$, $b = 4.0011 \cdot 10^7\,\text{s}^{-1}$.

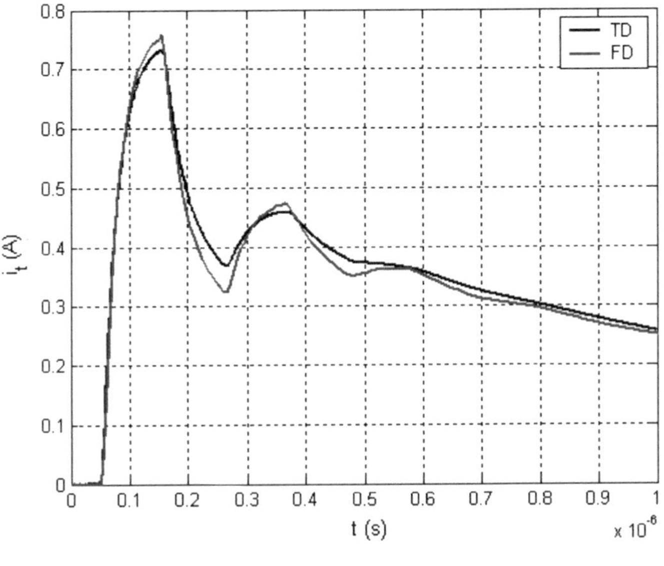

Fig. 1.4 Transient current induced at the centre of the grounding electrode

The transient current at the centre of the electrode obtained via the direct time domain approach and the indirect frequency domain approach GB-IBEM with Fast Fourier Transform (FFT) is shown in Fig. 1.4.

Satisfactory agreement between the results obtained via different approaches can be observed.

1.4 Concluding Remarks

The paper reviews electromagnetic modeling of grounding systems by means of the antenna theory (AT) approach in the frequency and time domain, respectively. The space-frequency and space-time Pocklington integro-differential equation, arising from the AT approach, are numerically solved by using the Galerkin–Bubnov scheme of the Indirect Boundary Element Method (GB-IBEM). The obtained numerical results for the current distribution and scattered voltage induced along the horizontal grounding electrode agree satisfactorily with the results calculated via other solution methods.

References

1. Ala, G., Di Silvestre, M.L.: A simulation model for electromagnetic transients in lightning protection systems. IEEE Trans. on Electromagn. Compat. **44**, 539–554 (2002)
2. Doric, V., Poljak, D., Roje, V.: Direct time domain analysis of a lightning rod based on the antenna theory. In: Proc. of 8th International Symposium on Electromagnetic Compatibility and Electromagnetic Biology, Saint-Petersburg, Russia (2009)
3. Grcev, L., Dawalibi, F.: An electromagnetic model for transients in grounding systems. IEEE Trans. on Power Deliv. **5**, 1773–1781 (1990)
4. Grcev, L.D., Menter, F.E.: Transient electromagnetic fields near large earthing systems. IEEE Trans. on Magn. **32**, 1525–1528 (1996)
5. Liu, Y., Zitnik, M., Thottappillil, R.: An improved transmission-line model of grounding system. IEEE Trans. on Electromagn. Compat. **43**, 348–355 (2001)
6. Lorentzou, M.I., Hatziargyriou, N.D., Papadias, B.C.: Time domain analysis of grounding electrodes impulse response. IEEE Trans. on Power Deliv. **18**, 517–524 (2003)
7. Miller, E.K., Landt, J.A.: Direct time-domain techniques for transient radiation and scattering from wires. In: Proceedings of the IEEE **68**, 1396–1423 (1980)
8. Olsen, R.G., Willis, M.C.: A comparison of exact and quasi-static methods for evaluating grounding systems at high frequencies. IEEE Trans. on Power Deliv. **11**, 1071–1081 (1996)
9. Poljak, D.: Advanced Modeling in Computational Electromagnetic Compatibility. Wiley, New Jersey (2007)
10. Poljak, D., Kovac, N.: Time domain modeling of a thin wire in a two-media configuration featuring a simplified reflection/transmission coefficient approach. Eng. Anal. with Bound. Elem. **33**, 283–293 (2009)
11. Poljak, D., Roje, V.: The Integral equation method for ground wire impedance. In: Constanda, C., Saranen, J., Seikkala, S. (eds.) Integral Methods in Science and Engineering, pp. 139–143. Longman, UK (1997)

12. Poljak, D., Rachidi, F., Tkachenko, S.V.: Generalized form of telegrapher's equations for the electromagnetic field coupling to finite-length lines above a lossy ground. IEEE Trans. on Electromagn. Compat. **49**, 689–697 (2007)
13. Poljak, D., Doric, V., Rachidi, F., Drissi, K., Kerroum, K., Tkachenko, S.V., et al.: Generalized form of telegrapher's equations for the electromagnetic field coupling to buried wires of finite length. IEEE Trans. on Electromagn. Compat. **51**, 331–337 (2009)
14. Tesche, F.M., Karlsson, T., Ianoz, M.: EMC Analysis Methods and Computational Models. Wiley, New Jersey (1997)

Chapter 2
On the Use of Analytical Methods in Electromagnetic Compatibility and Magnetohydrodynamics

Silvestar Šesnić and Dragan Poljak

Abstract The paper deals with the use of analytical methods for solving various integro-differential equations in electromagnetic compatibility, with the emphasis on the frequency and time domain solutions of the thin wire configurations buried in a lossy ground. Solutions in the frequency domain are carried out via certain mathematical manipulations with the current function appearing in corresponding integral equations. On the other hand, analytical solutions in the time domain are undertaken using the Laplace transform and Cauchy residue theorem. Obtained analytical results are compared to those calculated using the numerical solution of the frequency domain Pocklington equation, where applicable. Also, an overview of analytical solutions to the Grad–Shafranov equation for tokamak plasma is given.

Keywords Electromagnetic compatibility · Thin wire analysis · Integro-differential equations · Analytical methods · Magnetohydrodynamics

2.1 Introduction

The electromagnetic field coupling to thin wire scatterers can be treated either in frequency (FD) or time domain (TD) [18]. The principal advantage of the frequency domain approach is relative simplicity of both the formulation and the selected numerical treatment. However, time domain modeling ensures better physical insight, accurate modeling of highly resonant structures, possibility of calculating only early time period and easier implementation of nonlinearities [12, 21].

S. Šesnić (✉)
Department of Power Engineering, University of Split, FESB,
R. Boskovica 32, 21000 Split, Croatia
e-mail: ssesnic@fesb.hr

D. Poljak
Department of Electronics, University of Split, FESB,
R. Boskovica 32, 21000 Split, Croatia
e-mail: dpoljak@fesb.hr

S. Silvestrov and M. Rančić (eds.), *Engineering Mathematics I*,
Springer Proceedings in Mathematics & Statistics 178,
DOI 10.1007/978-3-319-42082-0_2

The formulation of the problem in thin wire analysis (FD or TD) is usually based on some variants of integral or integro-differential equation (Hallén or Pocklington type), respectively. Numerical modeling is widely used for solving various complex problems. On the other hand, analytical solution can be obtained when dealing with canonical problems, using a carefully chosen set of approximations [10, 26]. The advantage of analytical solutions over numerical ones is the ability to "follow up" the procedure with the complete control of adopted approximations. In this way, the insight into the physical characteristics of the problem is ensured, which is, when using numerical methods, rather complex task. Also, analytical solutions are readily implemented for benchmark purposes, as well as some fast engineering estimation of phenomena.

Valuable contributions in the area of analytical solutions of integral equations in electromagnetics are given by R.W.P. King *et al.* [9, 10]. S. Tkachenko derives the analytical solution for the current induced along the wire above perfectly conducting (PEC) ground using the transmission line modeling (TLM) for LF excitations [25]. On the other hand, time domain analytical modeling is not investigated to a greater extent and papers on the subject are rather scarce. A. Hoorfar and D. Chang give the solution for transient response of thin wire in free space using singularity expansion method [8]. R. Velazquez and D. Mukhedar derive analytical solution for the current induced along a grounding electrode, based on the TL model [27]. Analytical solutions in time domain have been reported by the authors in [20–22].

Analytical solutions pertaining to the mathematical model of fusion plasma, given by the set of magnetohydrodynamic (MHD) equations provide a satisfactory description of macroscopic plasma behavior. Combining MHD equations with Maxwell's equations of classical electrodynamics yields nonlinear second order differential equation known as Grad–Shafranov equation (GSE) [1]. Analytical solutions of the GSE are very useful for theoretical studies of plasma equilibrium, transport and MHD stability [28].

2.2 Thin Wire Models in Antenna Theory

2.2.1 Frequency Domain Formulation

Horizontal, perfectly conducting wire of length L and radius a, embedded in a lossy medium at depth d and excited by a plane wave is considered, as shown in Fig. 2.1. The medium is characterized with electric permittivity ε and conductivity σ. Dimensions of the structure satisfy the thin-wire approximation [12].

The current induced along the wire is governed by the inhomogeneous Pocklington integro-differential equation [17]

$$-\frac{1}{j4\pi\omega\varepsilon_{eff}}\left(\frac{\partial^2}{\partial x^2}-\gamma^2\right)\int_0^L I\left(x'\right)g\left(x,x'\right)dx'=E_x\left(\omega\right), \tag{2.1}$$

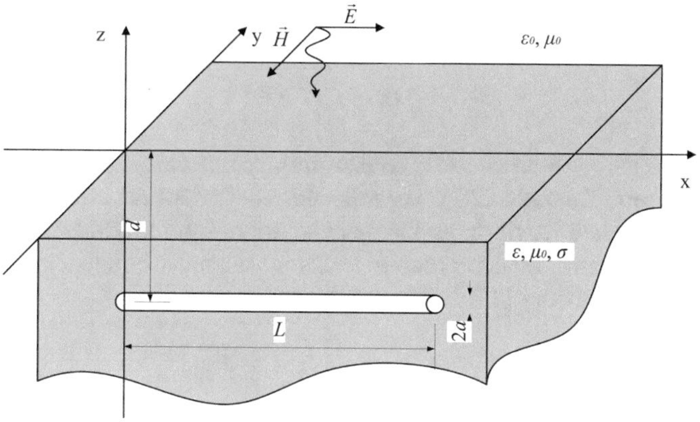

Fig. 2.1 Horizontal *straight thin wire* buried in a lossy medium

where $I(x')$ denotes current distribution along the wire. Complex permittivity of the medium is defined as

$$\varepsilon_{eff} = \varepsilon_r \varepsilon_0 - j\frac{\sigma}{\omega}, \tag{2.2}$$

where ε_r and σ represent relative electric permittivity and conductivity, respectively. Green's function $g(x, x')$ can be expressed as [14]

$$g(x, x') = g_0(x, x') - \Gamma_{ref}\, g_i(x, x'), \tag{2.3}$$

where $g_0(x, x')$ denotes the lossy medium Green's function

$$g_0(x, x') = \frac{e^{-\gamma R_1}}{R_1}, \tag{2.4}$$

and $g_i(x, x')$ is Green's function according to the image theory

$$g_i(x, x') = \frac{e^{-\gamma R_2}}{R_2}. \tag{2.5}$$

Propagation constant of the medium is defined in the following way

$$\gamma = \sqrt{j\omega\mu\sigma - \omega^2\mu\varepsilon}, \tag{2.6}$$

and distances R_1 and R_2 correspond to distances from the source and the image to the observation point, respectively

$$R_1 = \sqrt{(x - x')^2 + a^2},$$
$$R_2 = \sqrt{(x - x')^2 + 4d^2}. \tag{2.7}$$

Presence of the earth-air interface is taken into account via the reflection coefficient within the Green's function (2.3). The reflection coefficient can be taken in the form of Fresnel coefficient [3, 5] or, as a simpler solution, from Modified Image Theory (MIT) [23]. The Fresnel reflection coefficient is considered to be better approximation of the Sommerfeld theory [11] and is defined as

$$\Gamma_{ref}^{Fr} = \frac{\frac{1}{\underline{n}}\cos\theta - \sqrt{\frac{1}{\underline{n}} - \sin^2\theta}}{\frac{1}{\underline{n}}\cos\theta + \sqrt{\frac{1}{\underline{n}} - \sin^2\theta}},$$
$$\theta = arctg\frac{|x - x'|}{2d}, \ \underline{n} = \frac{\varepsilon_{eff}}{\varepsilon_0}. \tag{2.8}$$

On the other hand, the reflection coefficient that arises from MIT is defined as follows [23]

$$\Gamma_{ref}^{MIT} = -\frac{\varepsilon_{eff} - \varepsilon_0}{\varepsilon_{eff} + \varepsilon_0}. \tag{2.9}$$

The scattered voltage along the wire is defined as an integral of the vertical component of the scattered electric field and the Generalized Telegrapher's Equation for spatial distribution of the scattered voltage is given as [16]

$$V^{sct}(x) = -\frac{1}{j4\pi\omega\varepsilon_{eff}}\int_0^L \frac{\partial I(x')}{\partial x'} g(x, x')\, dx', \tag{2.10}$$

which can be easily determined, once the current distribution is known.

2.2.2 *Time Domain Formulation*

In the case of time domain formulation, the same configuration is considered as shown in Fig. 2.1 [7]. Governing equation for the unknown transient current flowing along the electrode is given in the form of time domain Pocklington integro-differential equation [21]

$$-\left(\frac{\partial^2}{\partial x^2} - \mu\sigma\frac{\partial}{\partial t} - \mu\varepsilon\frac{\partial^2}{\partial t^2}\right)\cdot\left[\frac{\mu}{4\pi}\int_0^L I\left(x', t - \frac{R}{v}\right)\frac{e^{-\frac{1}{\tau_g}\frac{R}{v}}}{R}dx' - \right.$$
$$\left. - \frac{\mu}{4\pi}\int_0^t\int_0^L \Gamma_{ref}^{MIT}(\tau) I\left(x', t - \frac{R^*}{v} - \tau\right)\frac{e^{-\frac{1}{\tau_g}\frac{R^*}{v}}}{R^*}dx'd\tau\right] = \left(\mu\varepsilon\frac{\partial}{\partial t} + \mu\sigma\right)E_x^{tr}(t), \tag{2.11}$$

where $I\left(x', t - \frac{R}{v}\right)$ represents the unknown transient current. Detailed derivation of (2.11) can be found in [21].

The distance from the source point in the wire axis to the observation point on the wire surface is given by

$$R = \sqrt{(x - x')^2 + a^2}, \tag{2.12}$$

while the distance from the source point on the image wire, according to the image theory is

$$R^* = \sqrt{(x - x')^2 + 4d^2}. \tag{2.13}$$

Time constant and propagation velocity in the lossy medium are defined as follows [21]

$$\tau_g = \frac{2\varepsilon}{\sigma}, \quad v = \frac{1}{\sqrt{\mu\varepsilon}}. \tag{2.14}$$

The reflection coefficient arising from the Modified Image Theory is given by inverse Laplace transform of (2.9) [23]

$$\Gamma_{ref}^{MIT}(t) = -\left[\frac{\tau_1}{\tau_2}\delta(t) + \frac{1}{\tau_2}\left(1 - \frac{\tau_1}{\tau_2}\right)e^{-\frac{t}{\tau_2}}\right], \tag{2.15}$$

where

$$\tau_1 = \frac{\varepsilon_0(\varepsilon_r - 1)}{\sigma}, \quad \tau_2 = \frac{\varepsilon_0(\varepsilon_r + 1)}{\sigma}. \tag{2.16}$$

Reflection coefficient (2.15) represents the simplest characterization of the earth-air interface, taking into account only medium properties. However, an extensive investigation of this coefficient applied to thin wires in two-media configuration has been carried out in [15].

2.3 Frequency Domain Applications of Analytical Methods

2.3.1 Horizontal Wire Below Ground

To solve the Pocklington equation (2.1) analytically, the integral on the left-hand side of (2.1) can be written in the following manner [20]:

$$\int_0^L I\left(x'\right) g\left(x, x'\right) dx' = I(x) \int_0^L g\left(x, x'\right) dx' + \int_0^L \left[I(x') - I(x)\right] g\left(x, x'\right) dx'. \quad (2.17)$$

The integral on the left hand side can be approximated by the first term on the right hand side of (2.17), thus neglecting the second integral. Furthermore, the characteristic integral term over the Green function is evaluated analytically. For the case of an imperfectly conducting ground the appropriate analytical integration of the first integral on the right hand side of (2.17) gives [25]

$$\int_0^L g\left(x, x'\right) dx' = \psi = 2\left(\ln\frac{L}{a} - \Gamma_{ref}^{MIT} \ln\frac{L}{2d}\right), \quad (2.18)$$

where reflection coefficient is given with (2.9).

After performing some mathematical manipulations, the analytical solution (2.1) can be obtained in the closed form and is given by

$$I\left(x, \omega\right) = \frac{4\pi e^{j\frac{a}{v}\omega}}{j\omega\mu\Psi\left(\omega\right)} E_x^{exc}\left(\omega\right)\left[1 - \frac{\cosh\left(\gamma\left(\frac{L}{2} - x\right)\right)}{\cosh\left(\gamma\frac{L}{2}\right)}\right]. \quad (2.19)$$

Figures 2.2 and 2.3 are related to horizontal wire of length L, radius $a = 0.01$ m, buried at depth $d = 2.5$ m in a lossy ground and illuminated by the plane wave of normal incidence transmitted into the ground with amplitude $E_0 = 1$ V/m at the interface between two media. Absolute value of spatial current distribution for lines $L = 5$ m and $L = 10$ m. The operating frequency of $f = 50$ MHz is shown. The conductivity

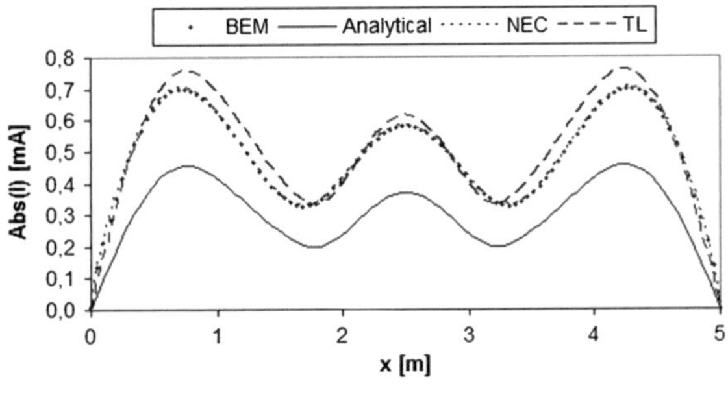

Fig. 2.2 Absolute value of current distribution along the single wire buried in a ground, $L = 5$ m

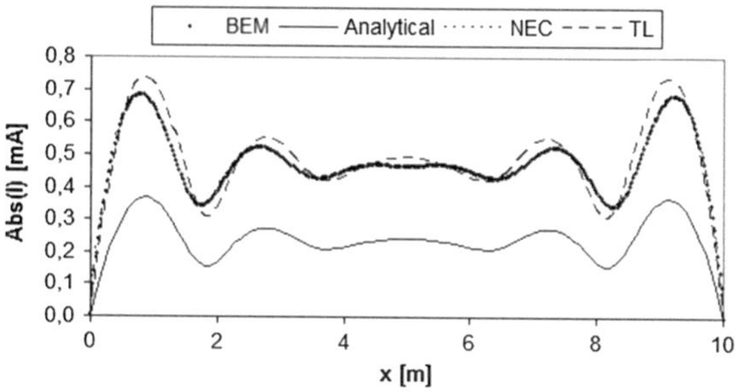

Fig. 2.3 Absolute value of current distribution along the single wire buried in a ground, $L = 10\,\text{m}$

of the ground is $\sigma = 0.01$ S/m and permittivity is $\varepsilon_r = 10$. The agreement between results obtained via different methods (analytical and numerical) is satisfactory.

2.3.2 Horizontal Grounding Electrode

When horizontal grounding electrode is considered, (2.1) can be written as a homogeneous equation, since source function is incorporated through the boundary condition [20]

$$-\frac{1}{j4\pi\omega\varepsilon_{eff}}\left(\frac{\partial^2}{\partial x^2}-\gamma^2\right)\int_0^L I\left(x'\right)g\left(x,x'\right)dx' = 0. \tag{2.20}$$

Now, the similar approach as in the case of horizontal wire can be adopted and (2.20) can be written as

$$-\frac{1}{j4\pi\omega\varepsilon_{eff}}\left(\frac{\partial^2}{\partial x^2}-\gamma^2\right)I\left(x\right)\int_0^L g\left(x,x'\right)dx' = 0. \tag{2.21}$$

Integral in (2.21) can be readily calculated as given in [20]

$$\int_0^L g\left(x,x'\right)dx' = 2\left(\ln\frac{L}{a}-\Gamma_{ref}\ln\frac{L}{2d}\right) = \Psi. \tag{2.22}$$

Now, the homogeneous Pocklington equation (2.21) simplifies into

$$\left(\frac{\partial^2}{\partial x^2}-\gamma^2\right)I\left(x\right) = 0. \tag{2.23}$$

Equation (2.23) is readily solved and the solution is given with

$$I\left(x\right) = I_g\frac{\sinh\left[\gamma\left(L-x\right)\right]}{\sinh\left(\gamma L\right)}. \tag{2.24}$$

The expression for scattered voltage can be obtained substituting (2.24) into (2.10), which yields

$$V^{sct}(x) = \frac{\gamma I_g}{j4\pi\omega\varepsilon_{eff}\sinh(\gamma L)} \int_0^L \cosh\left[\gamma(L-x)\right] g\left(x, x'\right) dx'. \qquad (2.25)$$

Integral in (2.25) is computed by means of standard numerical integration.

In Fig. 2.4, the current distribution along the electrode $L = 10\,\text{m}$, buried at $d = 0.3\,\text{m}$, with ground properties $\sigma = 0.01\,\text{S/m}$ and $\varepsilon_r = 10$ at the operating frequency $f = 10\,\text{MHz}$. The waveforms obtained via different approaches are very similar.

Figure 2.5 shows the results for transient impedance of an electrode of $L = 10\,\text{m}$, buried in a ground of conductivity $\sigma = 1\,\text{mS/m}$ for $1/10\,\mu\text{s}$ lightning pulse. It can be seen that the agreement between analytical and numerical results is very good, except for the early time response where discrepancy of around 10 % can be observed.

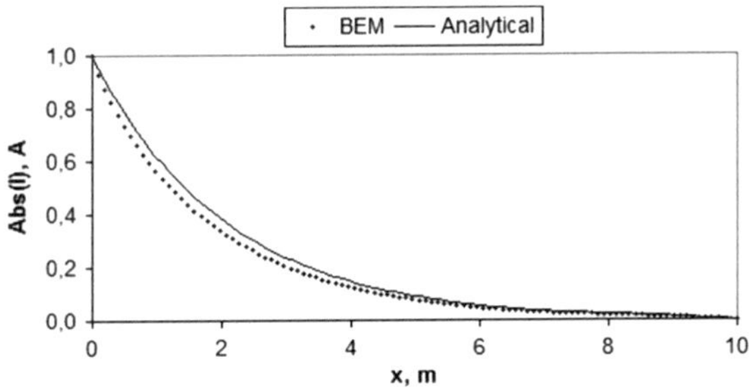

Fig. 2.4 Absolute value of a current distribution along the horizontal electrode

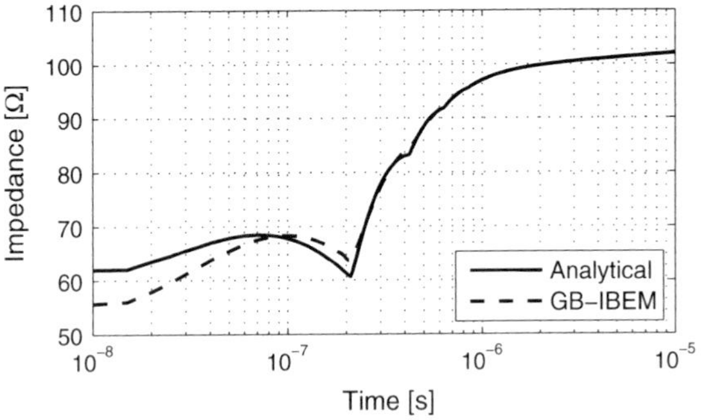

Fig. 2.5 Transient impedance of the grounding electrode

2.4 Time Domain Applications of Analytical Methods

2.4.1 Horizontal Wire Below Ground

To obtain analytical solution of (2.11), the integral operator is simplified, using addition and subtraction technique

$$\int_0^L I\left(x', t-\frac{R}{v}\right)\frac{e^{-\frac{1}{\tau_g}\frac{R}{v}}}{R}dx' = I\left(x, t-\frac{a}{v}\right)\int_0^L \frac{e^{-\frac{1}{\tau_g}\frac{R}{v}}}{R}dx'. \tag{2.26}$$

This approximation has proven to be valid in papers by Tijhuis *et al.* [4, 24]. Next step in solving the differential equation (2.11) is to apply the Laplace transform and obtain the following equation

$$\begin{aligned}(\mu\varepsilon s+\mu\sigma)\,E_x^{tr}(s) &= -\frac{\mu}{4\pi}\left(\frac{\partial^2}{\partial x^2}-\mu\sigma s-\mu\varepsilon s^2\right)\cdot\\ \cdot I(x,s)\,e^{-\frac{a}{v}s}&\left[\int_0^L \frac{e^{-\frac{1}{\tau_g}\frac{R}{v}}}{R}dx' - \Gamma_{ref}^{MIT}(s)\int_0^L \frac{e^{-\frac{1}{\tau_g}\frac{R^*}{v}}}{R^*}dx'\right].\end{aligned} \tag{2.27}$$

Integrals in (2.27) can be solved analytically as follows [25]

$$\Psi(s) = \int_0^L \frac{e^{-\frac{1}{\tau_g}\frac{R}{v}}}{R}dx' - \Gamma_{ref}^{MIT}(s)\int_0^L \frac{e^{-\frac{1}{\tau_g}\frac{R^*}{v}}}{R^*}dx' = 2\left(\ln\frac{L}{a}+\frac{s\tau_1+1}{s\tau_2+1}\ln\frac{L}{2d}\right). \tag{2.28}$$

Now, relation (2.27) can be written as

$$\frac{\partial^2 I(x,s)}{\partial x^2} - \gamma^2 I(x,s) = -\frac{4\pi}{\mu s\Psi(s)}e^{\frac{a}{v}s}\gamma^2 E_x^{tr}(s). \tag{2.29}$$

The solution of (2.29) can be readily obtained, prescribing the boundary conditions at the wire ends

$$\begin{aligned} I(0,s) &= 0,\\ I(L,s) &= 0. \end{aligned} \tag{2.30}$$

The solution of (2.29) is written as

$$I(x,s) = \frac{4\pi e^{\frac{a}{v}s}}{\mu s\Psi(s)}E_x^{tr}(s)\left[1-\frac{\cosh\left(\gamma\left(\frac{L}{2}-x\right)\right)}{\cosh\left(\gamma\frac{L}{2}\right)}\right]. \tag{2.31}$$

To obtain the solution for the current distribution in time domain, inverse Laplace transform has to be performed featuring the Cauchy residue theorem [19]

$$f(t) = \lim_{y \to \infty} \frac{1}{j2\pi} \int_{x-jy}^{x+jy} e^{ts} F(s)\, ds = \sum_{k=1}^{n} \operatorname{Res}(s_k). \tag{2.32}$$

Calculating all the residues of the function (2.31) and undertaking the inverse transform as in (2.32), the following expression is obtained

$$I(x,t) = \frac{4\pi}{\mu} \left\{ \begin{array}{l} R(s_\Psi) \left[1 - \frac{\cosh\left(\gamma_\Psi \left(\frac{L}{2} - x\right)\right)}{\cosh\left(\gamma_\Psi \frac{L}{2}\right)} \right] e^{\left(t + \frac{a}{v}\right) s_\Psi} - \\ -\frac{\pi}{\mu\varepsilon L^2} \sum_{n=1}^{\infty} \frac{2n-1}{\pm\sqrt{b^2 - 4c_n} s_{1,2n} \Psi\left(s_{1,2n}\right)} \sin \frac{(2n-1)\pi x}{L} e^{\left(t + \frac{a}{v}\right) s_{1,2n}} \end{array} \right\}, \tag{2.33}$$

where coefficients $R(s_\Psi)$ and s_Ψ represent physical properties of the system

$$R(s_\Psi) = \frac{1}{2 \ln \frac{L}{2d} \frac{s_\Psi}{s_\Psi \tau_2 + 1} \left(\tau_1 - \tau_2 \frac{s_\Psi \tau_1 + 1}{s_\Psi \tau_2 + 1} \right)}, \quad s_\Psi = -\frac{\ln \frac{L}{a} + \ln \frac{L}{2d}}{\tau_1 \ln \frac{L}{a} + \tau_2 \ln \frac{L}{2d}}. \tag{2.34}$$

Furthermore, other coefficients in relation (2.33) are given as follows

$$\begin{gathered} \gamma_\Psi = \sqrt{\mu\varepsilon \left(s_\Psi^2 + b s_\Psi \right)}, \\ s_{1,2n} = \frac{1}{2} \left(-b \pm \sqrt{b^2 - 4c_n} \right), \\ b = \frac{\sigma}{\varepsilon}, \quad c_n = \frac{(2n-1)^2 \pi^2}{\mu\varepsilon L^2}, \quad n = 1, 2, 3, \ldots . \end{gathered} \tag{2.35}$$

Expression (2.33) represents the space-time distribution of the current along the straight wire buried in a lossy medium excited by an impulse excitation.

Furthermore, the response to an arbitrary excitation can be obtained performing the corresponding convolution. The excitation function is plane wave in the form of double exponential electromagnetic pulse tangential to the wire [16]

$$E_x(t) = E_0 \left(e^{-\alpha t} - e^{-\beta t} \right). \tag{2.36}$$

In Fig. 2.6, transient current at the center of the straight wire with $L = 1$ m, $d = 30$ cm, $\sigma = 10$ mS/m is shown. Relatively good agreement between the results is achieved for a short wire and higher conductivity of a medium.

Figure 2.7 shows the transient current induced at the center of straight longer wires buried in a lossy medium with $\sigma = 1$ mS/m. For a 10 m–long wire the agreement between the results is rather satisfactorily.

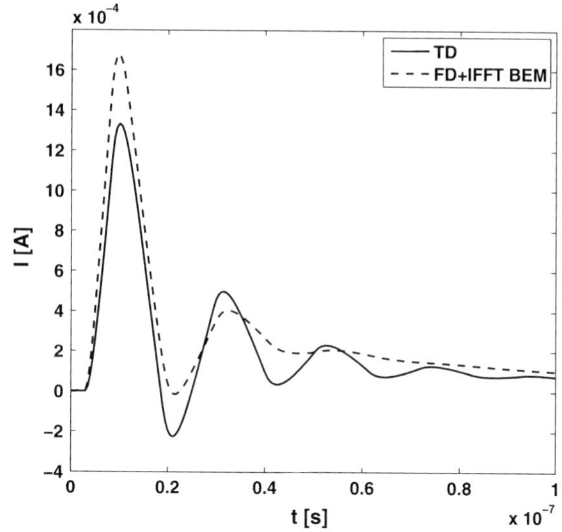

Fig. 2.6 Transient current at the center of the straight wire, $L = 1\,\text{m}$, $d = 30\,\text{cm}$, $\sigma = 10\,\text{mS/m}$

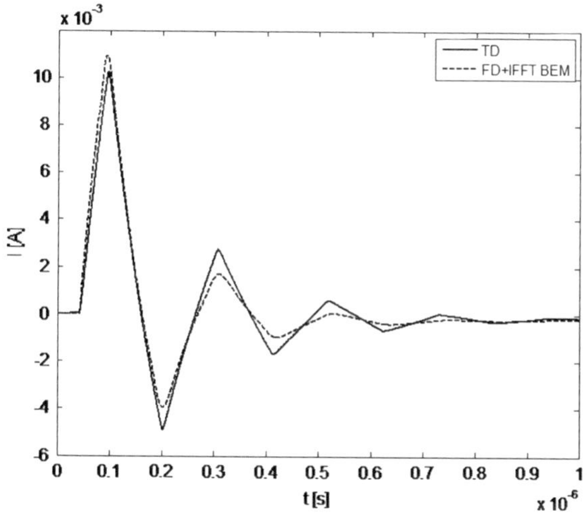

Fig. 2.7 Transient current at the center of the straight wire, $L = 10\,\text{m}$, $d = 4\,\text{m}$, $\sigma = 1\,\text{mS/m}$

2.4.2 *Horizontal Grounding Electrode*

Homogeneous variant of integro-differential equation (2.11), representing the governing equation for grounding electrode can be solved analytically, as it has been reported recently by the authors in [22]. The governing equation (2.11) is simplified using (2.26). Now (2.11) can be written as follows

$$\frac{\partial^2 I(x,s)}{\partial x^2} - \gamma^2 I(x,s) = 0. \tag{2.37}$$

Prescribing the boundary conditions at the wire ends

$$I(0, s) = I_g(s),$$
$$I(L, s) = 0, \tag{2.38}$$

the solution of (2.37) is readily obtained in the form

$$I(x, s) = I_g(s) \frac{\sinh\left[\gamma (L - x)\right]}{\sinh(\gamma L)}. \tag{2.39}$$

To obtain the solution for the current distribution in the time domain, inverse Laplace transform is performed and Cauchy residue theorem is applied [19] using (2.32). Having determined the residues of (2.39), the time domain counterpart is given

$$I(x, t) = \frac{2\pi}{\mu\varepsilon L^2} \sum_{n=1}^{\infty} \frac{(-1)^{n-1} n}{\pm\sqrt{b^2 - 4c_n}} \sin \frac{n\pi (L - x)}{L} e^{t s_{1,2n}}, \tag{2.40}$$

where corresponding coefficients are

$$s_{1,2n} = \frac{1}{2}\left(-b \pm \sqrt{b^2 - 4c_n}\right),$$
$$b = \frac{\sigma}{\varepsilon}, \quad c_n = \frac{n^2\pi^2}{\mu\varepsilon L^2}, \quad n = 1, 2, 3, \ldots . \tag{2.41}$$

Equation (2.40) represents an analytical expression for the space-time distribution of the current flowing along the grounding electrode excited by an equivalent current source in the form of the Dirac pulse. On the other hand, one of the functions most frequently used to represent the lightning current is the double exponential pulse, given with [13]

$$I_g(t) = I_0 \left(e^{-\alpha t} - e^{-\beta t}\right). \tag{2.42}$$

Analytical convolution is undertaken with (2.40) and (2.42), to obtain the expression for the current flowing along the electrode

$$I(x, t) = \frac{2\pi I_0}{\mu\varepsilon L^2} \sum_{n=1}^{\infty} \frac{(-1)^{n-1} n}{\pm\sqrt{b^2 - 4c_n}} \sin \frac{n\pi (L - x)}{L} \cdot$$
$$\cdot \left(\frac{e^{s_{1,2n} t} - e^{-\alpha t}}{s_{1,2n} + \alpha} - \frac{e^{s_{1,2n} t} - e^{-\beta t}}{s_{1,2n} + \beta} \right). \tag{2.43}$$

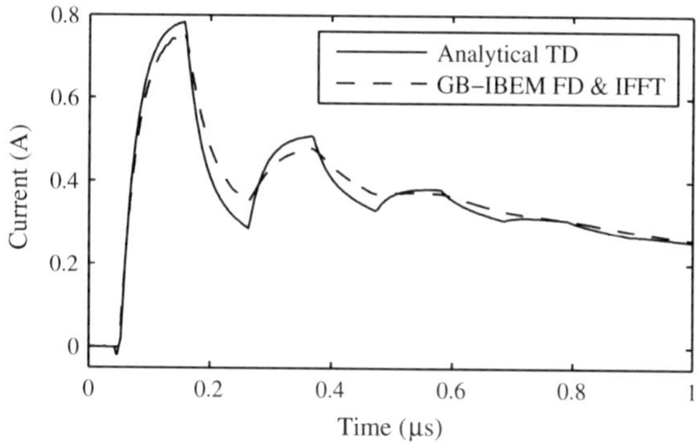

Fig. 2.8 Transient current at the center of the grounding electrode, 0.1/1 μs pulse

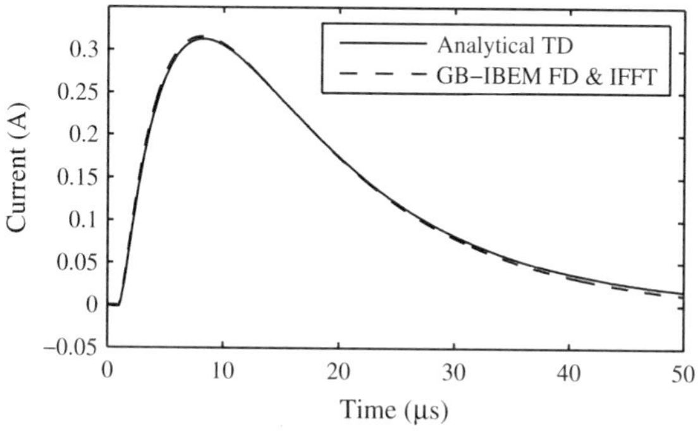

Fig. 2.9 Transient current at the center of the grounding electrode, 1/10 μs pulse

Equation (2.43) represents the expression for the space-time distribution of the current flowing along the electrode due to a double exponential current source excitation.

Analytical results for the transient current induced at the center of the electrode are calculated with (2.43) and are compared to the results obtained via numerical approach. The results shown in Fig. 2.8 are calculated for the grounding electrode with $L = 10\,\text{m}$, buried in a lossy ground with the conductivity $\sigma = 1\,\text{mS/m}$. The agreement between the results is very good.

The results shown in Fig. 2.9 are related to calculations performed for electric properties of the ground $\sigma = 0.833\,\text{mS/m}$ and $\varepsilon_r = 9$. It is worth emphasizing that low ground conductivity is considered. The electrode is buried at depth $d = 0.5\,\text{m}$. Length of the grounding electrode is $L = 200\,\text{m}$. The agreement between analytical and numerical results for the current induced at the center of the electrodes is very good, especially for the longer electrode.

2.5 Some Analytical Solutions to the Grad–Shafranov Equation

Grad–Shafranov equation describing the plasma equilibrium is given as [28]

$$\frac{\partial^2 \psi}{\partial r^2} - \frac{1}{r}\frac{\partial \psi}{\partial r} + \frac{\partial^2 \psi}{\partial z^2} = -f\frac{df}{d\psi} - \mu_0 r^2 \frac{dP}{d\psi}. \tag{2.44}$$

Various analytical solutions of GSE have been derived so far [6]. The analytical solutions are essential in describing various parameters that are involved in real tokamak scenarios as they are well suited for benchmarking various numerical codes. In this section, four different analytical solutions will be presented, with a short overview of their derivation as well as the emphasis to their applications.

2.5.1 *Solution of the Homogeneous Equation*

In order to obtain any solution corresponding to the realistic source functions that appear on the right-hand side of (2.44), it is necessary to define possible solutions of the homogeneous equation given by

$$\frac{\partial^2 \psi}{\partial r^2} - \frac{1}{r}\frac{\partial \psi}{\partial r} + \frac{\partial^2 \psi}{\partial z^2} = 0. \tag{2.45}$$

Solution of (2.45) can be obtained by variable separation and is given with

$$\psi_0(r,z) = \left(c_1 r J_1(kr) + c_2 r Y_1(kr)\right)\left(c_3 e^{kz} + c_4 e^{-kz}\right). \tag{2.46}$$

On the other hand, the solutions can also be based on the series expansion [28]

$$\psi_0 = \sum_{n=0,2,\ldots} f_n(r)\, z^n. \tag{2.47}$$

One of the possible solutions satisfying these conditions and suitable for further implementation is [28]

$$\psi_0(r,z) = c_1 + c_2 r^2 + c_3\left(r^4 - 4r^2 z^2\right) + c_4\left(r^2 \ln r - z^2\right). \tag{2.48}$$

2.5.2 The Solov'ev Equilibrium

The Solov'ev equilibrium is the simplest usable solution to the inhomogeneous GSE [6]. It has been widely used in studies of plasma equilibrium, transport and MHD stability analysis.

The source functions in Solov'ev equilibrium are linear in ψ and are given as [2]

$$P(\psi) = \frac{A}{\mu_0}\psi, \quad f^2(\psi) = 2B\psi + F_0^2, \tag{2.49}$$

with the corresponding solution

$$\psi(r, z) = \psi_0(r, z) - \frac{A}{8}r^4 - \frac{B}{2}z^2. \tag{2.50}$$

Wide variety of plasma shapes can be generated using (2.50). However, the current profile of this solution is restricted, since implementation of A and B allow choosing only two plasma parameters.

2.5.3 The Herrnegger–Maschke Solutions

The solution to the GSE for a parabolic source functions was reported in [6]

$$P(\psi) = \frac{C}{2\mu_0}\psi^2, \quad f^2(\psi) = D\psi^2 + F_0^2. \tag{2.51}$$

The solution of (2.51) can be given in the form of Coulomb wave functions as [2]

$$\psi = \alpha\left(F_0(\eta, x) + \gamma G_0(\eta, x)\right)\cos(kz). \tag{2.52}$$

As is the case for the Solov'ev equilibrium, the Herrnegger–Maschke solutions have only two free parameters, namely C and D, which allow independent specification of plasma current and pressure ratio.

2.5.4 Mc Carthy's Solution

Innovative source functions were introduced by Mc Carthy in [6]. These source functions are dissimilar in their nature and describes a linear dependence of pressure and quadratic dependence of the current profile

$$P(\psi) = \frac{S}{\mu_0}\psi, \quad f^2(\psi) = T\psi^2 + 2U\psi + F_0^2. \tag{2.53}$$

Equation (2.53) can be solved by the separation of variables where the following equations are obtained

$$\frac{\partial^2 H(z)}{\partial z^2} + k^2 H(z) = 0, \tag{2.54}$$

$$\frac{\partial^2 G(r)}{\partial r^2} - \frac{1}{r}\frac{\partial G(r)}{\partial r} - \left(k^2 - T\right) G(r) = 0. \tag{2.55}$$

The solution for $H(z)$ is readily obtained as

$$H(z) = c_1 e^{jkz} + c_2 e^{-jkz}, \tag{2.56}$$

while the solution of (2.55) is given with

$$G(r) = rB_1(ar), \tag{2.57}$$

where B_1 denotes the family of Bessel functions and parameter a satisfies the equation [6]

$$a^2 = \pm\left(T - k^2\right). \tag{2.58}$$

More mathematical details on these families can be found in [6].

To obtain exact solution of (2.56) and (2.57) for various real scenarios, the numerical solution of the free boundary problem (with a conventional equilibrium solver) and subsequent projection of the numerically obtained solution onto the exact solutions via a least squares fitting procedure is implemented [6]. The obtained solution can be written in the form

$$\begin{aligned}\psi = c_1 + c_2 r^2 + rJ_1(pr)(c_3 + c_4 z) + c_5 \cos pz + c_6 \sin pz + \\ +r^2 (c_7 \cos pz + c_8 \sin pz) + c_9 \cos p\sqrt{r^2 + z^2} + \\ +c_{10} \sin p\sqrt{r^2 + z^2} + rJ_1(vr)(c_{11} \cos qz + c_{12} \sin qz) + \\ +rJ_1(qr)(c_{13} \cos vz + c_{14} \sin vz) + \\ +rY_1(vr)(c_{15} \cos qz + c_{16} \sin qz) + \\ +rY_1(qr)(c_{17} \cos vz + c_{18} \sin vz),\end{aligned} \tag{2.59}$$

where corresponding vector of coefficients c_i can be found in [6].

Fig. 2.10 Exact GSE solution for ASDEX Upgrade discharge # 10 958, $t = 5.20\,\mathrm{s}$

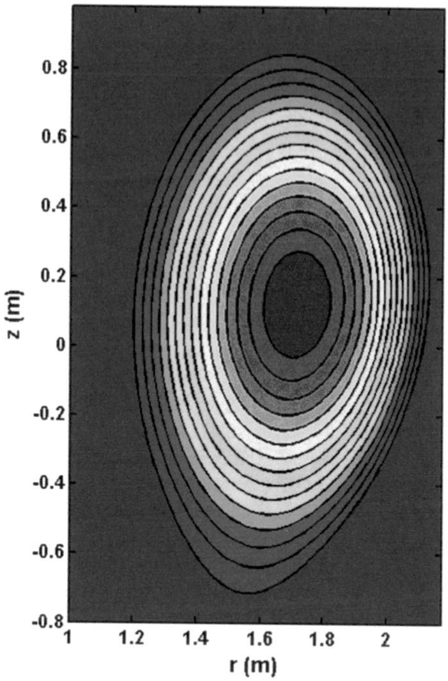

2.5.5 *Computational Example*

Computational example depicted in Fig. 2.10 corresponds to the results for tokamak equilibrium obtained using analytical solution (2.59). The highest value for the poloidal magnetic flux $\psi_{max} = 1.4\,\mathrm{Tm}^2$ is observed at the center of tokamak plasma, as it is expected, while the final contour (called separatrix) defines the area where the value of the magnetic flux is equal to zero.

2.6 Concluding Remarks

In the paper, some analytical methods for solving various integro-differential equationss in electromagnetic compatibility have been reviewed. Of particular interest are thin wire configurations buried in a lossy medium. Both frequency and time domain solutions are considered. Solutions in the frequency domain are obtained by performing certain mathematical manipulations with the unknown current function. On the other hand, solutions in the time domain are carried out using the Laplace transform and Cauchy residue theorem. The trade-off between the presented methods is given in this review paper, as well. Obtained analytical results are compared to those calculated by means of various numerical solutions, where applicable. Finally, an overview of well-established and widely used analytical solutions of the Grad–Shafranov equation is given and discussed.

References

1. Ambrosino, G., Albanese, R.: Magnetic control of plasma current, position, and shape in Tokamaks: A survey or modeling and control approaches. IEEE Control Syst. **25**, 76–92 (2005)
2. Atanasiu, C.V., Günter, S., Lackner, K., Miron, I.G.: Analytical solutions to the Grad–Shafranov equation. Phys. of Plasmas (1994-present) **11**, 3510–3518 (2004)
3. Barnes, P.R., Tesche, F.M.: On the direct calculation of a transient plane wave reflected from a finitely conducting half space. IEEE Trans. Electromagn. Compat. **33**, 90–96 (1991)
4. Bogerd, J.C., Tijhuis, A., Klaasen, J.J.A.: Electromagnetic excitation of a thin wire: A traveling-wave approach. IEEE Trans. Antennas Propag. **46**, 1202–1211 (1998)
5. Bridges, G.E.J.: Fields generated by bare and insulated cables buried in a lossy half-space. IEEE Trans. Geosci. Remote Sens. **30**, 140–146 (1992)
6. Carthy, P.J.M.: Analytical solutions to the Grad-Shafranov equation for tokamak equilibrium with dissimilar source functions. Phys. of Plasmas (1994-present) **6**, 3554–3560 (1999)
7. Grcev, L., Dawalibi, F.: An electromagnetic model for transients in grounding systems. IEEE Trans. Power Deliv. **5**, 1773–1781 (1990)
8. Hoorfar, A., Chang, D.: Analytic determination of the transient response of a thin-wire antenna based upon an SEM representation. IEEE Trans. Antennas Propag. **30**, 1145–1152 (1982)
9. King, R.W.P.: Embedded bare and insulated antennas. IEEE Trans. Biomed. Eng. **BME–24**, 253–260 (1977)
10. King, R.W.P., Fikioris, G.J., Mack, R.B.: Cylindrical Antennas and Arrays. Cambridge University Press, UK (2002)
11. Miller, E.K., Poggio, A.J., Burke, G.J.: An integro-differential equation technique for the time-domain analysis of thin wire structures. I. The numerical method. J. of Computational Phys. **12**, 24–48 (1973)
12. Poljak, D.: Advanced Modeling in Computational Electromagnetic Compatibility. Wiley, New Jersey (2007)
13. Poljak, D., Doric, V.: Wire antenna model for transient analysis of simple grounding systems, Part I: The vertical grounding electrode. Prog. In Electromagn. Res. **64**, 149–166 (2006)
14. Poljak, D., Doric, V.: Wire antenna model for transient analysis of simple grounding systems, Part II: The horizontal grounding electrode. Prog. In Electromagn. Res. **64**, 167–189 (2006)
15. Poljak, D., Kovac, N.: Time domain modeling of a thin wire in a two-media configuration featuring a simplified reflection/transmission coefficient approach. Eng. Anal. with Bound. Elem. **33**, 283–293 (2009)
16. Poljak, D., Doric, V., Rachidi, F., Drissi, K., Kerroum, K., Tkachenko, S.V., Sesnic, S.: Generalized form of telegrapher's equations for the electromagnetic field coupling to buried wires of finite length. IEEE Trans. on Electromagn. Compat. **51**, 331–337 (2009)
17. Poljak, D., Sesnic, S., Goic, R.: Analytical versus boundary element modelling of horizontal ground electrode. Eng. Anal. with Bound. Elem. **34**, 307–314 (2010)
18. Rao, S.M.: Time Domain Electromagnetics. Academic Press, Cambridge (1999)
19. Schiff, J.L.: The Laplace Transform: Theory and Applications. Springer, Heidelberg (1999)
20. Sesnic, S., Poljak, D.: Antenna model of the horizontal grounding electrode for transient impedance calculation: Analytical versus boundary element method. Eng. Anal. with Bound. Elem. **37**, 909–913 (2013)
21. Sesnic, S., Poljak, D., Tkachenko, S.V.: Time domain analytical modeling of a straight thin wire burried in a lossy medium. Prog. In Electromagn. Res. **121**, 485–504 (2011)

22. Sesnic, S., Poljak, D., Tkachenko, S.V.: Analytical modeling of a transient current flowing along the horizontal grounding electrode. IEEE Trans. on Electromagn. Compat. **55**, 1132–1139 (2013)
23. Takashima, T., Nakae, T., Ishibashi, R.: Calculation of complex fields in conducting media. IEEE Trans. on Electr. Insul. **EI–15**, 1–7 (1980)
24. Tijhuis, A., Zhongqiu, P., Bretones, A.: Transient excitation of a straight thin-wire segment: A new look at an old problem. IEEE Trans. on Antennas and Propag. **40**, 1132–1146 (1992)
25. Tkatchenko, S., Rachidi, F., Ianoz, M.: Electromagnetic field coupling to a line of finite length: theory and fast iterative solutions in frequency and time domains. IEEE Trans. on Electromagn. Compat. **37**, 509–518 (1995)
26. Tkatchenko, S., Rachidi, F., Ianoz, M.: High-frequency electromagnetic field coupling to long terminated lines. IEEE Trans. on Electromagn. Compat. **43**, 117–129 (2001)
27. Velazquez, R., Mukhedkar, D.: Analytical modelling of grounding electrodes transient behavior. IEEE Trans. Power Appar. Syst. **PAS–103**, 1314–1322 (1984)
28. Zheng, S.B., Wootton, A.J., Solano, E.R.: Analytical tokamak equilibrium for shaped plasmas. Phys. of Plasmas (1994-present) **3**, 1176–1178 (1996)

Chapter 3
Analysis of Horizontal Thin-Wire Conductor Buried in Lossy Ground: New Model for Sommerfeld Type Integral

Milica Rančić, Radoslav Jankoski, Sergei Silvestrov and Slavoljub Aleksić

Abstract A new simple approximation that can be used for modeling of one type of Sommerfeld integrals typically occurring in the expressions that describe sources buried in the lossy ground, is proposed in the paper. The ground is treated as a linear, isotropic and homogenous medium of known electrical parameters. Proposed approximation has a form of a weighted exponential function with an additional complex constant term. The derivation procedure of this approximation is explained in detail, and the validation is done applying it in the analysis of a bare conductor fed in the center and immersed in the lossy ground at arbitrary depth. Wide range of ground and geometry parameters of interest has been taken into consideration.

Keywords Current distribution · Horizontal conductor · Integral equation · Lossy ground · Point-matching method · Sommerfeld integral

3.1 Introduction

Significant effort has been put into evaluation of the influence of real ground parameters on the near- and far-field characteristics of wire conductors (or systems consisting of them) located in the air above lossy ground, or buried inside of it [1–8, 10–22,

---

M. Rančić (✉) · S. Silvestrov
Division of Applied Mathematics, The School of Education, Culture and Communication, Mälardalen University, Box 883, 721 23 Västerås, Sweden
e-mail: milica.rancic@mdh.se

S. Silvestrov
e-mail: sergei.silvestrov@mdh.se

R. Jankoski
FEIT, Ss. Cyril and Methodius University, Skopje, Macedonia
e-mail: radoslavjankoski@gmail.com

S. Aleksić
Faculty of Electronic Engineering, University of Niš, Niš, Serbia
e-mail: slavoljub.aleksic@elfak.ni.ac.rs

S. Silvestrov and M. Rančić (eds.), *Engineering Mathematics I*,
Springer Proceedings in Mathematics & Statistics 178,
DOI 10.1007/978-3-319-42082-0_3

24–37]. The methods applied in this research field range from simplified analytical to rigorous full-wave ones.

The one often used in cases of conductors buried in the ground is the transmission line model (TLM) [3, 16, 17, 20], which offers advantages of analytical approaches: simplicity and short calculation time. However, the TLM introduces calculation errors depending on the electrical properties of the ground, burial depth, and frequency range in question. More specifically, it is reliable for deep-buried long horizontal conductors at frequencies below MHz range, [16, 17, 20].

On the other hand, using the full-wave approach [1, 4–6, 12–15, 18, 19, 21], any kind of arbitrarily positioned wire system could be analyzed, at any frequency of interest with no restrictions to the electrical parameters of the ground. This approach is based on formulation of the electric field integral equation (EFIE) and its solution using an appropriate numerical method (e.g. method of moments, boundary element method). The influence of the ground parameters is taken into account through Sommerfeld integrals, which are a part of the kernel of the formulated integral equation (e.g. Pocklington, Hallén, etc.). Although the calculation accuracy that comes with this approach is high, greater computational costs also need to be paid, which depends on the numerical method used for EFIE solving, and the way Sommerfeld integrals are dealt with.

Basically, two approaches can be taken for the latter issue. The first, more time-consuming one, but also the one yielding most accurate results is any method of numerical integration of such integrals [4, 5, 21, 24, 33, 37]. A variety of methods have been proposed that could be roughly divided into a group of methods of direct integration (integration along the real axis), and a group of methods that consider changing of the integration path in the complex plane. The second approach considers approximate solving of these integrals using different methods, [6, 9, 11–15, 18, 19, 22, 25–32, 35, 36]. The reflection coefficient method, the method of images, methods considering approximation of the transformed reflection coefficient (spectral reflection coefficient - SRC) that is a part of the integrand, are some of the directions that researchers took in this area.

For the cases of wires buried in the ground, the influence of the air/ground boundary surface is usually taken into account using the reflection coefficient (or the transmission coefficient) approach (RC or TC, respectively, [5, 12–18, 20]), or the modified image theory (MIT, [15, 18, 19]). The latter one can only account for the electrical properties of the ground, not the burial depth, and its validity is frequency dependent (up to 1 MHz), [16]. On the other hand, the simplicity of the MIT and low computational cost that comes with it are also present in the RC or TC approaches; however, the plane wave incidence angle and wire depth are here taken into account. A drawback of these approaches is that they are valid for the far-field region, whereas the influence of the lossy ground is primarily noticeable in the close proximity of the sources. As an improvement, in [36] authors propose an approximation of the Sommerfeld integrals using a linear combination of 15 exponential functions with certain unknown constants obtained using the least-squares method. According to the authors, the maximum relative error of calculations is less than 0.1 per cent in a wide range of tested parameters describing the geometry and the ground, [36].

Similar solution, but applicable to static and quasi-static cases, is used in [11, 35]. The integrands are approximated by a set of exponential functions with unknown exponents and weight coefficients.

In this paper the authors propose a new model for approximation of one type of Sommerfeld integrals occurring in cases of conductors buried at arbitrary depth in the lossy ground parallel to the air/ground surface. This model is based on the procedure proposed by the first author in [25–32], which considers approximation of a part of the integrand using a weighted exponential function with an additional unknown complex constant term. This procedure has been successfully employed for approximation of two forms of Sommerfeld integrals appearing in expressions describing the Hertz's vector potential in the surroundings of sources positioned in the air above lossy ground. Proposed solutions have been applied to near- and far-field analysis of different wire antenna structures arbitrarily located in the air above lossy soil, [25–32], and modeling of the lightning discharge using an antenna model, [9].

Application of the newly proposed approximation of the integral in question is validated analyzing a centrally fed horizontal conductor immersed in the lossy ground. An integral equation of Hallén's type (HIE) is solved applying the point-matching method (PMM) as in [25–32], and adopting the polynomial current approximation as in [21, 22, 25–32]. Different burial depths of the wire, and different ground types, are considered at various frequencies. Obtained results are, were possible, compared to the TC approach in combination with the PMM solution to the HIE. Also, Partial Element Equivalent Circuit (PEEC) method applied in [10], and a so-called Hybrid Circuit Model (HCM) proposed in [7, 8], are also used for comparison purposes. Based on presented results, corresponding conclusions are given, and possibilities for further research are discussed.

3.2 Problem Formulation

Let us observe a centrally fed horizontal thin-wire conductor with lengths of conductor halves l, and cross-section radii a, buried in the lossy half-space (LHS) at depth h, as illustrated in Fig. 3.1. The LHS is considered a homogeneous, linear and isotropic medium of known electrical parameters. Electrical parameters of the air are:

- $\sigma_0 = 0$ - conductivity;
- ε_0 - permittivity;
- μ_0 - permeability,

and of the soil:

- σ_1 - conductivity;
- $\varepsilon_1 = \varepsilon_{r1}\varepsilon_0$ - permittivity (ε_{r1} - relative permittivity);
- $\mu_1 = \mu_0$ - permeability;
- $\underline{\sigma}_i = \sigma_i + j\omega\varepsilon_1$ - complex conductivity;

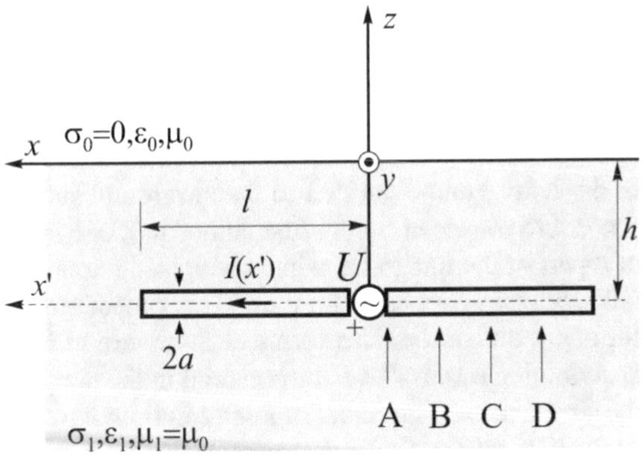

Fig. 3.1 Illustration of a horizontal conductor buried in the lossy half-space

- $\gamma_i = \alpha_i + \mathrm{j}\beta_i = (\mathrm{j}\omega\mu\underline{\sigma}_i)^{1/2},\ i = 0,\ 1$ - complex propagation constant ($i = 0$ for the air, and $i = 0$ for the LHS);
- $\omega = 2\pi f$ - angular frequency;
- $\underline{\varepsilon}_{r1}$ - complex relative permittivity;
- $\underline{n} = \gamma_1/\gamma_0 = \varepsilon_{r1}^{1/2} = (\varepsilon_{r1} - j60\sigma_1\lambda_0)^{1/2}$ - refractive index, and
- λ_0 - wavelength in the air.

The Hertz's vector potential has two components at an arbitrary point $M_1(x, y, z)$ in the ground in the vicinity of the conductor, i.e. $\mathbf{\Pi}_1 = \Pi_{x1}\hat{x} + \Pi_{z1}\hat{z}$, [5, 12–16, 18–20, 34]. Consequently, the tangential component of the scattered electric field can be expressed as:

$$E_{x1}^{sct}(x, x') = \left[\frac{\partial^2}{\partial x^2} - \underline{\gamma}_1^2\right]\Pi_{x1} + \frac{\partial^2\Pi_{z1}}{\partial x \partial z}, \tag{3.1}$$

where

$$\Pi_{x1} = \frac{1}{4\pi\underline{\sigma}_1}\int_{-l}^{l} I(x')\left[K_o(x, x') - K_i(x, x') + U_{11}\right]\mathrm{d}x', \tag{3.2}$$

$$\Pi_{z1} = \frac{1}{4\pi\underline{\sigma}_1}\int_{-l}^{l} I(x')\frac{\partial W_{11}}{\partial x}\mathrm{d}x', \tag{3.3}$$

with $I(x')$ - the current distribution along the conductor (x'- axis assigned to the wire);

$$K_o(x, x') = e^{-\underline{\gamma}_1 r_o},\ r_o = \sqrt{\rho^2 + a^2},\ \rho = \left|x - x'\right|, \tag{3.4}$$

$$K_i(x, x') = e^{-\underline{\gamma}_1 r_i},\ r_i = \sqrt{\rho^2 + (2h)^2},\ \rho = \left|x - x'\right|, \tag{3.5}$$

$$U_{11} = \int_{\alpha=0}^{\infty} \tilde{T}_{\eta 1}(\alpha)e^{-u_1(z+h)}\frac{\alpha}{u_1}J_0(\alpha\rho)\mathrm{d}\alpha, \tag{3.6}$$

$$W_{11} = \int_{\alpha=0}^{\infty} \tilde{T}_{\eta 2}(\alpha) e^{-u_1(z+h)} \frac{\alpha}{u_1} J_0(\alpha\rho) \mathrm{d}\alpha, \tag{3.7}$$

$$\tilde{T}_{\eta 1}(\alpha) = \frac{2u_1}{u_0 + u_1}, \; u_i = \sqrt{\alpha^2 + \underline{\gamma}_i^2}, \; i = 0, 1, \tag{3.8}$$

$$\tilde{T}_{\eta 2}(\alpha) = \frac{2u_1(u_0 - u_1)}{\underline{\gamma}_1^2 u_0 + \underline{\gamma}_0^2 u_1}, \; u_i = \sqrt{\alpha^2 + \underline{\gamma}_i^2}, \; i = 0, 1, \tag{3.9}$$

where $J_0(\alpha\rho)$ is the zero-order Bessel function of the first kind. Adopting (3.2) and (3.3), expression (3.1) can be written as

$$E_{x1}^{sct}(x, x') = \frac{1}{4\pi\underline{\sigma}_1} \int_{-l}^{l} I(x') G(x, x') \mathrm{d}x', \text{ and} \tag{3.10}$$

$$G(x, x') = \left[\frac{\partial^2}{\partial x^2} - \underline{\gamma}_1^2\right]\left[K_o(x, x') - K_i(x, x') + U_{11}\right] + \frac{\partial^2}{\partial x \partial z}\left[\frac{\partial W_{11}}{\partial x}\right]. \tag{3.11}$$

Since, according to [34], integral given by (3.7) can be rewritten as $\frac{\partial W_{11}}{\partial z} = -\underline{\gamma}_1^2 V_{11} - U_{11}$, where

$$V_{11} = \int_{\alpha=0}^{\infty} \tilde{T}_{\eta 3}(\alpha) e^{-u_1(z+h)} \frac{\alpha}{u_1} J_0(\alpha\rho) \mathrm{d}\alpha, \text{ and} \tag{3.12}$$

$$\tilde{T}_{\eta 3}(\alpha) = \frac{2u_1}{\underline{\gamma}_1^2 u_0 + \underline{\gamma}_0^2 u_1}, \; u_i = \sqrt{\alpha^2 + \underline{\gamma}_i^2}, \; i = 0, 1, \tag{3.13}$$

then the expressions (3.10) and (3.11) can be rewritten as

$$E_{x1}^{sct}(x, x') = \frac{1}{4\pi\underline{\sigma}_1} \begin{bmatrix} \frac{\partial^2}{\partial x^2} \int_{-l}^{l} I(x') \left[K_o(x, x') - K_i(x, x') - \underline{\gamma}_1^2 V_{11}\right] \mathrm{d}x' - \\ -\underline{\gamma}_1^2 \int_{-l}^{l} I(x') \left[K_o(x, x') - K_i(x, x') + U_{11}\right] \mathrm{d}x' \end{bmatrix}. \tag{3.14}$$

Boundary condition for the total tangential component of the electric field vector must be satisfied at any given point on the conductor's surface, and if the wire is perfectly conducting then

$$E_{x1}^{sct}(x, x') + E_{x1}^{tr}(x, x') = 0, \tag{3.15}$$

where $E_{x1}^{tr}(x, x')$ is the transmitted electric field. Now, the integral equation-IE (3.15) has the form

$$E_{x1}^{tr}(x, x') = -E_{x1}^{sct}(x, x'), \tag{3.16}$$

which, as a solution, gives the current distribution along the observed conductor. However, in order to do so, a group of improper integrals, referred to as integrals of Sommerfeld type, needs to be solved. Those would be integrals given by (3.6) and (3.7) if the formulation (3.10) and (3.11) is substituted in (3.16), or a set of integrals (3.6) and (3.12), if (3.14) is adopted.

3.3 Sommerfeld Integral Approximations

Different approaches have been applied in this field, but most of them start with a simplified version of the Green's function (3.11) having the following form:

$$G(x, x') = \left[\frac{\partial^2}{\partial x^2} - \underline{\gamma}_1^2\right]\left[K_o(x, x') - K_i(x, x') + U_{11}\right], \tag{3.17}$$

which means that only one Sommerfeld integral (the one given by (3.6)) needs to be solved. The following sub-sections will give an overview of a solution already proposed in the literature (Sect. 3.3.1), and also a newly developed one by the authors of this paper (Sect. 3.3.2).

3.3.1 Transmission Coefficient (TC) Approach

Transmission coefficient approach [3, 12], substitutes the part $-K_i(x,\ x') + U_{11}$ in (3.17) by

$$-K_i(x,\ x') + U_{11} = -K_i(x,\ x')\Gamma_{TM}^{trans.}, \tag{3.18}$$

i.e. approximates the U_{11} by

$$U_{11}^a \approx K_i(x,\ x')\left(1 - \Gamma_{TM}^{trans.}\right), \tag{3.19}$$

where

$$\Gamma_{TM}^{trans.} = \frac{2\underline{n}\cos\theta}{\underline{n}^2\cos\theta + \sqrt{\underline{n}^2 - \sin^2\theta}}, \theta = \arctan\frac{\rho}{2h}, \tag{3.20}$$

presents the transmission coefficient for TM polarization.

3.3.2 Two-Image Approximation - TIA

In this paper, the authors propose a new approximation for the integral (3.6), so-called two-image approximation (TIA) developed using the procedure applied in [25–32] for modeling two different forms of Sommerfeld integrals occurring in cases of sources located in the air above LHS.

1st case: Let us assume the expression (3.8) in the following form:

$$\tilde{T}_{\eta 1}^{a} = \underline{B} + \underline{A} e^{-(u_1 - \underline{\gamma}_1)\underline{d}}, \tag{3.21}$$

where $\underline{B}$, $\underline{A}$ and $\underline{d}$ are unknown complex constants. When (3.21) is substituted into (3.6), taking into account the identity, [23],

$$\int_{\alpha=0}^{\infty} \frac{e^{-|c|\sqrt{\alpha^2+\underline{\gamma}_1^2}}}{\sqrt{\alpha^2+\underline{\gamma}_1^2}} \alpha \, J_0(\alpha\rho) \mathrm{d}\alpha = K_c(x, x') = \frac{e^{-\underline{\gamma}_1\sqrt{\rho^2+|c|^2}}}{\sqrt{\rho^2+|c|^2}}, \tag{3.22}$$

the following general TIA approximation of (3.6) is obtained:

$$U_{11}^{a}(x, x') = \underline{B} K_{zh}(x, x') + \underline{A} e^{\underline{\gamma}_1 |\underline{d}|} K_{zhd}(x, x'), \tag{3.23}$$

where

$$K_{zh}(x, x') = e^{-\underline{\gamma}_1 r_{zh}}, \; r_{zh} = \sqrt{\rho^2 + (z+h)^2}, \tag{3.24}$$

$$K_{zhd}(x, x') = e^{-\underline{\gamma}_1 r_{zhd}}, \; r_{zhd} = \sqrt{\rho^2 + (z + h + |\underline{d}|)^2}. \tag{3.25}$$

Constants $\underline{B}$, $\underline{A}$ and $\underline{d}$ are evaluated matching the expressions (3.8) and (3.21), as well as their first derivative, at certain characteristic points in the range of integration of (3.6). One possibility is as follows,

1. Matching point 1: $u_1 \to \infty$

$$\tilde{T}_{\eta 1}(u_1 \to \infty) = 1, \tag{3.26}$$

$$\tilde{T}_{\eta 1}^{a}(u_1 \to \infty) = \underline{B}. \tag{3.27}$$

2. Matching point 2: $u_1 = \underline{\gamma}_0$

$$\tilde{T}_{\eta 1}(u_1 = \underline{\gamma}_0) = \frac{2}{1 + \sqrt{2 - \underline{n}^2}}, \tag{3.28}$$

$$\tilde{T}_{\eta 1}^{a}(u_1 = \underline{\gamma}_0) = \underline{B} + \underline{A} e^{-\underline{\gamma}_0 (1-\underline{n})\underline{d}}. \tag{3.29}$$

3. Matching point for the first derivative: $u_1 = \underline{\gamma}_0$

$$\tilde{T}'_{\eta 1}(u_1 = \underline{\gamma}_0) = \frac{-2(\underline{n}^2 - 1)}{\underline{\gamma}_0\sqrt{2-\underline{n}^2}\left(1+\sqrt{2-\underline{n}^2}\right)^2}, \tag{3.30}$$

$$\tilde{T}'^{a}_{\eta 1}(u_1 = \underline{\gamma}_0) = -\underline{d}\underline{A}e^{-\underline{\gamma}_0(1-\underline{n})\underline{d}}. \tag{3.31}$$

Equating (3.26) and (3.27), (3.28) and (3.29), and (3.30) and (3.39), a system of three equations over three unknown constants $\underline{B}$, $\underline{A}$ and $\underline{d}$ is formed, and the solution is given in the first row of Table 3.1.

2nd case: The same approximation of (3.6) can be achieved if we assume (3.8) as

$$\tilde{T}^{a}_{\eta 1} = \underline{B} + \underline{A}e^{-u_1\underline{d}}, \tag{3.32}$$

then (3.6) gets the form

$$U^{a}_{11}(x, x') = \underline{B}K_{zh}(x, x') + \underline{A}K_{zhd}(x, x'), \tag{3.33}$$

For the same matching points as previously, we get

1. Matching point 1: $u_1 \to \infty$

$$\tilde{T}_{\eta 1}(u_1 \to \infty) = 1, \tag{3.34}$$

$$\tilde{T}^{a}_{\eta 1}(u_1 \to \infty) = \underline{B}. \tag{3.35}$$

2. Matching point 2: $u_1 = \underline{\gamma}_0$

$$\tilde{T}_{\eta 1}(u_1 = \underline{\gamma}_0) = \frac{2}{1+\sqrt{2-\underline{n}^2}}, \tag{3.36}$$

$$\tilde{T}^{a}_{\eta 1}(u_1 = \underline{\gamma}_0) = \underline{B} + \underline{A}e^{-\underline{\gamma}_0\underline{d}}. \tag{3.37}$$

3. Matching point for the first derivative: $u_1 = \underline{\gamma}_0$

$$\tilde{T}'_{\eta 1}(u_1 = \underline{\gamma}_0) = \frac{-2(\underline{n}^2 - 1)}{\underline{\gamma}_0\sqrt{2-\underline{n}^2}\left(1+\sqrt{2-\underline{n}^2}\right)^2}, \tag{3.38}$$

$$\tilde{T}'^{a}_{\eta 1}(u_1 = \underline{\gamma}_0) = -\underline{d}\underline{A}e^{-\underline{\gamma}_0\underline{d}}. \tag{3.39}$$

The values obtained for $\underline{B}$, $\underline{A}$ and $\underline{d}$ are listed in the second row of Table 3.1.

Table 3.1 Obtained values of constants describing proposed TIA model

TIA Model	$\underline{B}$	$\underline{A}$	$\underline{d}$
1st *case*	1	$\frac{1-\sqrt{2-\underline{n}^2}}{1+\sqrt{2-\underline{n}^2}} e^{\underline{\gamma}_0(1-\underline{n})\underline{d}}$	$\frac{2}{\underline{\gamma}_0\sqrt{2-\underline{n}^2}}$
2nd *case*	1	$\frac{1-\sqrt{2-\underline{n}^2}}{1+\sqrt{2-\underline{n}^2}} e^{\underline{\gamma}_0\underline{d}}$	$\frac{2}{\underline{\gamma}_0\sqrt{2-\underline{n}^2}}$

3.4 Solution of the Integral Equation

In order to validate the application of the proposed approximation for the integral (3.6), the integral equation (3.16) will be solved for the current. First, the form of the Hallén's IE (HIE) is obtained as a solution of the partial differential equation that arises from (3.16). For the case of a thin-wire conductor centrally-fed by a Dirac's δ-generator, $E_{x1}^{tr}(x, x') = U\delta(x)$, $U = 1$ V, and taking into account the simplified Green's function given by (3.17), the HIE becomes:

$$\int_{-l}^{l} I(x')\left[K_o(x,x') - K_i(x,x') + U_{11}\right]\mathrm{d}x' - C\cos(\mathrm{j}\underline{\gamma}_1 x) = \mathrm{j}\frac{\underline{n}}{60}U\sin(\mathrm{j}\underline{\gamma}_1 x), \tag{3.40}$$

where C is an integration constant.

In order to solve (3.40), the point-matching method (PMM) is applied, giving us a system of linear equations with current distribution and integration constant C as unknowns. In this paper, we adopt the entire domain polynomial current approximation for the current as in [21, 22, 25–32]:

$$I(u' = x'/l) = \sum_{m=0}^{M} I_m u'^m, \quad 0 \le u' \le 1, \tag{3.41}$$

where $I_m, m = 0,\ 1, ..., \ M$, are complex current coefficients. This amounts to a total of $(M + 1) + 1$ unknowns, which calls for as much linear equations. The matching is done at $(M + 1)$ points that are chosen as $x_i = il/M$, $i = 0,\ 1,\ 2, ...,\ M$. This way, a system of $(M + 1)$ linear equations is formed, lacking one additional equation to account for the unknown integration constant C. This remaining linear equation is obtained applying the condition for vanishing of the current at the conductor's end, which corresponds to $I(-l) = I(l) = 0$. If we adopt TC or TIA model for (3.6), the system of equations becomes:

$$\sum_{m=0}^{M} I_m \int_{-l}^{l} \left(\frac{x'}{l}\right)^m \left[K_o(x,x') - K_i(x,x') + U_{11}^a(x,x')\right]\mathrm{d}x' - C\cos(\beta_0\underline{n}x_i) = $$

$$= -\mathrm{j}\frac{\underline{n}}{60}U\sin(\beta_0\underline{n}x_i),\ i = 0,\ 1,\ 2, ...,\ M, \tag{3.42}$$

$$\sum_{m=0}^{M} I_m = 0. \qquad (3.43)$$

3.5 Numerical Results

First we observed the convergence of the PMM method when the TIA approach to solving Sommerfeld integral (3.6) is adopted. The current magnitude is calculated for different values of the order of the polynomial current distribution M. Obtained results along a half of the conductor, which correspond to burial depth of $h = 0.1$ m, can be observed from Figs. 3.2 and 3.3 for two frequency values: 1 and 10 MHz, respectively. Each figure includes two diagrams corresponding to two different values of the ground conductivity: (a) $\sigma_1 = 0.001$ S/m and (b) $\sigma_1 = 0.01$ S/m. The analysis is performed for the case of the conductor's half-length $l = 5$ m, cross-section radius $a = 5$ mm, and electric permittivity of the ground $\varepsilon_{r1} = 10$.

The conductor with the same geometry parameters is considered again, for two cases of burial depths: (a) $h = 1.0$ m and (b) $h = 5.0$ m. The variable parameter in all figures is the specific conductivity of the ground, and it takes three values: $\sigma_1 = 0.001, 0.01$ and 0.1 S/m. The analysis is performed for the case of the electric permittivity of the ground $\varepsilon_{r1} = 10$. The results obtained by the PMM method and both the TC and newly proposed TIA approach are compared to the corresponding ones obtained by the methods from [7, 8, 10]. In [10] the authors employ the Partial Element Equivalent Circuit (PEEC) method, while in [7, 8] a so-called Hybrid Circuit Model (HCM) is proposed. Satisfying accordance of the results can be observed from the presented results. This is especially noticeable for more deeply buried conductors, and lower frequencies.

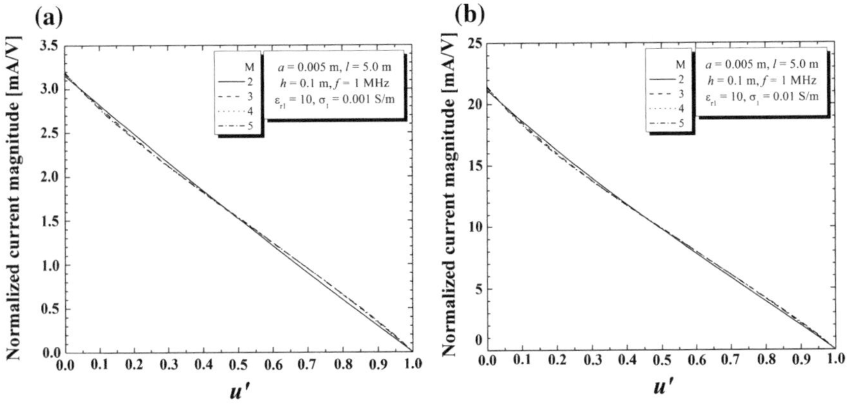

Fig. 3.2 Current magnitude along a half of the conductor for burial depth of 0.1 m and two values of ground conductivities: **a** 0.001 S/m, **b** 0.01 S/m. Order of polynomial current approximation M is taken as a parameter. Frequency is 1 MHz

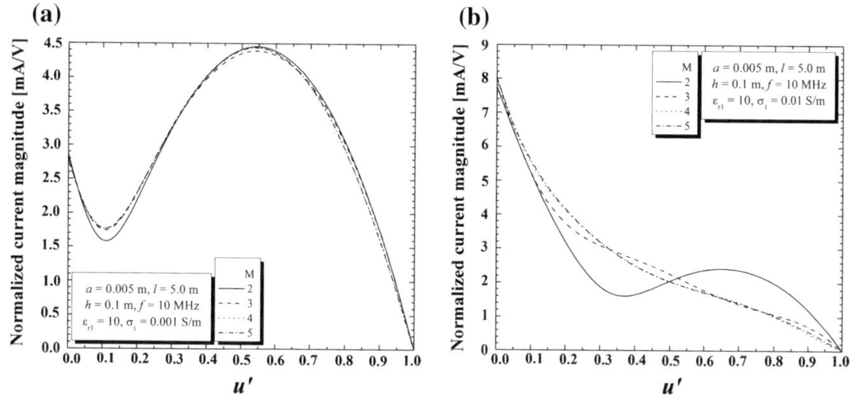

Fig. 3.3 Current magnitude along a half of the conductor for burial depth of 0.1 m and two values of ground conductivities: **a** 0.001 S/m, **b** 0.01 S/m. Order of polynomial current approximation M is taken as a parameter. Frequency is 10 MHz

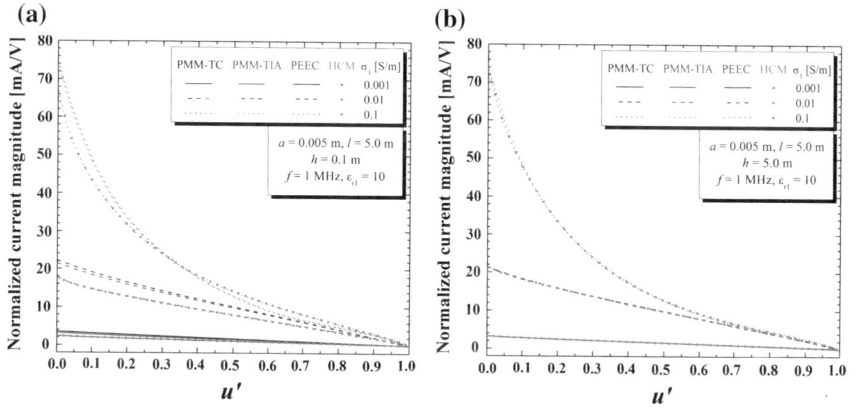

Fig. 3.4 Current magnitude along a half of the conductor buried at **a** 0.1 m, **b** 5 m for frequency of 1 MHz. Conductivity of the ground is taken as a parameter. Comparison of different methods

Next, results for the current magnitude are given in Figs. 3.4 and 3.5 for two frequency values, 1 and 10 MHz, respectively.

Figures 3.6, 3.7 and 3.8 illustrate the current magnitude distribution for three different frequencies: 1, 5 and 10 MHz. Two cases of burial depth are considered: (a) $h = 1.0$ m and (b) $h = 5.0$ m. The conductor's geometrical parameters are the same as previously.

Each figure corresponds to the same electrical permittivity ($\varepsilon_{r1} = 10$), while the ground's conductivity is varied ($\sigma_1 = 0.001, 0.01$ and 0.1 S/m). Again, the results obtained by different methods, PMM-TC, PMM-TIA, PEEC, and HCM, are compared.

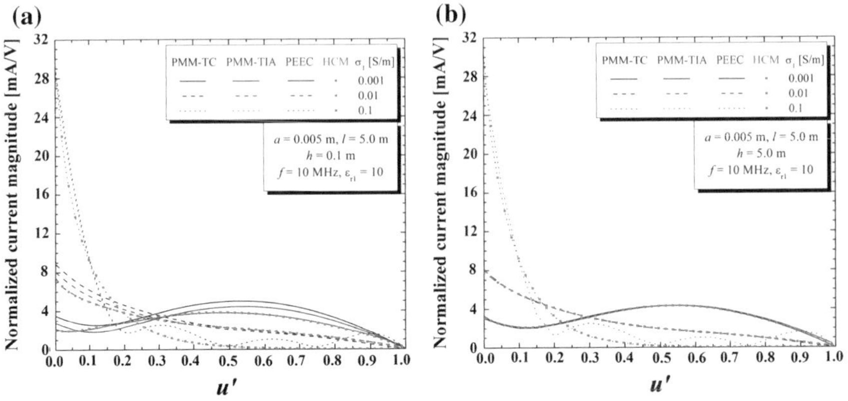

Fig. 3.5 Current magnitude along a half of the conductor buried at **a** 0.1 m, **b** 5 m for frequency of 10 MHz. Conductivity of the ground is taken as a parameter. Comparison of different methods

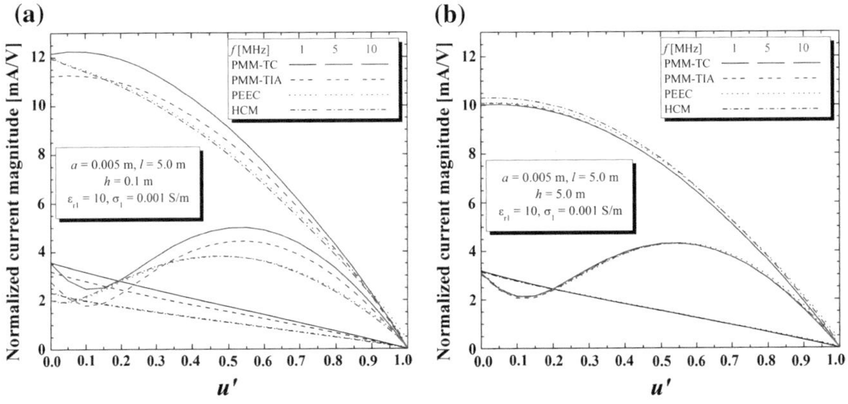

Fig. 3.6 Current magnitude along a half of the conductor buried at **a** 0.1 m and **b** 5 m for ground conductivity of 0.001 S/m. Frequency is taken as a parameter. Comparison of different methods

Final set of numerical results illustrates the influence of different values of the electrical permittivity at two frequencies: (a) 1 MHz and (b) 5 MHz. Figures 3.9, 3.10 and 3.11 correspond to three cases of specific ground conductivity $\sigma_1 = 0.001$ S/m, 0.01 S/m, and 0.1 S/m. The conductor has the same geometry as previously, and is positioned at $h = 5.0$ m below the boundary surface air/LHS. The results obtained by the PMM-TIA (solid squares), and the PEEC method (continual lines) are presented. Observed values of the electrical permittivity are: $\varepsilon_{r1} = 1, 2, 5, 10, 20, 36$ and 81. This influence is most noticeable at higher frequencies and for lower values of the ground conductivity.

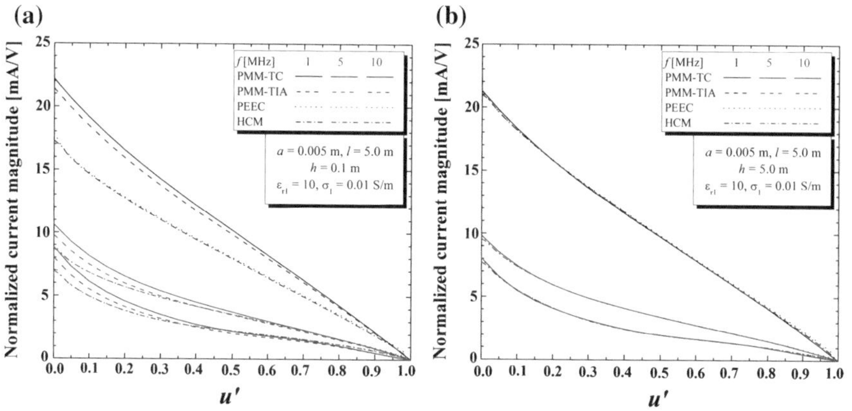

Fig. 3.7 Current magnitude along a half of the conductor buried at **a** 0.1 m and **b** 5 m for ground conductivity of 0.01 S/m. Frequency is taken as a parameter. Comparison of different methods

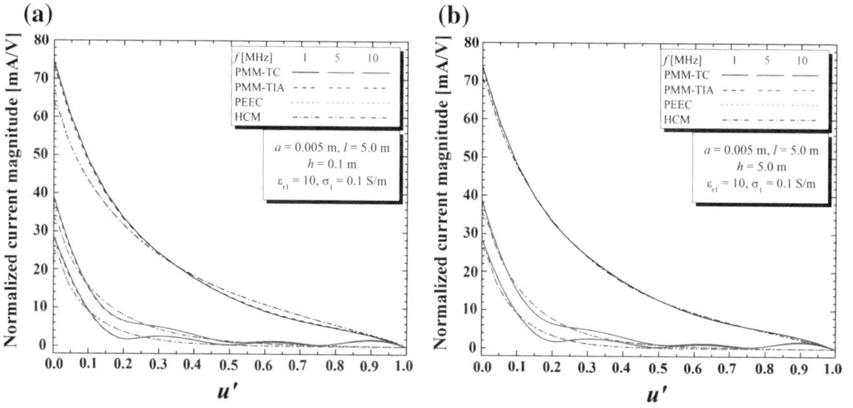

Fig. 3.8 Current magnitude along a half of the conductor buried at **a** 0.1 m and **b** 5 m for ground conductivity of 0.1 S/m. Frequency is taken as a parameter. Comparison of different methods

3.6 Conclusion

The aim of the paper to effectively approximate one form of Sommerfeld integrals has been achieved developing a simple approximation in a form of a weighted exponential function with an additional constant term, denoted here as two-image approximation (TIA). Proposed approximation is valid over a wide range of parameters (electrical parameters of the ground and geometry parameters). Presented numerical results show that the proposed model in combination with the PMM method can be successfully applied to frequency analysis of conductors buried in the lossy medium.

Furthermore, presented results indicate a possibility of effective application of the proposed procedure to other forms of Sommerfeld integrals that also appear

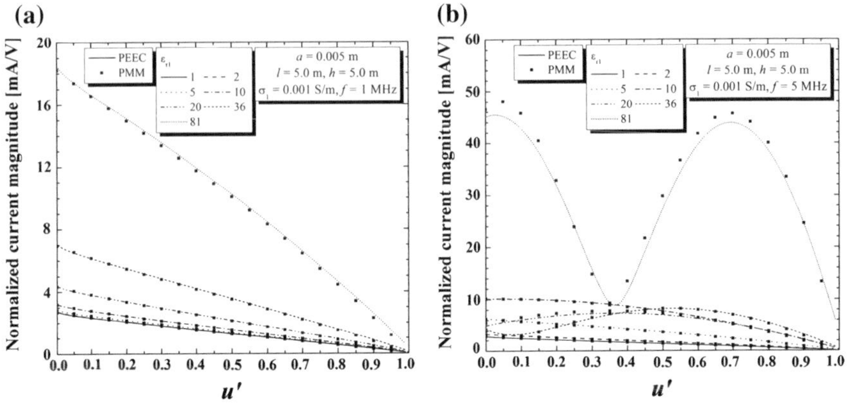

Fig. 3.9 Current magnitude along a half of the conductor for frequencies **a** 1 MHz, **b** 5 MHz. Electrical permittivity of the ground is taken as a parameter. Ground conductivity is 0.001 S/m

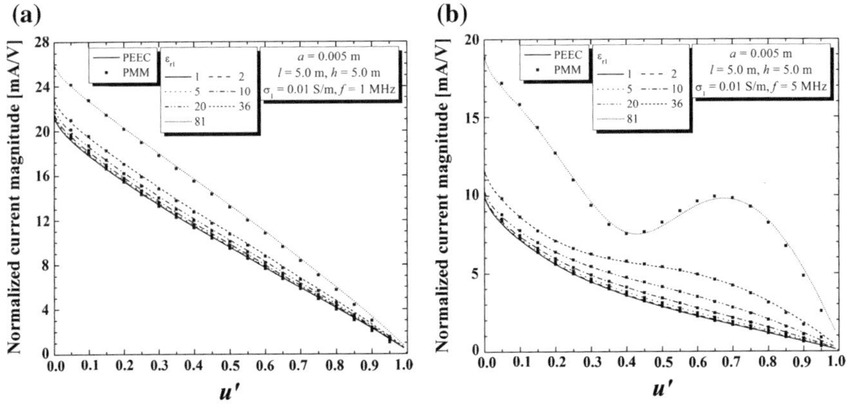

Fig. 3.10 Current magnitude along a half of the conductor for frequencies **a** 1 MHz, **b** 5 MHz. Electrical permittivity of the ground is taken as a parameter. Ground conductivity is 0.01 S/m

in the observed case of sources buried in the lossy ground (the ones given by (3.7) and (3.12)), which are usually neglected [3, 5, 6, 12–20, 24]. This would yield a more stringent analysis, and also a more accurate one, of not only antennas immersed in the lossy ground, but also wire grounding systems in such soil, buried telecommunication cables exposed to electromagnetic interferences, submarine dipoles, bare or isolated antennas embedded in dissipative media, etc.

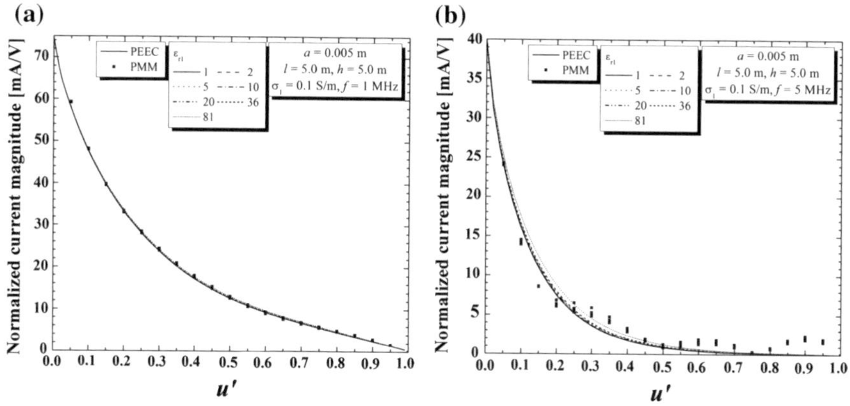

Fig. 3.11 Current magnitude along a half of the conductor for frequencies **a** 1 MHz, **b** 5 MHz. Electrical permittivity of the ground is taken as a parameter. Ground conductivity is 0.1 S/m

References

1. Arnautovski-Toseva, V., Drissi, K.E.K., Kerroum, K.: HF comparison of image and TL models of a horizontal thin-wire conductor in finitely conductive earth. In: Proceedings of SOFTCOM 2011, Split, Croatia, pp. 1–5 (2011)
2. Banos, A.: Dipole Radiation in the Presence of a Conducting Half-Space. Pergamon Press, New York (1966)
3. Bridges, G.E.: Transient plane wave coupling to bare and insulated cables buried in a lossy half-space. IEEE Trans. EMC **37**(1), 62–70 (1995)
4. Burke, G.J., Miller, E.K.: Modeling antennas near to and penetrating a lossy interface. IEEE Trans. AP **32**(10), 1040–1049 (1984)
5. Grcev, L., Dawalibi, F.: An electromagnetic model for transients in grounding systems. IEEE Trans. Power Deliv. **5**(4), 1773–1781 (1990)
6. Grcev, L.D., Menter, F.E.: Transient electro-magnetic fields near large earthing systems. IEEE Trans. Magn. **32**(3), 1525–1528 (1996)
7. Jankoski, R., Kuhar, A., Grcev, L.: Frequency domain analysis of large grounding systems using hybrid circuit model. In: Proceedings of 8th International Ph.D. Seminar on Computational and Electromagnetic Compatibility - CEMEC 2014, Timisora, Romania, pp. 1–4 (2014)
8. Jankoski, R., Kuhar, A., Markovski, B., Kacarska, M., Grcev, L.: Application of the electric circuit approach in the analysis of grounding conductors. In: Proceedings of 5th International Symposium on Applied Electromagnetics - SAEM 2014, Skopje, Macedonia, pp. 1–7 (2014)
9. Javor, V., Rančić, P.D.: Electromagnetic field in the vicinity of lightning protection rods at a lossyground. IEEE Trans. EMC **51**(2), 320–330 (2009)
10. Kuhar, A., Jankoski, R., Arnautovski-Toseva, V., Ololoska-Gagoska, L., Grcev, L.: Partial element equivalent circuit model for a perfect conductor excited by a current source. CD Proceedings of 11th International Conference on Applied Electromagnetics - PES 2013,Niš, Serbia, pp. 1–4 (2013)
11. Poljak, D.: New numerical approach in the analysis of a thin wire radiating over a lossy half-space. Int. J. Numer. Methods Eng. **38**(22), 3803–3816 (1995)
12. Poljak, D.: Electromagnetic modeling of finite length wires buried in a lossy half-space. Eng. Anal. Bound. Elem. **26**(1), 81–86 (2002)
13. Poljak, D., Doric, V.: Time domain modeling of electromagetic filed coupling to finite length wires embedded in a dielectric half-space. IEEE Trans. EMC **47**(2), 247–253 (2005)

14. Poljak, D., Doric, V.: Wire antenna model for transient analysis of simple grounding systems, Part II: The horizontal grounding electrode. Progress Electromagn. Res. **64**, 167–189 (2006)
15. Poljak, D., Gizdic, I., Roje, V.: Plane wave coupling to finite length cables buried in a lossy ground. Eng. Anal. Bound. Elem. **26**(9), 803–806 (2002)
16. Poljak, D., Doric, V., Drissi, K.E.K., Kerroum, K., Medic, I.: Comparison of wire antenna and modified transmission line approach to the assessment of frequency response of horizontal grounding electrodes. Eng. Anal. Bound. Elem. **32**(8), 676–681 (2008)
17. Poljak, D., Doric, V., Rachidi, F., Drissi, K.E.K., Kerroum, K., Tkatchenko, S., et al.: Generalized form of telegrapher's equations for the electromagnetic field coupling to buried wires of finite length. IEEE Trans. EMC. **51**(2), 331–337 (2009)
18. Poljak, D., Sesnic, S., Goic, R.: Analytical versus boundary element modelling of horizontal ground electrode. Eng. Anal. Bound. Elem. **34**(4), 307–314 (2010)
19. Poljak, D., Drissi, K.E.K., Kerroum, K., Sesnic, S.: Comparison of analytical and boundary element modeling of electromagnetic field coupling to overhead and buried wires. Eng. Anal. Bound. Elem. **35**(3), 555–563 (2011)
20. Poljak, D., Lucic, R., Doric, V., Antonijevic, S.: Frequency domain boundary element versus time domain finite element model for the transient analysis of horizontal grounding electrode. Eng. Anal. Bound. Elem. **35**(3), 375–382 (2011)
21. Popović, B.D., Djurdjević, D.: Entire-domain analysis of thin-wire antennas near or in lossy ground. IEE Proc. Microw. Antennas Propag. **142**(3), 213–219 (1995)
22. Popović, B.D., Petrović, V.V.: Horizontal wire antenna above lossy half-space: simple accurate image solution. Int. J. Numer. Model.: Electron. Netw. Devices Fields **9**(3), 194–199 (1996)
23. Prudnikov, A.P., Brychkov, YuA, Marichev, O.I.: Integrals and Series: Special Functions, vol. 2, p. 189. CRC Press, Florida (1998)
24. Rahmat-Samii, Y., Parhami, P., Mittra, R.: Loaded horizontal antenna over an imperfect ground. IEEE Trans. AP **26**(6), 789–796 (1978)
25. Rančić, M. P., Aleksić, S.: Simple numerical approach in analysis of horizontal dipole antennas above lossy half-space. IJES - Int. J. Emerg. Sci. 1(4), 586–596 [PES 2011, Serbia, 2011, paper no. O4–3] (2011)
26. Rančić, M.P., Aleksić, S.: Simple approximation for accurate analysis of horizontal dipole antenna above a lossy half-space. Proceedings of 10th International Conference on Telecommunications in Modern Satellite. Cable and Broadcasting Services - TELSIKS 2011, pp. 432–435. Niš, Serbia (2011)
27. Rančić, M. P., Aleksić, S.: Horizontal dipole antenna very close to lossy half-space surface. Electrical Review. **7b**, 82–85 (2012) [ISEF 2011, Portugal, 2011, PS.4.19]
28. Rančić, M. P., Aleksić, S.: Analysis of wire antenna structures above lossy homogeneous soil. In: Proc. of 21st Telecommunications Forum (TELFOR), pp. 640–647, Belgrade, Serbia, (2013)
29. Rančić, M.P., Rančić, P.D.: Vertical linear antennas in the presence of a lossy half-space: An improved approximate model. Int. J. Electron. Commun. AEUE **60**(5), 376–386 (2006)
30. Rančić, M.P., Rančić, P.D.: Vertical dipole antenna above a lossy half-space: efficient and accurate two-image approximation for the Sommerfeld integral. In: CD Proceedings of EuCAP'06, paper No121, Nice, France (2006)
31. Rančić, M.P., Rančić, P.D.: Horizontal linear antennas above a lossy half-space: a new model for the Sommerfeld's integral kernel. Int. J. Electron. Commun. AEUE **65**(10), 879–887 (2011)
32. Rančić, M.P., Aleksić, S., Khavanova, M.A., Petrov, R.V., Tatarenko, A.S., Bichurin, M.I.: Analysis of symmetrical dipole antenna on the boundary between air and lossy half-space. Mod. Probl. Sci. Educ. **6**, 1–7 (2012). (in Russian)
33. Sarkar, T.K.: Analysis of arbitrarily oriented thin wire antennas over a plane imperfect ground. Int. J. Electron. Commun. (AEUE) **31**, 449–457 (1977)
34. Siegel, M., King, R.W.P.: Radiation from linear antennas in a dissipative half-space. IEEE Trans. AP **19**(4), 477–485 (1971)
35. Vujević, S., Kurtović, M.: Efficient use of exponential approximation of the kernel function in interpretation of resistivity sounding data. Int. J. Eng. Model. **5**(1–2), 45–52 (1992)

36. Vujević, S., Sarajcev, I.: A new algorithm for exponential approximation of Sommerfeld integrals. In: Proceedings of 11th Int. DAAM Symp. on Intelligent Manufacturing and Automation: Man - Machine - Nature, pp. 485–486, Vienna, Austria, (2000)
37. Zou, J., Zhang, B., Du, X., Lee, J., Ju, M.: High-efficient evaluation of the lightning electromagnetic radiation over a horizontally multilayered conducting ground with a new complex integration path. IEEE Trans. on EMC. **PP(99)**, 1–9 (2014)

Chapter 4
Comparison of TL, Point-Matching and Hybrid Circuit Method Analysis of a Horizontal Dipole Antenna Immersed in Lossy Soil

Radoslav Jankoski, Milica Rančić, Vesna Arnautovski-Toseva and Sergei Silvestrov

Abstract HF analysis of a horizontal dipole antenna buried in lossy ground has been performed in this paper. The soil is treated as a homogenous half-space of known electrical parameters. The authors compare the range of applicability of two forms of transmission line model, a hybrid circuit method, and a point-matching method in such analysis.

Keywords Transmission line model · Point matching method · Hybrid circuit method

4.1 Introduction

Modeling of wire conductors buried in finitely conducting soil has been a subject of great amount of research, [1, 2, 4–10, 13–15]. This problem has been dealt with in different ways, from application of rigorous full-wave approaches to simplified ones more suitable for practical engineering studies. In this paper the authors compare three different concepts.

R. Jankoski (✉) · V. Arnautovski-Toseva
Ss. Cyril and Methodius University, FEIT, Skopje, Macedonia
e-mail: radoslavjankoski@gmail.com

V. Arnautovski-Toseva
e-mail: atvesna@feit.ukim.edu.mk

M. Rančić · S. Silvestrov
Division of Applied Mathematics, The School of Education, Culture and Communication, Mälardalen University, Box 883, 721 23 Västerås, Sweden
e-mail: milica.rancic@mdh.se

S. Silvestrov
e-mail: sergei.silvestrov@mdh.se

S. Silvestrov and M. Rančić (eds.), *Engineering Mathematics I*,
Springer Proceedings in Mathematics & Statistics 178,
DOI 10.1007/978-3-319-42082-0_4

The first one considers the Transmission Line derived for the case of the centrally fed horizontal dipole antenna and buried in lossy ground. Two considered variants take into account the properties of the ground through ground impedances given by Sunde in [13], and Theethayi et al. in [15].

The second one, the integral equation (IE) approach considers solution of the Hallén's IE using the point-matching method, and the entire domain polynomial representation of the current distribution along the observed antenna, [11, 12].

Finally, the modification of the well-known PEEC (Partial Element Equivalent Circuit) method successfully applied in [4, 5], and denoted as the Hybrid Circuit Method (HCM) is used also in this paper. The HCM, unlike PEEC, assumes that capacitive and inductive coupling between different parts of a thin wire conductor can be modelled without cell shifting. This method has been validated against the full wave approach in [4, 5], and will be used here as the reference one.

The analysis is performed in a wide frequency range for different ground conductivities and geometry parameters. Corresponding comments are given in the conclusion.

4.2 Geometry Layout

Let us consider a symmetrical horizontal dipole antenna (HDA) with conductor length l, and cross-section radius a, buried in the lossy half-space (LHS) at depth h, Fig. 4.1. The LHS is idealized as a homogeneous, linear and isotropic medium of known electrical parameters (air: $\sigma_0 = 0$ - conductivity; ε_0 - permittivity; μ_0 - permeability; soil: σ_1 - conductivity; $\varepsilon_1 = \varepsilon_{r1}\varepsilon_0$ - permittivity (ε_{r1} - relative permittivity); $\mu_1 = \mu_0$ - permeability; $\underline{\sigma}_i = \sigma_i + j\omega\varepsilon_1$ - complex conductivity; $\gamma_i = \alpha_i + \mathrm{j}\beta_i = (\mathrm{j}\omega\mu\underline{\sigma}_i)^{1/2}$, $i = 0,\ 1$ - complex propagation constant ($i = 0$ for the air, and $i = 0$ for the LHS); $\omega = 2\pi f$ - angular frequency; $\underline{\varepsilon}_{r1}$ - complex relative permittivity; $\underline{n} = \gamma_1/\gamma_0 = \underline{\varepsilon}_{r1}^{1/2} = (\varepsilon_{r1} - j60\sigma_1\lambda_0)^{1/2}$ - refractive index, and λ_0 - wavelength in the air). The HDA is fed in the center by harmonic voltage generator $U = 1$ V in range 1 kHz–10 MHz.

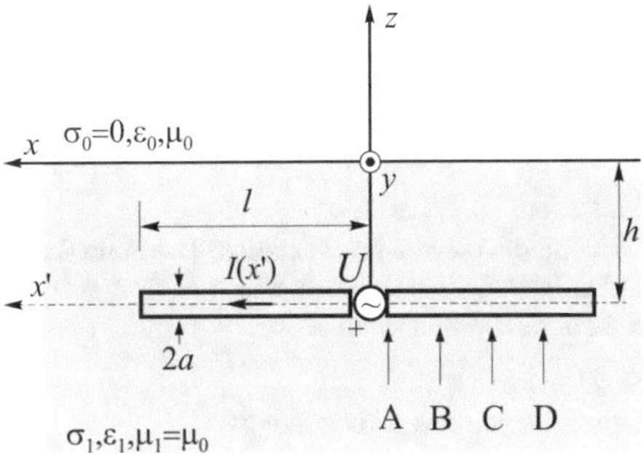

Fig. 4.1 Illustration of a *horizontal conductor* buried in the lossy half-space

4.3 Transmission Line Model (TLM)

The frequency domain formulation of the transmission line equations, [1], is as follows:

$$\frac{\mathrm{d}U(x')}{\mathrm{d}x'} + ZI(x') = 0, \tag{4.1}$$

$$\frac{\mathrm{d}I(x')}{\mathrm{d}x'} + YI(x') = 0, \tag{4.2}$$

where $U(x')$ and $I(x')$ are line voltage and current, Z and Y are ground impedance and admittance, respectively, and $\underline{\gamma}_1^2 = ZY$. Substituting voltage from (4.2) into (4.1), we get

$$\frac{\mathrm{d}^2 I(x')}{\mathrm{d}x'^2} - \underline{\gamma}_1^2 I(x') = 0, \tag{4.3}$$

whose solution presents the current distribution along the HDA

$$I(x') = \begin{cases} \dfrac{U \sinh \underline{\gamma}_1 x'}{Z_0 \sinh \underline{\gamma}_1 L} \sinh \underline{\gamma}_1 l, & -l < x' < 0, \\[2ex] \dfrac{U \sinh \underline{\gamma}_1 l}{Z_0 \sinh \underline{\gamma}_1 L} \sinh \underline{\gamma}_1 (L - x'), & 0 < x' < l, \end{cases} \tag{4.4}$$

where $Z_0 = Z/\underline{\gamma}_1$ - characteristic impedance, and $L = 2l$.

According to Sunde's expression [13], Z can be expressed as:

$$Z^S = \frac{\mathrm{j}\omega\mu_0}{2\pi}\left[K_0(\underline{\gamma}_1 a) - K_0(\underline{\gamma}_1\sqrt{a^2+4h^2}) + 2\int_0^\infty \frac{e^{-2h\sqrt{\alpha^2+\underline{\gamma}_1^2}}\cos\alpha a}{\alpha+\sqrt{\alpha^2+\underline{\gamma}_1^2}}\,\mathrm{d}\alpha\right], \tag{4.5}$$

$K_0(*)$ being the Bessel function of the second kind and order zero. Authors in [15] propose a more simplified formula:

$$Z^{Log} = \frac{\mathrm{j}\omega\mu_0}{2\pi}\ln\left[\frac{1+\underline{\gamma}_1 a}{\underline{\gamma}_1 a}\right]. \tag{4.6}$$

Depending on the chosen expression for Z, two variants of the TLM are obtained: TLM-Sunde and TLM-Log corresponding to (4.5) and (4.6), respectively.

4.4 Point-Matching Method (PMM)

The point-matching method (PMM) is applied to solve the Hallén's integral equation (HIE) that arises from the partial differential equation obtained satisfying the boundary condition for the tangential component of the electric field on the surface of the HDA, [6, 11, 12]. For the case of the HDA centrally-fed by a Dirac's δ-generator of voltage U, and taking into account the simplified form of the Green's function [7–10], the HIE becomes:

$$\int_{-l}^{l} I(x')\left[K_o(x,x') - K_i(x,x') + U_{11}\right]\mathrm{d}x' - C\cos(\mathrm{j}\underline{\gamma}_1 x) = \mathrm{j}\frac{n}{60}U\sin(\mathrm{j}\underline{\gamma}_1 x), \tag{4.7}$$

$$K_o(x,x') = e^{-\underline{\gamma_1} r_o},\ r_o = \sqrt{\rho^2 + a^2},\ \rho = \left|x - x'\right|, \tag{4.8}$$

$$K_i(x,x') = e^{-\underline{\gamma_1} r_i},\ r_i = \sqrt{\rho^2 + (2h)^2},\ \rho = \left|x - x'\right|, \tag{4.9}$$

$$U_{11} = \int_{\alpha=0}^{\infty} \tilde{T}_{\eta 1}(\alpha) e^{-u_1(z+h)} \frac{\alpha}{u_1} J_0(\alpha\rho)\mathrm{d}\alpha, \tag{4.10}$$

$$\tilde{T}_{\eta 1}(\alpha) = \frac{2u_1}{u_0 + u_1},\ u_i = \sqrt{\alpha^2 + \underline{\gamma}_i^2},\ i = 0, 1, \tag{4.11}$$

where $J_0(\alpha\rho)$ is the zero-order Bessel function of the first kind, and C is an integration constant.

Applying the PMM, the HIE is transformed into a system of linear equations with current distribution $I(x')$ and constant C being the unknowns. The entire domain polynomial current approximation is adopted as in [6, 11, 12]:

$$I(u' = x'/l) = \sum_{m=0}^{M} I_m u'^m,\ 0 \le u' \le 1, \tag{4.12}$$

where $I_m, m = 0, 1, ..., M$, are complex current coefficients. The matching is done at $(M+1)$ points that are chosen as $x_i = il/M, i = 0, 1, 2, ..., M$. The $(M+2) - th$ equation needed for evaluation of C is derived from (4.12) with the assumption that the current vanishes at HDA's ends. The system of equations becomes:

$$\sum_{m=0}^{M} I_m \int_{-l}^{l} \left(\frac{x'}{l}\right)^m \left[K_o(x,x') - K_i(x,x') + U_{11}^a(x,x')\right]\mathrm{d}x' - C\cos(\beta_0 \underline{n} x_i) = \tag{4.13}$$

$$= -\mathrm{j}\frac{n}{60}U\sin(\beta_0 \underline{n} x_i),\ i = 0, 1, 2, ..., M,$$

$$\sum_{m=0}^{M} I_m = 0. \tag{4.14}$$

where U_{11}^{a} denotes the approximation of the Sommerfeld integral (4.10), which is here obtained applying three approaches:

1. Modified Image Theory (MIT) approach, models (4.10) by

$$U_{11}^{a} \approx K_i(x,\ x')\left(1-\Gamma_{E}^{ref.}\right), \tag{4.15}$$

   where $\Gamma_{E}^{ref.} = -(\underline{n}^2-1)/(\underline{n}^2+1)$ is the reflection coefficient due to earth/air interface, [2, 8–10, 14].
2. Transmission coefficient approach (TC), [7],

$$U_{11}^{a} \approx K_i(x,\ x')\left(1-\Gamma_{TM}^{trans.}\right), \tag{4.16}$$

   where $\Gamma_{TM}^{trans.} = \frac{2\underline{n}\cos\theta}{\underline{n}^2\cos\theta+\sqrt{\underline{n}^2-\sin^2\theta}}$ - the transmission coefficient for TM polarization, and $\theta = \arctan\frac{\rho}{2h}$.
3. Two-image approximation (TIA), [6], models (4.10) by:

$$U_{11}^{a}(x,x') = \underline{B}K_{zh}(x,x') + \underline{A}e^{\gamma_1|\underline{d}|}K_{zhd}(x,x'), \tag{4.17}$$

   where

$$K_{zh}(x,x') = e^{-\underline{\gamma}_1 r_{zh}},\ r_{zh} = \sqrt{\rho^2+(z+h)^2},$$
$$K_{zhd}(x,x') = e^{-\underline{\gamma}_1 r_{zhd}},\ r_{zhd} = \sqrt{\rho^2+(z+h+|\underline{d}|)^2},$$
$$\underline{B}=1,\ \underline{A} = \frac{1-\sqrt{2-\underline{n}^2}}{1+\sqrt{2-\underline{n}^2}}e^{\underline{\gamma}_0(1-\underline{n})\underline{d}},\ \underline{d}\frac{2}{\underline{\gamma}_0\sqrt{2-\underline{n}^2}}.$$

4.5 Hybrid Circuit Method (HCM)

The HCM will be explained based on a single cell used for representing the observed structure, Fig. 4.2. The following notations are used: V_l and V_m are nodal voltages; J_l and J_m are nodal leakage currents; U_k is the voltage of the k-th segment; I_k is the total leakage current of the k-th segment. The voltage of the k-th segment is assumed to be an average value of node voltages:

$$U_k = \frac{V_l+V_m}{2}. \tag{4.18}$$

Consequently, the following matrix representation of the voltage distribution is obtained:

$$[U] = [Q][V], \tag{4.19}$$

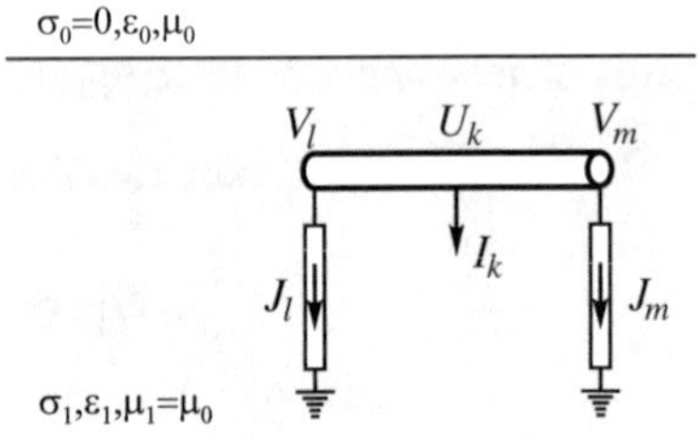

Fig. 4.2 Illustration of a *HCM cell*

where the elements of $[V]$ matrix are node voltages, and the elements of $[Q]$ matrix are evaluated as follows:

$$q_{i,j} = \begin{bmatrix} 1/2, & i - \text{th branch is connected to the } j - \text{th node,} \\ 0, & i - \text{th branch is not connected to the } j - \text{th node.} \end{bmatrix} \quad (4.20)$$

Each leakage current I_k is broken down into two currents $(I_k/2)$. This assumption leads to the following matrix equation describing node leakage currents $[J]$ (with $[I_l]$ - matrix of leakage currents):

$$[J] = [Q]^T[I_l]. \quad (4.21)$$

The inductive coupling between wire segments is represented by self and mutual partial inductances:

$$L_{mn} = \frac{\mu_0 e^{-\underline{\gamma}_1 r_{mn}} \cos\theta_{mn}}{4\pi} \int_{l_m}\int_{l_n} \frac{\mathrm{d}l_m\,\mathrm{d}l_n}{r_{mn}}, \quad (4.22)$$

where $\mathrm{d}l_m$ and $\mathrm{d}l_n$ are elementary lengths of analyzed segments, while the θ_{mn} is the angle between them. Conductive and capacitive coupling is represented by means of complex resistivities evaluated as follows:

$$R_{mn} = \frac{1}{4\pi l_m l_n \underline{\sigma}_1}\left[e^{-\underline{\gamma}_1 r_{rm}} \int_{l_m}\int_{l_n} \frac{\mathrm{d}l_m\,\mathrm{d}l_n}{r_{mn}} + e^{-\underline{\gamma}_1 r''_{rm}} R_t \int_{l_m}\int_{l_n} \frac{\mathrm{d}l_m\,\mathrm{d}l_n}{r''_{mn}}\right], \quad (4.23)$$

$$R_t = \frac{\sigma_1 + \mathrm{j}\omega\varepsilon_0(\varepsilon_{r1} - 1)}{\sigma_1 + \mathrm{j}\omega\varepsilon_0(\varepsilon_{r1} + 1)}, \quad (4.24)$$

r_{rm} is the distance between the m-th and n-th segment while r''_{rm} is the distance between the image of the m-th segment and n-th segment.

Now, the relationship between the voltage and the leakage currents of segments can be written as:

$$[I_l] = [G][U], \quad (4.25)$$

where $[G]$ - matrix of conductances.

Combining (4.19), (4.21), and (4.25) we get the relation between the node leakage currents and node voltages:

$$[J] = [Q]^T [G][Q][V]. \tag{4.26}$$

Using the Modified Nodal Analysis (MNA), [3], a compact form of the Conventional Nodal Analysis (CNA) is obtained:

$$[I_s] - [A]^T [Z]^{-1} [V_s] = [Y][V], \tag{4.27}$$

where $[A]$ is the incidence matrix, $[Z]$ is the longitudinal impedance matrix, $[Y] = [Q]^T [G][Q] + [A]^T [Z]^{-1} [A]$ is the matrix of admittances, and $[I_s]$ and $[V_s]$ are matrices of external current and voltage sources.

Once the voltage distribution is determined based on (4.27), the longitudinal current distribution is obtained as:

$$[I] = [Z]^{-1}([V_s] + [A][V]). \tag{4.28}$$

4.6 Numerical Results

Dependence of the current magnitude at specific points along the HDA arm in the observed frequency range from 1 kHz–10 MHz, is presented in Figs. 4.3, 4.4 and 4.5. The HDA's conductors have a cross-section radius of $a = 0.001$ m, and the antenna is buried at depth $h = 0.5$ m. Considered electrical parameters of the ground are: electrical permittivity $\varepsilon_{r1} = 10$, and specific conductivities a) $\sigma_1 = 0.001$ S/m, b) $\sigma_1 = 0.01$ S/m, and c) $\sigma_1 = 0.1$ S/m. Current is calculated at four different points along one arm of the HDA (A, B, C and D from Fig. 4.1).

Figures 4.3a, 4.4a and 4.5a illustrate the dependence of the current magnitude on frequency for a case of a short dipole $l = 10$ m, while corresponding results obtained for a long HDA ($l = 50$ m) are illustrated in Figs. 4.3b, 4.4b and 4.5b.

We have compared results obtained by methods described in previous sections, which are denoted as:

- TLM-Sunde - transmission line model, expression (4.5);
- TLM-Log - transmission line model, expression (4.6);
- PMM-MIT - point-matching method, MIT approach;
- PMM-TC - point-matching method, TC approach;
- PMM-TIA - point-matching method, TIA approach;
- HCM - hybrid circuit method.

For the sake of evaluating limitations of mentioned methods, the results obtained by a full-wave approach are also presented in the figures.

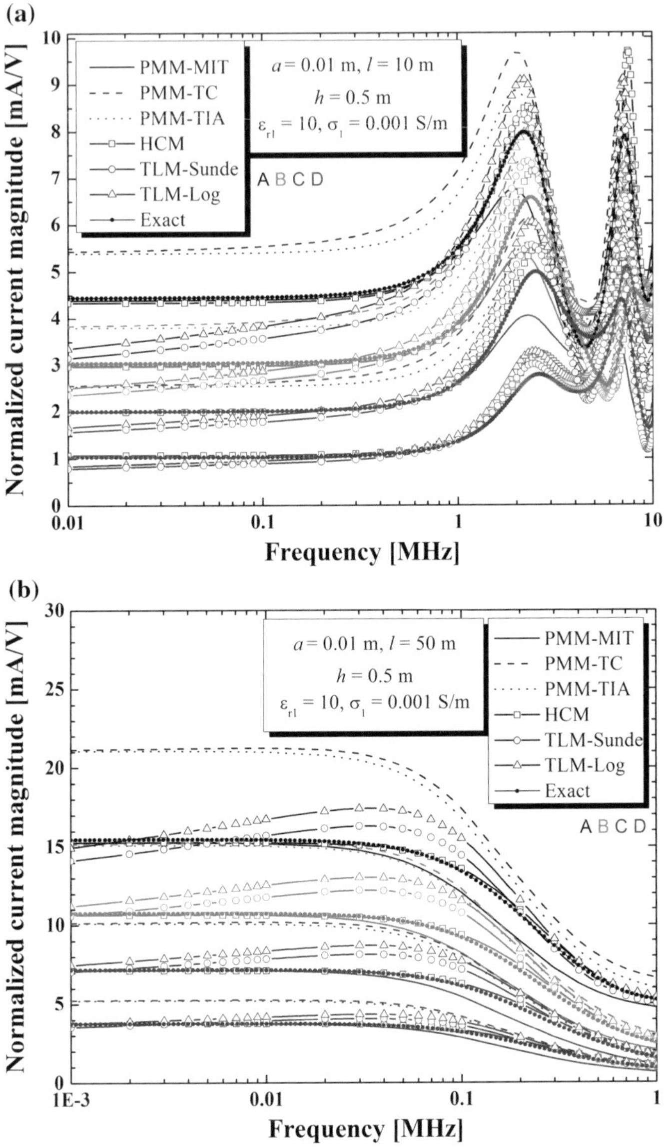

Fig. 4.3 Current magnitude at different points along an arm of a **a** short and **b** long HDA versus frequency for ground conductivity of 0.001 S/m. Comparison of different methods

The comparison of the same methods, for a case of more deeply buried HDA ($h = 5.0$ m) is given in Table 4.1. The results correspond to current magnitude calculated at point A (feeding point of the HDA) of HDA's arm with length of $l = 10$ m.

Table 4.1 Current magnitude at feeding point of a short HDA versus frequency for different ground conductivities. HDA is buried at $h = 0.5$ m

σ_1 (S/m)	f (MHz)	Current magnitude (mA)					
		TLM-Sunde	TLM-Log	PMM-MIT	PMM-TC	PMM-TIA	HCM
0.001	0.01	3.161	3.356	5.300	5.393	5.390	5.207
	0.1	3.613	3.833	5.309	5.411	5.401	5.219
	1	5.391	5.624	6.438	6.673	6.621	6.438
	2	8.766	8.964	8.437	8.810	8.632	8.677
	5	2.688	2.651	3.906	4.081	4.122	3.935
	10	3.128	3.286	5.573	5.487	5.421	4.985
0.01	0.01	36.037	38.224	52.931	53.855	53.850	52.036
	0.1	40.734	42.489	49.804	50.896	50.884	49.463
	1	19.050	19.333	21.808	21.671	21.656	21.169
	2	13.833	14.054	16.164	16.122	16.130	15.652
	5	9.706	9.922	11.519	11.524	11.525	11.025
	10	7.747	7.924	9.339	9.338	9.338	8.824
0.1	0.01	406.829	424.353	497.472	507.974	508.195	494.045
	0.1	190.461	193.416	218.053	216.921	216.736	211.808
	1	72.102	73.753	84.638	84.640	84.640	81.365
	2	54.806	56.122	62.652	62.652	62.652	61.393
	5	38.453	39.439	41.807	41.807	41.807	42.256
	10	29.634	30.428	31.511	31.511	31.507	31.776

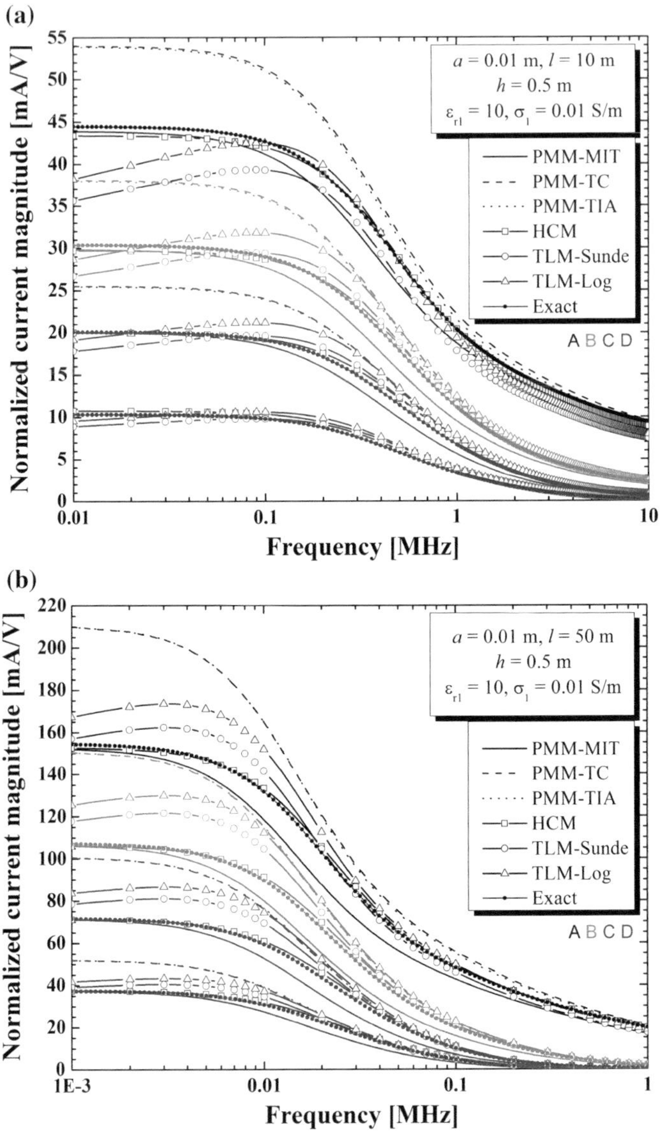

Fig. 4.4 Current magnitude at different points along an arm of a **a** short and **b** long HDA versus frequency for ground conductivity of 0.01 S/m. Comparison of different methods

4.7 Conclusion

Exact modeling of wire conductors buried in lossy soil at arbitrary depth calls for tedious numerical integration of Sommerfeld's integrals, which appear in the

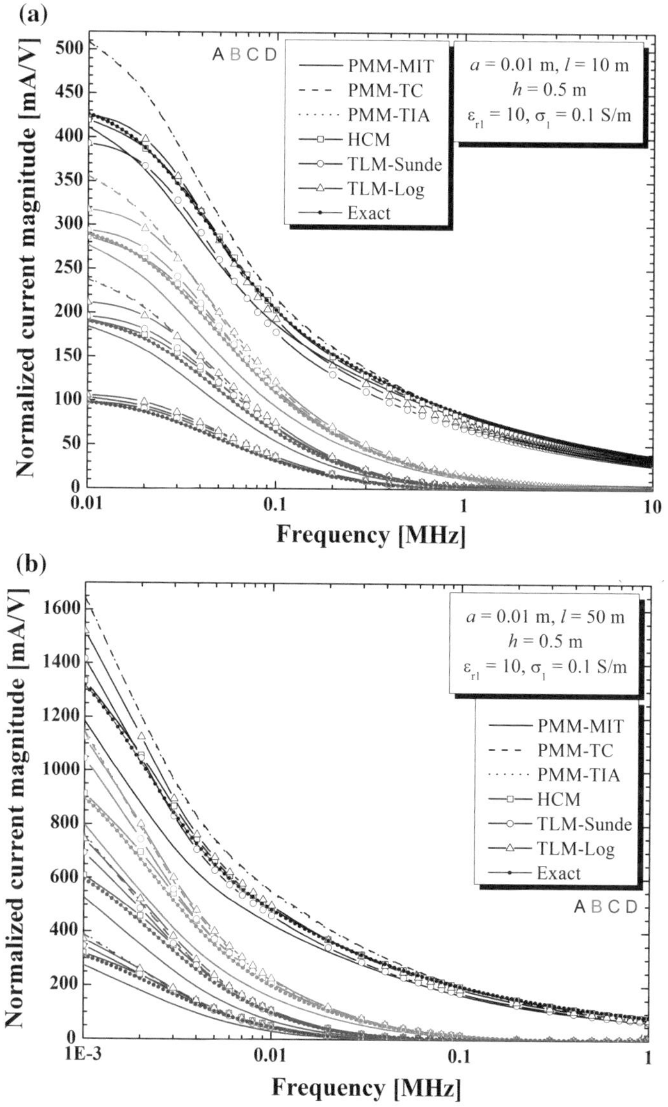

Fig. 4.5 Current magnitude at different points along an arm of a **a** short and **b** long HDA versus frequency for ground conductivity of 0.1 S/m. Comparison of different methods

expression describing the electromagnetic field in its surroundings. In that sense, an approximate approach that can be applied in the analysis of such structures in desired frequency spectrum, and for wide range of ground and geometry parameters, is welcomed. This paper deals with three such methods used for HDA analysis.

Observed behavior in the considered frequency range can be summarized as follows:

- The biggest difference between the methods can be observed at the HDA's feeding point A regardless of the antenna length, burial depth, or conductivity of the ground. The best accordance with the exact solution is obtained by the HCM method at all points along the HDA.
- For the lower frequency spectrum, the HCM method best agrees with the PMM-MIT approach for the lowest conductivity value (Fig. 4.3), but with its increase becomes closer to PMM-TC and PMM-TIA values, especially for higher frequencies regardless of HDA length. For $h = 5$ m, differences between the HCM and all variants of PMM method are barely noticeable.
- Both TL models are in the best accordance with the reference HCM results for the cases of higher ground conductivities, longer antenna, and higher range of frequencies (Figs. 4.4b and 4.5b). The same characteristics can be also observed for more deeply buried HDA (Table 4.1).

Some improvements are possible for the PMM approach, taking into account the actual Green's function ((4.19) in [7]) instead of the simplified one used in this paper, which will be explored in our future work.

References

1. Arnautovski-Toseva, V., Drissi, K. E. K., Kerroum, K.: HF comparison of image and TL models of a horizontal thin-wire conductor in finitely conductive earth. In: Proceedings of SOFTCOM 2011, pp. 1–5. Split, Croatia (2011)
2. Grcev, L.D., Menter, F.E.: Transient electromagnetic fields near large earthing systems. IEEE Trans. Magn. **32**(3), 1525–1528 (1996)
3. Ho, C.W., Ruehli, A.E., Brennan, P.A.: The modified nodal approach to network analysis. IEEE Trans. on Circuits Syst. **14**(3), 873–878 (1975)
4. Jankoski, R., Kuhar, A., Markovski, B., Kacarska, M., Grcev, L.: Application of the electric circuit approach in the analysis of grounding conductors. In: Proceedings of 5th International Symposium on Applied Electromagnetics - SAEM 2014, pp. 1–7. Skopje, Macedonia (2014)
5. Jankoski, R., Kuhar, A., Grcev, L.: Frequency domain analysis of large grounding systems using hybrid circuit model. In: Proceedings of 8th International PhD Seminar on Computational and Electromagnetic Compatibility - CEMEC 2014, pp. 1–4. Timisora, Romania (2014)
6. Monsefi, F., Jankoski, R., Rančić, M., Silvestrov, S.: Evaluating parameters of approximate functions for representation of Sommerfeld integrals. Paper Presented at 16th Conference of the Applied Stochastic Models and Data Analysis - ASMDA 2015, Piraeus, Greece (2015)
7. Poljak, D.: Electromagnetic modeling of finite length wires buried in a lossy half-space. Eng. Anal. with Bound. Elem. **26**(1), 81–86 (2002)
8. Poljak, D., Gizdic, I., Roje, V.: Plane wave coupling to finite length cables buried in a lossy ground. Eng. Anal. with Bound. Elem. **26**(9), 803–806 (2002)
9. Poljak, D., Sesnic, S., Goic, R.: Analytical versus boundary element modelling of horizontal ground electrode. Eng. Anal. with Bound. Elem. **34**(4), 307–314 (2010)
10. Poljak, D., Drissi, K.E.K., Kerroum, K., Sesnic, S.: Comparison of analytical and boundary element modeling of electromagnetic field coupling to overhead and buried wires. Eng. Anal. with Bound. Elem. **35**(3), 555–563 (2011)

11. Rančić, M.P., Rančić, P.D.: Vertical linear antennas in the presence of a lossy half-space: An improved approximate model. Int. J. Electron. Commun. AEUE. **60**(5), 376–386 (2006)
12. Rančić, M.P., Rančić, P.D.: Horizontal linear antennas above a lossy half-space: A new model for the Sommerfeld's integral kernel. Int. J. El. Commun. AEUE. **65**(10), 879–887 (2011)
13. Sunde, E.D.: Earth Conduction Effects in Transmission Systems. Dover publication, New York (1968)
14. Takashima, T., Nakae, T., Ishibasi, R.: Calculation of complex fields in conducting media. IEEE Trans. on Electr. Insul. **EI–15**(1), 1–7 (1980)
15. Theethayi, N., Thottappillil, R., Paolone, M., Nucci, C., Rachidi, F.: External impedance and admittance of buried horizontal wires for transient studies using transmission line analysis. IEEE Trans. Dielectr. Electr. Insul. **14**(3), 751–761 (2007)

Chapter 5
Theoretical Study of Equilateral Triangular Microstrip Antenna and Its Arrays

Pushpanjali G. Metri

Abstract Novel design of equilateral triangular microstrip antenna is proposed at X-band frequency. The antenna is designed, fabricated and tested for single and multiband operation. This study presents the theory developed with respect to the experimental work carried out on design and development of equilateral triangular microstrip array antenna (ETMAA). The experimental impedance bandwidth of single element conventional equilateral triangular microstrip antenna (CETMA) is found to be 5.02 %. The two, four and eight elements of ETMAA have been designed and fabricated using low cost glass epoxy substrate material. The array elements are excited using corporate feed technique. The effect of slot in enhancing the impedance bandwidth is studied by placing the slot in the radiating elements of ETMAA. The study is also made by exciting the array element of ETMAA through aperture coupling. For eight elements, maximum 33.8 % impedance bandwidth is achieved, which is 6.73 times more than the impedance bandwidth of conventional single element CETMA. The experimental impedance bandwidths are verified theoretically and they are in good agreement. The obtained experimental results and theoretical study of the proposed antennas are given and discussed in detail.

Keywords Theoretical study of ETMAA · Gap-coupled feeding technique · Slot loading technique

5.1 Introduction

A microstrip or patch antenna is a low-profile antenna that has a number of advantages over other antennas: it is light weight, low volume, low profile, planar configurations which can be made conformal, low fabrication cost, readily amenable to mass production and electronics like LNA's and SSPA's, and can be integrated with these antennas quite easily [27]. While the antenna can be a 3-D structure (wrapped

P.G. Metri (✉)
Department of Physics, Sangameshwar College, Solapur, Maharashtra, India
e-mail: pushpa22metri@gmail.com

S. Silvestrov and M. Rančić (eds.), *Engineering Mathematics I*,
Springer Proceedings in Mathematics & Statistics 178,
DOI 10.1007/978-3-319-42082-0_5

around a cylinder, for example), it is usually flat and that is why patch antennas are sometimes referred to as planar antennas.

Global demand for voice, data and video related services continues to grow faster than the required infrastructure can be deployed. Despite huge amount of money that has been spent in attempts to meet the need of the world market, the vast majority of people on Earth still do not have access to quality communication facilities. The greatest challenge faced by governments and service providers is the "last-mile" connection, which is the final link between the individual user or business users and worldwide network [35]. Copper wires, traditional means of providing this "last-mile" connection is both costly and inadequate to meet the needs of the bandwidth intensive applications. Coaxial cable and power line communications all have technical limitations. And fiber optics, while technically superior but is extremely expensive to install to every home or business user. To overcome this wireless connection is being seen as an alternative to quickly and cost effectively meet the need for flexible broadband links.

Basically an antenna can be considered as the connecting link between free-space and transmitter or receiver [2]. Presently there exist different types of antennas. The design and development of microwave antennas are the most important task in microwave communication systems to achieve the desired radiation requirements. Among the various types of microwave antennas, the microstrip antennas (MSAs) have found one of the important classes within the broad field of microwave antennas because of their diversified applications in microwave communication.

Microstrip antenna technology has been the most rapidly developing topic in the antenna field receiving the creative attentions of academic, industrial and professional engineers and researchers throughout the world [29]. As the microstrip antenna is planar in configuration, it enjoys all the advantages of printed circuit technology. A microstrip antenna in its simplest form consists of a radiating patch, power dividers, matching networks and phasing circuits photoetched on one side of the dielectric substrate board. The other side of the board is a metallic ground plane [3].

G.A. Deschamps first proposed the concept of the microstrip antenna in 1953 [13]. However, practical antennas were developed by Robert E. Munson [26] and John Q. Howell [16] in the 1970s. As a result, microstrip antennas have quickly evolved from academic novelty to commercial reality, with applications in a wide variety of microwave systems.

The radiating patch of microstrip antenna can be of any geometry viz: rectangular, triangular, circular, square, elliptical, sectoral, annular ring etc. Among the various types of microstrip antenna configurations, the rectangular geometry is commonly used. But, one of the most attractive features of the equilateral triangular microstrip antenna is that, the area necessary for the patch becomes about half as large as that of a nearly rectangular or square microstrip antenna designed for the same frequency [36].

Microstrip antennas despite their potential advantages also have some drawbacks compared to conventional microwave antennas. One of the major drawbacks is its narrow impedance bandwidth i.e. 1–2 %. Increasing the impedance bandwidth of microstrip antennas has become an important task and is the major thrust of research in microwave communication. Various techniques have been reported in the literature for enhancing the impedance bandwidth of microstrip antennas. But equilateral triangular microstrip antenna found handful of investigations. The literature study shows that, there is still void and hence requires further investigation to enhance its impedance bandwidth.

In the present investigation, work is mainly concentrated on the enhancement of impedance bandwidth of equilateral triangular microstrip array antennas using:

i. Corporate feed technique,
ii. Slot-loading technique,
iii. Aperture-coupled feeding technique, and
iv. Gap-coupled technique.

The conventional equilateral triangular microstrip antenna (CETMA) is designed and fabricated using glass-epoxy substrate material. The two, four and eight elements equilateral triangular microstrip array antennas (ETMAAs) are constructed using the same substrate material. The antenna elements are excited through corporate feed technique. The change in impedance bandwidth is studied for these antennas. Further, by loading slot in the radiating elements, the effect of slot in enhancing the impedance bandwidth is studied. The elements of slot loaded antennas are excited through aperture coupled feeding. The comparative study of impedance bandwidth is made between corporate fed slot loaded and aperture coupled equilateral triangular microstrip array antennas. The impedance bandwidths of these antennas are studied comparatively.

5.2 Types of Microstrip Antennas

The approaching maturity of microstrip antenna technology, coupled with the increasing demand and applications these antennas are mainly classified into three basic categories [3]:

1. Microstrip patch antennas,
2. Microstrip traveling wave antennas,
3. Microstrip slot antennas.

The present study is carried out experimentally for microstrip patch antennas and microstrip slot antennas. Theory for the proposed antennas is also developed for validation of experimental results of the antennas. This chapter presents the theoretical calculation of impedance bandwidth of conventional equilateral triangular microstrip antenna (CETMA) and equilateral triangular microstrip array antennas (ETMAAs). The impedance bandwidth is determined separately for corporate and aperture coupled fed ETMAAs.

Several techniques have been developed and found quite useful to enhance the impedance bandwidth of microstrip antennas (MSAs), such as use of stacked technique [5–7, 22], use of corporate feed technique [12, 15], use of parasitic elements [1], use of thick dielectric substrate [4, 6, 12], use of additional resonator [18], use of aperture-coupled technique [11, 28], use of electromagnetically coupled technique [17], use of L-shaped probe [25], use of a resonant slot inserted in the main patch [21, 23, 31, 33], etc.

Various theoretical analyses are available in the literature to validate the experimental impedance bandwidth of MSAs. Joseph Helszajn and David S. James [14] have described theoretical and experimental results on planar resonators of equilateral triangular resonators having magnetic sidewalls. The TM fields in such resonators with magnetic boundary conditions are obtained by duality from the TE modes with electric boundaries. The theoretical description includes the cutoff numbers of the first few modes. The performance of a microstrip circulator using a triangular resonator is also described by them.

Girish Kumar et al. [18] have verified the theoretical bandwidth of MSA consisting of additional resonator gap-coupled to the radiating edges of resonant patch. They have verified the theoretical bandwidth using Green's function approach and the segmentation method.

Kai Fong Lee et al. [20] have analyzed the equilateral triangular patch antenna by means of the cavity model. They have given the theoretical formulas and the characteristics obtained from the theory including radiation patterns, percentages of power radiated, total Q factors, input impedances and their variations with feed position. They opined that, the equilateral triangular patch can be designed to function as a triple frequency antenna.

Wei Chen et al. [10] have presented a critical study of the resonant frequency of the equilateral triangular patch antenna. They compared their results with experimental results reported by other scientists using the moment method and the Gang's hypothesis analysis.

Qing Song and Xue-Xia Zhang [34] have verified the theoretical bandwidth of two element gap-coupled microstrip array antenna by using the spectral dyadic Green's function for a grounded dielectric slab and the moment method.

Girish Kumar and K.P. Ray [19] have given the expressions for calculating the percentage impedance bandwidth of rectangular microstrip antenna (RMSA). In this study theoretical determination of impedance bandwidth of ETMA and ETMAAs has been made with the help of the equations given by Girish Kumar and K.P. Ray [8: pp.13] by replacing [3, 8] W/L ratio with $(n \times Se)$ where, S_e is the effective side length of the equilateral triangular radiating patch and n is the number of equilateral triangular radiating patches. The impedance bandwidth of single element conventional equilateral triangular microstrip antenna (CETMA) is determined by this method.

The equations given by Girish Kumar and K.P. Ray have been extended to determine the impedance bandwidth of ETMAAs. The extended equations are applied to determine the impedance bandwidth of two, four and eight element ETMAAs. As the number of array elements increases in ETMAAs the multiplying factor S_e also increases accordingly.

The theoretical impedance bandwidth of ETMA and ETMAAs computed on the basis of the above theory is compared to the experimental impedance bandwidth for the validation.

5.2.1 Theoretical Impedance Bandwidth

The conventional equilateral triangular microstrip antenna (CETMA) and equilateral triangular microstrip array antennas (ETMAAs) have been designed for TE10 mode.

The expression derived by Girish Kumar and K.P. Ray [19] for the calculation of percentage impedance bandwidth are in terms of patch dimensions and substrate parameters. The expressions are, [19],

$$Impedence\ bandwidth(\%) = \left[\frac{A \times h}{\lambda_0 \sqrt{\varepsilon_r}}\right] \times \sqrt{\frac{W}{L}}, \tag{5.1}$$

where:
h - thickness of the substrate,
ε_r - relative permittivity of the substrate,
W - width of the patch,
L - length of the patch,
λ_0 - free-space wavelength,
A - correction factor.

The correction factor A changes as the value of $\left[\frac{A \times h}{\lambda_0 \sqrt{\varepsilon_r}}\right]$ changes [19], which is given by:

$$A = 180 \text{ for } \left[\frac{A \times h}{\lambda_0 \sqrt{\varepsilon_r}}\right] \leq 0.045,$$
$$A = 200 \text{ for } 0.045 \leq \left[\frac{A \times h}{\lambda_0 \sqrt{\varepsilon_r}}\right] \leq 0.075,$$
$$A = 220 \text{ for } \left[\frac{A \times h}{\lambda_0 \sqrt{\varepsilon_r}}\right] \geq 0.075.$$

In the present investigation the value of correction factor A is taken as 180 because the calculated value of $\left[\frac{A \times h}{\lambda_0 \sqrt{\varepsilon_r}}\right]$ for CETMA and ETMAAs is less than 0.045 determined for the known value of h, λ_0 and ε_r.

The expression (5.1) given by Girish Kumar et al. [19] is for RMSA. But in the present study the geometry of radiating elements are equilateral triangular in shape. Therefore the Eq. (5.1) is converted for equilateral triangular microstrip antenna and arrays by replacing [3, 14, 19] the W/L ratio with $(n \times S_e)$. The modified equation for CETMA is given by

$$Impedence\ bandwidth(\%) = \left[\frac{A \times h}{\lambda_0 \sqrt{\varepsilon_r}}\right] \times \sqrt{n \times S_e}, \tag{5.2}$$

where:
S_e - effective side length of the equilateral triangular radiating patch, and
n - number of equilateral triangular radiating patches.

In Eq. (5.2) the value of S_e is given by the formula, [19],

$$S_e = S + \frac{4h}{\varepsilon_e}, \tag{5.3}$$

where:
S - side length of the equilateral triangular microstrip patch,
ε_e - effective dielectric constant.

The value of S and ε_r are determined using the following Eqs. (5.4), [30], and (5.5), respectively.

$$S = \left(\frac{S_{eff1} + S_{eff2}}{2}\right), \tag{5.4}$$

$$\varepsilon_e = \frac{\varepsilon_r + 1}{2} + \frac{\varepsilon_r - 1}{2}\left(1 + 12\frac{h}{s}\right)^{-\frac{1}{2}}. \tag{5.5}$$

The Eq. (5.2) is extended to calculate the impedance bandwidth of corporate fed two, four and eight element array antennas. Further, (5.2) is also used to determine the impedance bandwidth of aperture-coupled fed four and eight element array antennas by considering the aperture coupled parameters. During the calculations of percentage impedance bandwidth of array elements the patch dimensions [19] in terms of area i.e. the area of the equilateral triangular radiating patch (A_t), area of slot loaded equilateral triangular microstrip radiating patch (A_{sp}) and capacitance of the slot (C_s) are taken into consideration. The capacitance of the slot C_s is calculated with the help of the transmission line model [3]. This analytical technique [19] is based on equivalent magnetic current distribution around the patch edges (similar to slot antennas).

5.2.1.1 Calculation of Impedance Bandwidth of CETMA

The CETMA is fed by 50 microstripline, which is connected at the center point C_p of the side length of the equilateral triangular microstrip patch. Between the equilateral triangular microstrip patch and 50 feed line a matching transform is used to avoid the mismatch. For CETMA, n is taken as 1 in Eq. (5.2) as CETMA consists of only one radiating element. Hence Eq. (5.2) reduces to,

$$Impedence\ bandwidth(\%) = \left[\frac{A \times h}{\lambda_0\sqrt{\varepsilon_r}}\right]\sqrt{S_e}. \tag{5.6}$$

Therefore, the theoretical impedance bandwidth of CETMA is calculated using the above equation which is found to be 5.35 %. This impedance bandwidth is recorded in Table 5.1. From this table it is seen that, the theoretical impedance bandwidth is in close agreement with the experimental value.

5.2.1.2 Calculation of Impedance Bandwidth of T-ETMAA

The two element equilateral triangular microstrip array antenna i.e. T-ETMAA is fed by corporate feed arrangement. The area of equilateral triangular radiating element (A_t) is taken into consideration [19] for calculating the impedance bandwidth of T-ETMAA. The value of A_t is determined using the basic formula of equilateral triangular element, which is given by

$$A_t = \frac{\sqrt{3}}{4} S^2, \tag{5.7}$$

where S is the side length of the equilateral triangular radiating element. The value of A_t is multiplied to Eq. (5.2) to find impedance bandwidth of T-ETMAA.

The value of n is taken as 2 in Eq. (5.2) as T-ETMAA consists of two radiating elements. Hence the extended formula for the determination of impedance bandwidth of T-ETMAA is given by

$$Impedence\ bandwidth(\%) = \left[\frac{A \times h}{\lambda_0 \sqrt{\varepsilon_r}}\right] \times (\sqrt{2 \times S_e}) A_t. \tag{5.8}$$

Table 5.1 shows the theoretical and experimental impedance bandwidths of T-ETMAA.

5.2.1.3 Calculation of Impedance Bandwidth of TS-ETMAA

The rectangular slots are loaded at the center of the radiating elements of two element slot-loaded equilateral triangular microstrip array antenna i.e. TS-ETMAA. Therefore the area of the slot loaded patch (A_{sp}) [19] and capacitance of the slot (C_s) [3] are considered during the calculation of impedance bandwidth of TS-ETMAA. Slot also resonates along with patch, which enhances the impedance bandwidth. The capacitance parameter C_s associated to the slot is responsible for its resonance. This C_s is evaluated using the transmission line model [3]. According to transmission line model, the C_s is given by

$$C_s = \frac{\Delta l \sqrt{\varepsilon_{eff}}}{c \times Z_0}, \tag{5.9}$$

where Δl is the extension length and ε_{eff} is the effective dielectric constant. The value of Δl and ε_{eff} are evaluated from (5.10) and (5.11), respectively. For these calculations ε_e is taken from (5.5).

$$\Delta l = 0.412h \left[\frac{(\varepsilon_e + 0.3)\left(\frac{W}{h} + 0.264\right)}{(\varepsilon_e - 0.258)\left(\frac{W}{h}\right) + 0.8}\right], \tag{5.10}$$

$$\varepsilon_{eff} = \varepsilon_r - \frac{\varepsilon_r - \varepsilon_e}{1 + G\left(\frac{f_r}{f_p}\right)^2}, \tag{5.11}$$

where

$$G = \left(\frac{Z_0 - 5}{60}\right)^{\frac{1}{2}} + (0.004 \times Z_0), \tag{5.12}$$

$$f_p = \frac{Z_0}{2\mu h}, \tag{5.13}$$

$$\mu_0 = 4\pi\, 10^{-9}, \tag{5.14}$$

$$Impedence\ \ bandwidth(\%) = \left[\left(\frac{A \times h}{\lambda_0 \sqrt{\varepsilon_r}}\right) \times \sqrt{2 \times S_e \times A_{sp}}\right] + C_s, \tag{5.15}$$

where:
A_{sp} - area of the slot loaded patch excluding the area of rectangular slot, and
C_s - capacitance of the slot.

Hence the value of A_{sp} in (5.15) is calculated by the formula

$$A_{sp} = A_t - A_s. \tag{5.16}$$

In the above equation the value of A_t is calculated with the help of (5.7) and A_s is the area of rectangular slot which is given by

$$A_s = L_s \times W_s, \tag{5.17}$$

where:
L_s - length of the rectangular slot,
W_s - width of the rectangular slot.

The impedance bandwidth of TS-ETMAA is calculated using (5.15) and is recorded in Table 5.1.

5.2.1.4 Calculation of Impedance Bandwidth of FS-ETMAA

The radiating elements of TS-ETMAA are increased from two to four to construct four element slot-loaded equilateral triangular microstrip array antenna i.e. FS-ETMAA. The value of n is taken as 4 in (5.2) for FS-ETMAA as FS-ETMAA consists of four radiating elements and a slot at their centre. The total capacitance effect caused by slots in two radiating elements of FS-ETMAA is minimized by the capacitance effect of slots produced by the remaining two elements of FS-ETMAA. The slot in the elements acts as series capacitances and hence C_s decreases. The two set of elements in FS-ETMAA are resonating independently and gives two operating

Table 5.1 Verification of impedance bandwidth of corporate fed and aperture-coupled fed ETMA and ETMAAs

Antennas	Impedence bandwidth (%)		Error(%)
	Theoretical	Experimental	
CETMA	5.35	5.02	6.57
T-ETMAA	4.76	4.50	5.77
TS-ETMAA	7.12	7.35	3.12
FS-ETMAA	6.62	6.68	0.89
F-ETMAA	9.44	9.11	3.62
ES-ETMAA	9.96	10.20	2.35
FA-ETMAA	13.47	13.75	2.03
FAS-ETMAA	22.98	23.74	3.20
EAS-ETMAA	32.50	33.80	3.84

bands [30]. Therefore the total C_s due to slot in FS-ETMAA is subtracted as shown in the following Eq. (5.18)

$$Impedance\ bandwidth(\%) = \left[\left(\frac{A \times h}{\lambda_0\sqrt{\varepsilon_r}}\right) \times \sqrt{4 \times S_e \times A_{sp}}\right] - C_s. \quad (5.18)$$

The values of A_{sp} and C_s in (5.18) are determined with the help of (5.16) and (5.9), respectively. The obtained theoretical impedance bandwidth of FS-ETMAA is tabulated in Table 5.1.

5.2.1.5 Calculation of Impedance Bandwidth of F-ETMAA

The F-ETMAA is the extension of T-ETMAA. The radiating elements of T-ETMAA are increased from two to four to construct four element equilateral triangular microstrip array antenna i.e. F-ETMAA. The (5.8) used for the calculation of impedance bandwidth of T-ETMAA, is also used here for the impedance bandwidth calculation of F-ETMAA. But the value of n is taken as 4 in this case (i.e. 2 is replaced by 4) as F-ETMAA consists of four radiating elements. From the experimental results of the proposed antennas [30], it is clear that the antenna is resonating for four bands of frequencies. This indicates that each element of F-ETMAA is resonating independently. The coupling effect caused by the total radiating area (A_t) of the two adjacent elements in F-ETMAA if subtracted as shown in the following Eq. (5.19), and the obtained overall impedance bandwidth now becomes equal to the experimental impedance bandwidth of F-ETMAA. Hence (5.8) becomes

$$Impedence\ bandwidth(\%) = \left[\left(\frac{A \times h}{\lambda_0\sqrt{\varepsilon_r}}\right) \times \sqrt{4 \times S_e}\right] - 2A_t. \quad (5.19)$$

The value of A_t is calculated using (5.7). The impedance bandwidth of F-ETMAA obtained from (5.19) is recorded in Table 5.1.

5.2.1.6 Calculation of Impedance Bandwidth of ES-ETMAA

The eight element slot-loaded equilateral triangular microstrip array antenna i.e. ES-ETMAA is the extension of FS-ETMAA. The number of radiating elements is increased from four to eight. The value of n is taken as 8 in (5.18) (i.e. 4 is replaced by 8) for ES-ETMAA as this antenna consists eight radiating elements. By comparing the graphs of the experimental results of the said antennas [30], it is clear that, ES-ETMAA resonates for four bands of frequencies by increasing elements from four to eight. But the overall impedance bandwidth is more in this case when compared to FS-ETMAA. The capacitance effect due to slot is similar in this case also as explained in Sect. 5.2.1.4. The equation used to determine the impedance bandwidth of ES-ETMAA is given by

$$Impedance\ bandwidth(\%) = \left[\left(\frac{A \times h}{\lambda_0 \sqrt{\varepsilon_r}}\right) \times \sqrt{8 \times S_e \times A_{sp}}\right] - C_s. \quad (5.20)$$

The value of A_{sp} is calculated using (5.16) and the value of C_s is calculated with the help of (5.9). The impedance bandwidth of ES-ETMAA is determined using (5.20) and is recorded in Table 5.1.

5.2.1.7 Calculation of Impedance Bandwidth of FA-ETMAA

The F-ETMAA is fed by aperture-coupling to construct four element aperture-coupled equilateral triangular microstrip array antenna i.e. FA-ETMAA. The radiating elements of FA-ETMAA are excited through coupling slots. The coupling slot resonates nearer to the patch resonance [33]. The total area A_t of radiating elements becomes virtually twice the actual area. Hence the basic Eq. (5.2), if multiplied by $2A_t$, now predicts the impedance bandwidth of FA-ETMAA. Hence (5.2) becomes

$$Impedence\ bandwidth(\%) = \left[\left(\frac{A \times h}{\lambda_0 \sqrt{\varepsilon_r}}\right) \times (\sqrt{4 \times S_e})\right] \times (2A_t). \quad (5.21)$$

The value of n in (5.2) is taken as 4 as FA-ETMAA consists of four radiating elements. The value of A_t is calculated using (5.7). The impedance bandwidth of FA-ETMAA is calculated using (5.21) and is given in Table 5.1.

5.2.1.8 Calculation of Impedance Bandwidth of FAS-ETMAA

The four element aperture-coupled slot-loaded equilateral triangular microstrip array antenna i.e. FAS-ETMAA is the extension of FS-ETMAA. The radiating elements of FS-ETMAA are fed by aperture-coupling to construct FAS-ETMAA. The equation used for calculating the impedance bandwidth of FS-ETMAA is taken in this case to determine the impedance bandwidth of FAS-ETMAA. But in FAS-ETMAA the coupling slots are kept exactly below the slot etched in the radiating elements of FAS-ETMAA separated by a substrate material. The coupling slot and slot in the radiating elements are resonating independently [8]. Capacitance C_s associates for each slot is in parallel, and hence C_s doubles the actual value. Therefore, $2C_s$ is multiplied to modified equation of (5.2) as in (5.18) to now predict the impedance bandwidth of FAS-ETMAA. Hence, the basic Eq. (5.6) for FAS-ETMAA becomes,

$$Impedance\ \ bandwidth(\%) = \left[\left(\frac{A \times h}{\lambda_0 \sqrt{\varepsilon_r}}\right) \times (\sqrt{4 \times S_e \times A_{sp}})\right] \times 2C_s. \quad (5.22)$$

The impedance bandwidth of FAS-ETMAA is calculated using the above formula (5.22) and is tabulated in Table 5.1.

5.2.1.9 Calculation of Impedance Bandwidth of EAS-ETMAA

The eight element aperture-coupled slot-loaded equilateral triangular microstrip array antenna i.e. EAS-ETMAA is the extension of ES-ETMAA. The radiating elements of ES-ETMAA are fed by aperture-coupling to construct EAS-ETMAA. The impedance bandwidth of this antenna is calculated by using (5.22). But the value of n is taken as 8 (i.e. 4 is replaced by 8 in (5.22)) as EAS-ETMAA consists of 8 radiating elements. The obtained impedance bandwidth of EAS-ETMAA is given in Table 5.1.

Figure 5.1 gives the clear view of the comparison of theoretical and experimental impedance bandwidth of proposed antennas. From the graph it is clear that, theoretical and experimental results are with close agreement. From Table 5.1 it is clear that, the theoretical impedance bandwidth of CETMA, corporate and aperture coupled fed ETMAAs are in good agreement with the experimental results. The last column of the Table 5.1 clearly tells that, the percentage errors between experimental and theoretical results are minimum, which validate that the developed theory is with good agreement with the designed equilateral triangular microstrip antennas and its arrays.

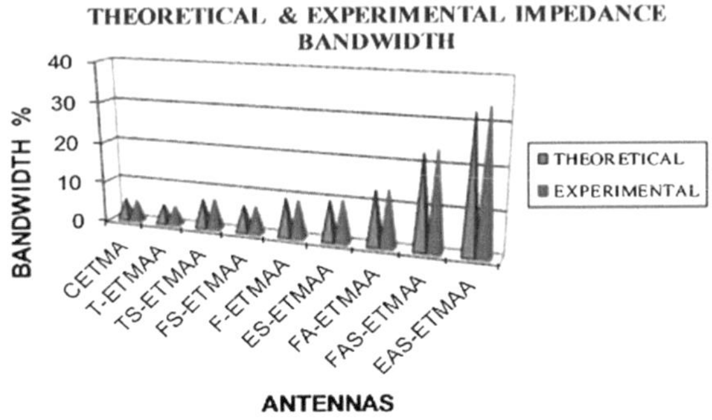

Fig. 5.1 Graph showing the comparison of theoretical and experimental impedance bandwidth of proposed antennas

5.3 Conclusion

The antennas reported in this study have been designed at X-band frequency of 9.4 GHz and are fabricated using low-cost glass epoxy dielectric substrate material. From the return loss [32] graph of CETMA, it is clear that, antenna resonating very close to the designed frequency of 9.4 GHz. This validates the design of CETMA. The dual band operation of antenna is easily achieved by simply increasing the array elements of TS-ETMAA from two to four. This newly obtained antenna FS-ETMAA is more useful for SAR application [24]. The multiband operation of antenna is achieved by increasing the array elements in T-ETMAA from two to four (F-ETMAA) with wider impedance bandwidth of 9.11 %. This multiband operation can be used in mobile computing network applications [32]. Further, the multiband impedance bandwidth is enhanced from 9.11 % to 10.20 % by increasing array elements of F-ETMAA from four to eight and by using slot in the radiating elements i.e. ES-ETMAA. This show that by increasing number of array elements, use of optimum slot in array elements and use of corporate feed arrangement is more effective in enhancing the impedance bandwidth and for converting single band into dual and multiband operation of antenna.

The triple band operation of FA-ETMAA can be converted into dual wide bands by inserting the slot at the center of radiating elements i.e. FAS-ETMAA which is 1.73 times more when compared to the impedance bandwidth of FA-ETMAA. The dual impedance bandwidth of FAS-ETMAA can be converted into a single wide band of magnitude 33.80 % by increasing array elements of FAS-ETMAA from four to eight i.e. EAS-ETMAA. This shows the effect of slot and aperture coupling is quite effective in enhancing the impedance bandwidth of ETMAAs. The impedance bandwidth of EAS-ETMAA is 3.07 times more than found earlier [9]. Further, this antenna is simple in design, fabrication uses low cost substrate material and is compact as it uses only eight array elements when compared to similar study in which the antenna consisted of sixteen array elements arranged in four rows and four columns to get nearly 11 % impedance bandwidth by [9]. From Table 5.1 it is seen that the experimental impedance bandwidths of CETMA, corporate and aperture coupled fed ETMAAs are in good agreement with the theoretical impedance bandwidths.

References

1. Au, T.M., Luk, K.M.: Effect of parasitic element on the characteristics of microstrip antenna. IEEE Trans. Antennas Propagat. **39**, 1247–1251 (1991)
2. Balanis, C.A.: Antenna Theory Analysis and Design. Willey, New York (1982)
3. Bhal, I.J., Bhartia, P.: Microstrip Antennas. Dedham, Massachusetts (1981)
4. Bhatnagar, P.S., Daniel, J.P., Mahdjoubi, K., Terret, C.: Hybrid edge, gap and directly coupled triangular microstrip antenna. Electron. Lett. **22**, 853–855 (1986)
5. Bhatnagar, P.S., Daniel, J.P., Mahdjoubi, K., Terret, C.: Experimental study on stacked triangular microstrip antennas. Electron. Lett. **22**, 864–865 (1986)
6. Bhatnagar, P.S., Daniel, J.P., Mahdjoubi, K., Terret, C.: Displaced multilayer triangular elements widen antenna bandwidth. Electron. Lett. **24**, 962–964 (1988)
7. Bhatnagar, P. S., Adimo, M., Mahdjoubi, K., Terret, C.: Experimental study on stacked aperture fed triangular microstrip antenna. In: Proceedings APSYM-CUSAT 92, pp. 209–212. Kerala, India (1992)
8. Buerkle, A., Sarabandi, K., Mosallaei, H.: Compact-slot and dielectric resonator antenna with dual-resonance, broadband characteristic. IEEE Trans. Antennas Propagat. **53**, 1020–1027 (2005)
9. Chakraborty, S., Gupta, B., Poddar, D.R.: Development of closed form design formulae for aperture coupled microstrip antenna. J. of Sci. Ind. Res. **64**, 482–486 (2005)
10. Chen, W., Lee, K.-F., Dahele, J.S.: Theoretical and experimental studies of the resonant frequencies of the equilateral triangular microstrip antenna. IEEE Trans. Antennas Propagat. **40**, 1253–1256 (1992)
11. Croq, F., Pozar, D.M.: Multifrequency operation of microstrip antennas using aperture coupled parallel resonators. IEEE Trans. Antennas Propagat. **40**, 1367–1374 (1992)
12. Derneryd, A.G., Karlsson, I.: Broadband microstrip antenna element and array. IEEE Trans. Antennas Propagat. **29**, 140–141 (1981)
13. Deschamps, G. A.: Microstrip microwave antennas. In: 3rd USAF, Symposium on Antennas, (1953)
14. Helszajn, J., James, D.S.: Planar triangular resonators with magnetic walls. IEEE Trans. Microw. Theory Tech. **26**, 95–100 (1978)
15. Horng, T.-S., Alexopoulos, N.G.: Corporate feed design for microstrip arrays. IEEE Trans. Antennas Propagat. **41**, 1615–1624 (1993)
16. Howell, J.Q.: Microstrip antennas. IEEE Trans. Antennas Propagat. **23**, 90–93 (1975)
17. Katehi, P.B., Alexopoulos, N.G.: A bandwidth enhancement method for microstrip antennas. IEEE Trans. Antennas Propagat. **35**, 5–12 (1987)
18. Kumar, G., Gupta, K.C.: Broad-band microstrip antennas using additional resonators gap-coupled to the radiating edges. IEEE Trans. Antennas Propagat. **32**, 1375–1379 (1984)
19. Kumar, G., Ray, K.P.: Broadband Microstrip Antennas. Norwood, Massachusetts (2003)
20. Lee, K.-F., Luk, K.-M., Dahele, J.S.: Characteristics of the equilateral triangular patch antenna. IEEE Trans. Antennas Propagat. **36**, 1510–1517 (1988)
21. Liu, W.C., Liu, H.J.: Compact triple-band slotted monopole antenna with asymmetrical CPW grounds. Electron. Lett. **42**, 840–842 (2006)
22. Long, S.A., Walton, M.D.: A dual-frequency stacked circular-disc antenna. IEEE Trans. Antennas Propagat. **27**, 270–273 (1979)
23. Lu, J.-H., Tang, C.-L., Wong, K.-L.: Novel dual-frequency and broad-band designs of slot-loaded equilateral triangular microstrip antennas. IEEE Trans. Antennas Propagat. **48**, 1048–1053 (2000)
24. Maci, S., Gentili, G.B.: Dual-frequency patch antennas. IEEE Trans. Antennas Propagat. Mag. **39**, 13–19 (1997)
25. Mak, C.L., Luk, K.M., Lee, K.F.: Wideband triangular patch antenna. IEE Proc. Microw. Antennas Propagat. **146**, 167–168 (1999)
26. Munson, R.E.: Conformal microstrip antennas and microstrip phased arrays. IEEE Trans. Antennas Propagat. **22**, 74–78 (1974)

27. Orban, D., Moernaut, G. J. K.: The basics of patch antennas, updated. RF Globalnet (www.rfglobalnet.com) newsletter. (2009)
28. Pozar, D.M., Jackson, R.W.: An aperture coupled microstrip antenna with a proximity feed on a perpendicular substrate. IEEE Trans. Antennas Propagat. **35**, 728–731 (1987)
29. Pozar, D.M., Schaubert, D.H.: Microstrip Antennas: The Analysis and Design of Microstrip Antennas and Arrays. Wiley, New York (1995)
30. Pushpanjali, G. M.: Design, fabrication and evaluation of microstrip antennas. PhD Thesis, Gulbarga University, Gulbarga, India (2008)
31. Rafi, GhZ, Shafai, L.: Wideband V-slotted diamond-shaped microstrip patch antenna. Electron. Lett. **40**, 1166–1167 (2004)
32. Rhee, S., Yun, G.: CPW fed slot antenna for triple-frequency band operation. Electron. Lett. **42**, 952–953 (2006)
33. Row, J.-S.: Dual-frequency triangular planar inverted-F antenna. IEEE Trans. Antennas Propagat. **53**, 874–876 (2005)
34. Song, Q., Zhang, X.-X.: A Study on wide band gap-coupled microstrip antenna array. IEEE Trans. Antennas Propagat. **43**, 313–317 (1995)
35. Stevanovic, I., Skrivervik, A., Mosig, J. R.: Smart Antenna Systems for Mobile Communications. Final Report (2003)
36. Suzuki, Y., Miyano, N., Chiba, T.: Circularly polarized radiation from singly fed equilateral-triangular microstrip antenna. IEEE. Proc. **134**, 194–197 (1987)

Chapter 6
Green Function of the Point Source Inside/Outside Spherical Domain - Approximate Solution

Nenad Cvetković, Miodrag Stojanović, Dejan Jovanović, Aleksa Ristić, Dragan Vučković and Dejan Krstić

Abstract A brief review of derivation of two groups of approximate closed form expressions for the electrical scalar potential (ESP) Green functions that originates from the current of the point ground electrode (PGE) in the presence of a spherical ground inhomogeneity are presented in this paper. The PGE is fed by a very low frequency periodic current through a thin isolated conductor. One of approximate solutions is proposed in this paper. Known exact solutions that have parts in a form of infinite series sums are also given in this paper. In this paper, the exact solution is solely reorganized in order to facilitate comparison to the closed form solutions, and to estimate the error introduced by the approximate solutions. Finally, error estimation is performed comparing the results for the electrical scalar potential obtained applying the approximate expressions and the accurate calculations. This is illustrated by a number of numerical experiments.

Keywords Electrical scalar potential · Green's function · Spherical inhomogeneity

---

N. Cvetković (✉) · M. Stojanović · D. Jovanović · A. Ristić · D. Vučković
Faculty of Electronic Engineering, University of Niš,
Aleksandra Medvedeva 14, 18000 Niš, Serbia
e-mail: nenad.cvetkovic@elfak.ni.ac.rs

M. Stojanović
e-mail: miodrag.stojanovic@elfak.ni.ac.rs

D. Jovanović
e-mail: dejan.jovanovic@elfak.ni.ac.rs

A. Ristić
e-mail: aleksa.ristic@elfak.ni.ac.rs

D. Vučković
e-mail: dragan.vuckovic@elfak.ni.ac.rs

D. Krstić
Faculty of Occupational Safety, University of Niš, Čarnojevića 10A,
18000 Niš, Serbia
e-mail: dejan.krstic@znrfak.ni.ac.rs

S. Silvestrov and M. Rančić (eds.), *Engineering Mathematics I*,
Springer Proceedings in Mathematics & Statistics 178,
DOI 10.1007/978-3-319-42082-0_6

6.1 Introduction

Problems of potential fields related to the influence of spherical material inhomogeneities, have a rather rich history of over 150 years in different fields of mathematical physics. In the fields of electrostatic field, stationary and quasi-stationary current field, and magnetic field of stationary currents, problems of a point source in the presence of a spherical material inhomogeneity are gathered in the book by Stratton [19] and all later authors that have treated this matter quote this reference as the basic one. The authors of this paper will also consider the results from [19] as the referent ones.

The exact solution shown in [19, pp. 201–205] related to the point charge in the presence of the dielectric sphere, is obtained solving the Poisson, i.e. Laplace partial differential equation expressed in the spherical coordinate system, using the method of separated variables. Unknown integration constants are obtained satisfying the boundary conditions for the electrical scalar potential continuity and normal component of the electric displacement at the boundary of medium discontinuity, i.e. at the dielectric sphere surface. The obtained general solution for the electrical scalar potential, besides a number of closed form terms, also consists of a part in a form of an infinite series sum that has to be numerically summed.

Among work of other authors that have dealt with this problem, the following will be cited in this paper: Hannakam [10, 11], Reiß [17], Lindell et al. [13, 14, 18] and Velickovic [20, 21]. The last cited ones, according to the authors of this paper, gave an approximate closed form solution of the problem. This is also characterized in this paper. In [10], author through detail analysis manages to express a part of general solution in a form of infinite sums by a class of integrals whose solutions can not be given in a closed form, i.e. general solution of these integrals have to be obtained numerically. In [17], the author considers a problem of this kind with an aim to calculate the force on the point charge in the presence of a dielectric sphere. For this problem solving he uses Kelvin's inversion factor and introduces a line charge image, which coincides with results from [10]. Starting with the general solution from [19], authors in [13, 18] using different mathematical procedures practically obtain the same solutions as in [10], considering separately the case of a point charge outside the sphere [13] and the case of a point charge inside the sphere [18].

In [20, pp. 97–98] and [21] author deduces the closed form solution for the electrical scalar potential of the point source in the presence of sphere inhomogeneity in two steps. In the first one the author assumes a part of the solution that corresponds to images in the spherical mirror and approximately satisfies boundary conditions on the sphere surface. In the second step, assumed solutions are broadened by infinite sums that approximately correspond to the ones that occur as an exact general solution in [19], i.e. in other words, approximately satisfy the Laplace partial differential equation. Afterwards, unknown constants under the sum symbol are obtained satisfying the boundary condition of continuity of the normal component of total current density on the sphere surface. These solutions enable summing of infinite sums and presenting the general solution in a closed form. Well known mathematical tools from Legendre polynomial theory were used for the summing procedure.

Finally, the author of this paper has, analysing problems of this kind, started from the general solution from [19] and primarily, reorganized certain parts in the following way. A number of terms that correspond to images in the spherical mirror with unknown weight coefficients are singled out. Remaining parts of the general solution are infinite sums whose general, n-th term presents a product of an unknown integration constant, factored function of radial sphere coordinate $r^{-1(n+1)}$ and r^n, and Legendre polynomial of the first kind $P_n(\cos\theta)$. Afterwards, all unknown constants are determined satisfying mentioned boundary conditions, but in such a way that the condition for the electrical scalar potential continuity is completely satisfied, while the condition for the normal component of total current density can be fulfilled approximately. Approximate satisfying of this boundary condition is done in a way to sum a part of the general solution expressed by infinite sums in a closed form. This technique is well known and was very successfully, although under certain assumptions, used by many authors especially in the high frequency domain. For example, one of them is explicitly considered in [12], and one is implicitly given in [15, 16].

Among five quoted solutions, three will be analyzed in this paper, i.e. the accurate one from [19] as the referent one, approximate one from [20, 21] (which will be characterized in detail since this was not done in [20, 21]) and the second one, also approximate model, proposed in this paper.

In the second section of the paper three groups of cited expressions for the ESP distribution will be given with minimal remarks about their deduction. In this part of the paper, general expressions for the evaluation of error of the ESP calculation will be also presented.

In the third section of the paper, a part of numerical experiments whose results justify the use of approximate solutions and also present the error level done along the way will be presented.

Finally, based on the presented theory and performed numerical experiments, corresponding conclusion will be made and a list of used references will be given.

6.2 Theoretical Background

6.2.1 Description of the Problem

Spherical inhomogeneity of radius r_s is considered. Sphere domain is considered a linear, isotropic and homogenous semi-conducting medium of known electrical parameters σ_s, ε_s and $\mu_s = \mu_0$ (σ_s - specific conductivity, $\varepsilon_s = \varepsilon_0\varepsilon_{rs}$ - permittivity and μ_0 - permeability). The remaining space is also a linear, isotropic and homogenous semi-conducting medium of known electrical parameters (σ_1, $\varepsilon_1 = \varepsilon_0\varepsilon_{r1}$ and $\mu_1 = \mu_0$).

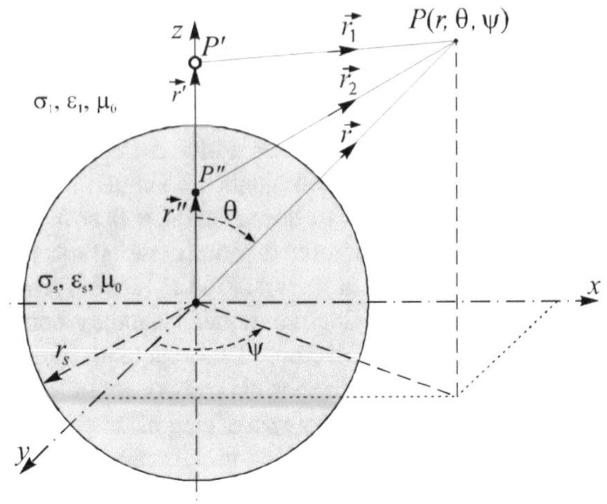

Fig. 6.1 The PGE outside the sphere

The spherical, i.e. Descartes' coordinate systems with their origins placed in the sphere centre are associated to the problem. At the arbitrary point P', defined by the position vector $\mathbf{r}' = r'\hat{z}$, a point current source is placed (so-called Point ground electrode - PGE), and is fed through a thin isolated conductor by a periodic current of intensity I_{PGE} and very low angular frequency ω, $\omega = 2\pi f$.

The location of the PGE can be outside the sphere, $r' \leq r_s$, or inside of it, $r' \geq r_s$, which also goes for the observed point P, defined by the vector $\mathbf{r}$, at which the potential and quasi-stationary current and electrical field structure are determined, i.e. for $r \geq r_s$ the point P is outside the sphere, and for $r \leq r_s$ inside of it.

In accordance with the last one, the electrical scalar potential $\varphi_{..}(\mathbf{r})$, total current density vector $\mathbf{J}_{..}(\mathbf{r})$ and electrical field vector $\mathbf{E}_{..}(\mathbf{r})$, will be denoted by two indexes $i, j = 1, s$ where the first one "i" denotes the medium where the quantity is determined, and the other one "j" the medium where the PGE is located. For example: $\varphi_{s1}(r)$ presents the potential calculated inside the sphere, $r \leq r_s$, when the PGE is located outside the sphere, $r' \leq r_s$. Also, in order to systemize text and ease its reading, the solutions that correspond to references [19–21], will be denoted in the exponent as follows: S-Stratton, V-Velickovic and R-Rancic, respectively. For example: $\varphi_{11}^{s}(r)$ presents the solution for the potential according to [19]—Stratton outside the sphere, $r \geq r_s$, when the PGE is located at point P' that is also outside the sphere, $r' \geq r_s$.

Problem geometry is illustrated graphically in Figs. 6.1 and 6.2, where Fig. 6.1 corresponds to the case when the PGE is placed outside the sphere, whilst Fig. 6.2 refers to its location inside of the sphere. Singled out images in the spherical mirror, i.e. P'' points with corresponding position vector $\mathbf{r}'' = r''\hat{z}$, where $r'' = r_s^2/r'$ is Kelvin's inversion factor of the spherical mirror, are also given in figures. Distance from the PGE, point P', to the observed point P is denoted by $r_1, r_1 = \sqrt{r^2 + r'^2 - 2rr'\cos\theta}$, and distance from the image in the spherical mirror, point P', to the point P by r_2, $r_2 = \sqrt{r^2 + r''^2 - 2rr''\cos\theta}$.

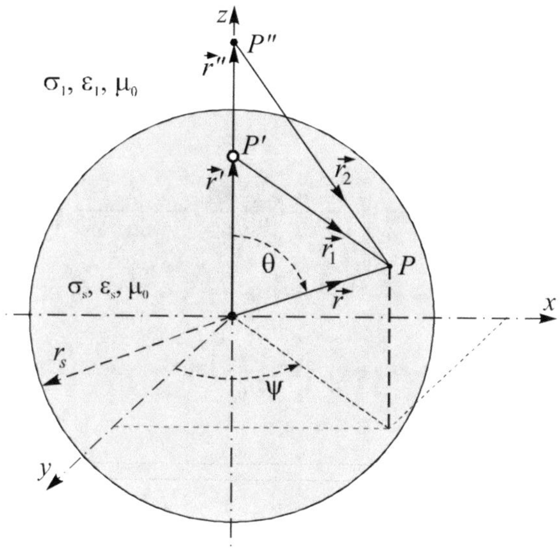

Fig. 6.2 The PGE inside the sphere

Finally, the following labels were used in the paper: $\bar{\sigma}_i = \sigma_i + j\omega\varepsilon_i$ - complex conductivity of i-th medium, $i = 1, s$; $\bar{\varepsilon}_{ri} = \varepsilon_{ri} - j\varepsilon_{ii} = \varepsilon_{ri} - j60\sigma_i\lambda_0$ - complex relative permittivity of i-th medium, $i = 1, s$ and λ_0—wave-length in the air; $\bar{\gamma}_i = (j\omega\mu_i\bar{\sigma}_i)^{1/2}$—complex propagation constant of i-th medium, $i = 1, s$; $\bar{n}_{ij} = \bar{\gamma}_i/\bar{\gamma}_j$ - complex refraction index of i-th and j-th medium, $i, j = 1, s$ and R_{1s}, T_{1s}, T_{s1}—quasi-stationary reflection and transmission coefficients defined by the following expression:

$$R_{1s} = \frac{\bar{\sigma}_1 - \bar{\sigma}_s}{\bar{\sigma}_1 + \bar{\sigma}_s} = \frac{\bar{n}_{1s}^2 - 1}{\bar{n}_{1s}^2 + 1} = T_{1s} - 1 = -R_{s1} = 1 - T_{s1}.$$

The time factor $\exp(j\omega t)$ is omitted in all relations.

6.2.2 *Exact ESP Solution According to [19]*

6.2.2.1 Electrical Scalar Potential (ESP)

The ESP function for any position of the PGE must satisfy the Poisson, i.e. Laplace partial differential equation, which are, in accordance with introduced labels for the spherical coordinate system, as follows:

- The PGE outside the sphere, $i = 1, s$, $r' \geq r_s$, Fig. 6.1,

$$\Delta\varphi_{i1} = \frac{1}{r^2}\frac{\partial}{\partial r}\left(r^2\frac{\partial\varphi_{i1}}{\partial r}\right) + \frac{1}{\sin\theta}\frac{\partial}{\partial\theta}\left(\sin\theta\frac{\partial\varphi_{i1}}{\partial\theta}\right) = = -\frac{I_{PGE}}{2\pi\bar{\sigma}_s}\frac{\delta(r-r')\delta(\theta)}{r^2\sin\theta}, r \geq r_s, \tag{6.1}$$

$$\Delta\varphi_{i1} = \frac{1}{r^2}\frac{\partial}{\partial r}(r^2\frac{\partial\varphi_{i1}}{\partial r}) + \frac{1}{\sin\theta}\frac{\partial}{\partial\theta}\left(\sin\theta\frac{\partial\varphi_{i1}}{\partial\theta}\right) = 0, r \leq r_s, \tag{6.2}$$

- The PGE inside the sphere, $i = 1, s, r' \leq r^s$, Fig. 6.2,

$$\Delta\varphi_{is} = \frac{1}{r^2}\frac{\partial}{\partial r}\left(r^2\frac{\partial\varphi_{i1}}{\partial r}\right) + \frac{1}{\sin\theta}\frac{\partial}{\partial\theta}\left(\sin\theta\frac{\partial\varphi_{i1}}{\partial\theta}\right) = 0, r \geq r_s, \tag{6.3}$$

$$\Delta\varphi_{is} = \frac{1}{r^2}\frac{\partial}{\partial r}\left(r^2\frac{\partial\varphi_{i1}}{\partial r}\right) + \frac{1}{\sin\theta}\frac{\partial}{\partial\theta}\left(\sin\theta\frac{\partial\varphi_{i1}}{\partial\theta}\right) = = -\frac{I_{PGE}}{2\pi\bar{\sigma}_1}\frac{\delta(r-r'\delta(\theta)}{r^2\sin\theta}, r \leq r_s, \tag{6.4}$$

where $\delta(r-r')$ and $\delta(\theta)$ are Dirac's δ -functions.
After the differential equations, for example (6.1) and (6.2), are solved applying the method of separating variables, the unknown integration constants are determined so the obtained solution satisfies the condition for the finite value of the potential at all points $r \in [0, \infty)$, except at $\mathbf{r} = \mathbf{r}'$. Remaining integration constants are determined from the electrical scalar potential boundary condition,

$$\varphi_{11}(r = r_s, \theta) = \varphi_{s1}(r = r_s, \theta), \tag{6.5}$$

and the one for the normal component of the total current density on the discontinuity surface, i.e.,

$$\bar{\sigma}_1\frac{\partial\varphi_{11}(r,\theta)}{\partial r}|_{r=r_s} = \bar{\sigma}_s\frac{\partial\varphi_{s1}(r,\theta)}{\partial r}|_{r=r_s}. \tag{6.6}$$

Finally, according to [19, Sect. 3.23, pp. 204, Eqs. (20)–(21)], the exact solution for the potential distribution is for $r' \geq r_s$:

$$\varphi_{11}^s(\mathbf{r}) = \frac{I_{PGE}}{4\pi\bar{\sigma}_1}\left[\frac{1}{r_1} + \sum_{n=0}^{\infty}\frac{n(\bar{\sigma}_1 - \bar{\sigma}_s)}{n\bar{\sigma}_s + (n+1)\bar{\sigma}_1}\frac{r_s^{2n+1}}{r'^{n+1}}\frac{P_n(\cos\theta)}{r^{n+1}}\right], r \geq r_s, \tag{6.7}$$

$$\varphi_{s1}^s(\mathbf{r}) = \frac{I_{PGE}}{4\pi\bar{\sigma}_1}\sum_{n=0}^{\infty}\frac{(2n+1)\bar{\sigma}_1 r^n}{n\bar{\sigma}_s + (n+1)\bar{\sigma}_1}\frac{P_n(\cos\theta)}{r'^{n+1}}, r \leq r_s, \tag{6.8}$$

where $P_n(\cos\theta)$ is Legendre polynomial of the first kind.

Keeping in mind the duality of electrostatic and quasi-stationary very low frequency current fields, in relation to the solution from [19, Eqs. (20)–(21)]:

- Labels introduced in expressions (6.7) and (6.8) fit the described geometry and used labels;
- q/ε_2 is substituted by $I_{PGE}/\bar{\sigma}_1$; and
- instead of permittivity, corresponding indexed complex conductivities are used, i.e. $\bar{\sigma}_s$ instead of ε_1, and $\bar{\sigma}_1$ instead of ε_2.

When expressions (6.7) and (6.8) are reorganized in such a way so they can be compared to approximate expressions, and additionally labelled with S - Stratton, the following exact solution is obtained:

$$\varphi_{11}^{s}(\mathbf{r}) = V_s\left[\frac{r_s}{r_1} + R_{1s}\frac{r_s}{r'}\left(\frac{r_s}{r_2} - \frac{r_s}{r}\right) - \right.$$
$$\left. - \frac{R_{1s}T_{1s}}{2}\sum_{n=1}^{\infty}\frac{1}{n+T_{1s}/2}\left(\frac{r''}{r}\right)^{n+1}P_n(\cos\theta)\right], r \geq r_s, \tag{6.9}$$

$$\varphi_{s1}^{s}(\mathbf{r}) = V_s\left[T_{1s}\frac{r_s}{r_1} - R_{1s}\frac{r_s}{r'} - \right.$$
$$\left. - \frac{R_{1s}T_{1s}}{2}\frac{r_s}{r'}\sum_{n=1}^{\infty}\frac{1}{n+T_{1s}/2}\left(\frac{r}{r'}\right)^{n}P_n(\cos\theta)\right], r \leq r_s, \tag{6.10}$$

where $V_s = I_{PGE}/(4\pi\bar{\sigma}_1 r_s)$, and R_{1s}, T_{1s}, are reflection and transmission coefficients, respectively.

In the same way, final solutions for Eqs. (6.3) and (6.4), for $r' \leq r_s$, that satisfy conditions (6.5) and (6.6) are:

$$\varphi_{1s}^{s}(\mathbf{r}) = V_s\left[T_{1s}\frac{r_s}{r_1} - R_{1s}\frac{r_s}{r} - \right.$$
$$\left. - \frac{R_{1s}T_{1s}}{2}\frac{r_s}{r'}\sum_{n=1}^{\infty}\frac{1}{n+T_{1s}/2}\left(\frac{r'}{r}\right)^{n+1}P_n(\cos\theta)\right], r \geq r_s, \tag{6.11}$$

$$\varphi_{ss}^{s}(\mathbf{r}) = V_s\left[\frac{T_{1s}}{T_{s1}}\frac{r_s}{r_1} - R_{1s}\frac{T_{1s}}{T_{s1}}\frac{r_s}{r'}\frac{r_s}{r_2} - R_{1s} - \right.$$
$$\left. - \frac{R_{1s}T_{1s}}{2}\sum_{n=1}^{\infty}\frac{1}{n+T_{1s}/2}\left(\frac{r}{r''}\right)^{n}P_n(\cos\theta)\right], r \leq r_s, \tag{6.12}$$

Comment: The last two expressions are not explicitly given in [19], as (6.7) and (6.8).

6.2.2.2 Quasi-Stationary Electrical and Current Field Structure

Once the potential distributions (6.9)–(6.10) and (6.11)–(6.12) are determined, the structure of the quasi-stationary field vectors are:

- Electrical field vector:

$$\mathbf{E}_{ij} \cong -grad\varphi_{ij} = -\frac{\partial \varphi_{ij}}{\partial r}\hat{r} - \frac{1}{r}\frac{\partial \varphi_{ij}}{\partial \theta}\hat{\theta}, i, j = 1, s; \tag{6.13}$$

- Total current density vector:

$$\mathbf{J}_{ij}^{tot} = \bar{\sigma}_i \mathbf{E}_{ij}, i, j = 1, s; \tag{6.14}$$

- Conduction current density vector:

$$\mathbf{J}_{ij} = \sigma_i \mathbf{E}_{ij}, i, j = 1, s. \tag{6.15}$$

6.2.3 ESP Solution According to [20, pp. 97–98] and [21]

The ESP solution proposed in [21] considers the following. Firstly, for the case $r' \geq r_s$, solution is proposed in a form:

$$\varphi_{11} \cong \frac{I_{PGE}}{4\pi\bar{\sigma}_1}\left[\frac{1}{r_1} + C_1\frac{1}{r_2} + C_2\frac{1}{r}\right], r \geq r_s, \tag{6.16}$$

$$\varphi_{s1} \cong \frac{I_{PGE}}{4\pi\bar{\sigma}_1}\left[C_3\frac{1}{r_1} + C_4\right], r \leq r_s, \tag{6.17}$$

where $C_1 - C_4$ are unknown constants that are determined satisfying the condition (6.5) and the one that the solution ((6.16), (6.17)) is also valid for the case of sphere with great radius. This solution is identical to the first three terms of the exact solution (6.9) and the first two of (6.10).

Since, this way obtained solution ((6.16), (6.17)) does not satisfy the boundary condition (6.6), the author broadened solutions (6.16) and (6.17) with two infinite series of general form:

$$\sum_{n=1}^{\infty} C_{5n}\left(\frac{r_s}{r}\right)^{\pm n} P_n(\cos\theta), \tag{6.18}$$

where C_{5n} are unknown constants, for "$+n$" in (6.18) the (6.16) is broadened and (6.17) for "$-n$". Unknown constants C_{5n} are determined using the condition (6.6), having the ESP final solution:

$$\phi_{11}^{V}(\mathbf{r}) \cong V_s \left[\frac{r_s}{r_1} + R_{1s}\frac{r_s}{r'}\left(\frac{r_s}{r_2} - \frac{r_s}{r}\right) + \right.$$
$$\left. + \frac{R_{1s}T_{1s}}{2} \ln \frac{r - r'' \cos\theta + r_2}{2r} \right], r \geq r_s, \tag{6.19}$$

$$\phi_{s1}^{V}(\mathbf{r}) \cong V_s \left[T_{1s}\frac{r_s}{r_1} - R_{1s}\frac{r_s}{r'} + \right.$$
$$\left. + \frac{R_{1s}T_{1s}}{2}\left(\frac{r_s}{r'}\right) \ln \frac{r' - r\cos\theta + r_1}{2r'} \right], r \leq r_s. \tag{6.20}$$

Label V-Velickovic in the exponent denotes that solutions (6.19) and (6.20) correspond to the ones from [21], and V_s is previously introduced constant that appears also in (6.9) and (6.10).

It should be noted that the introduced extension (6.18) for "$+n$", approximately satisfies the general solution of the Laplace equation, i.e. (6.7), where r_s/r is factored by $(n+1)$.

Similarly, the solutions for the potential when $r' \leq r_s$, i.e. the PGE is located inside the sphere, are also given in [21]. The solutions are as follows:

$$\phi_{1s}^{V}(\mathbf{r}) \cong V_s \left[T_{1s}\frac{r_s}{r_1} - R_{1s}\frac{r_s}{r} + \right.$$
$$\left. + \frac{R_{1s}T_{1s}}{2}\left(\frac{r_s}{r'}\right) \ln \frac{r - r'\cos\theta + r_1}{2r} \right], r \geq r_s, \tag{6.21}$$

$$\phi_{ss}^{V}(\mathbf{r}) \cong V_s \left[\frac{T_{1s}}{T_{s1}}\frac{r_s}{r_1} - R_{1s}\frac{T_{1s}}{T_{s1}}\frac{r_s}{r'}\frac{r_s}{r_2} - R_{1s} + \right.$$
$$\left. + \frac{R_{1s}T_{1s}}{2} \ln \frac{r'' - r\cos\theta + r_2}{2r''} \right], r \leq r_s. \tag{6.22}$$

6.2.4 *ESP Solution Proposed in This Paper*

If the general solution from [19] is reorganized under the sum symbol into a form that is for $r' \geq r_s$ given by (6.9) and (6.10), the following is obtained:

$$\varphi_{11}(\mathbf{r}) \cong \frac{I_{PGE}}{4\pi\bar{\sigma}_1}\left[\frac{1}{r_1} + C_1\frac{r_s}{r'}\frac{1}{r_2} + B_0\frac{r_s}{r} + \sum_{n=1}^{\infty} B_n \left(\frac{r_s}{r}\right)^{n+1} P_n(\cos\theta) \right], r \geq r_s, \tag{6.23}$$

$$\varphi_{s1}(\mathbf{r}) \cong \frac{I_{PGE}}{4\pi\bar{\sigma}_1}\left[D_1\frac{1}{r_1} + A_0 + \sum_{n=1}^{\infty} A_n \left(\frac{r}{r_s}\right)^n P_n \cos\theta\right], r \leq r_s, \tag{6.24}$$

where C_1, D_1, B_n and $A_n, n = 0, 1, \ldots$, are unknown constants. Starting from the boundary condition (6.5) we have $1 + C_1 = D_1$ and $B_n = A_n, n = 0, 1, \ldots$ The other boundary condition gives $C_1 = R_{1s}$, so $D_1 = T_{1s}$. If the condition (6.6) is approximately satisfied, we also have $A_0 = B_0 = -R_{1s}/r'$ and constants $B_n, n = 1, 2, \ldots$, related to (6.6) are determined from the condition

$$-\frac{R_{1s}T_{1s}}{2r'}\sum_{n=1}^{\infty}\left(\frac{r_s}{r'}\right)^n P_n(\cos\theta) = \sum_{n=1}^{\infty} n B_n P_n(\cos\theta). \tag{6.25}$$

In (6.6) remains a term in a form of a sum, i.e. the error "e" of satisfying the boundary condition (6.6) for the radial component of total current density is

$$\begin{aligned} e\{J_{11r}^{tot}\} &= \frac{I_{PGE}}{4\pi r_s}\sum_{n=1}^{\infty} B_n P_n(\cos\theta) = \\ &= -\bar{\sigma}_1 V_s \frac{R_{1s}T_{1s}}{2r'}\sum_{n=1}^{\infty}\frac{1}{n}\left(\frac{r_s}{r'}\right)^n P_n(\cos\theta) = \\ &= \bar{\sigma}_1 V_s \frac{R_{1s}T_{1s}}{2r'}\ln\frac{r' - r_s\cos\theta + r_{1s}}{2r'}, \end{aligned} \tag{6.26}$$

where r_{1s} is r_1 for $r = r_s$, and $B_n = A_n, n = 1, 2, \ldots$, from (6.25).

If we substitute the solution for $B_n = A_n, n = 1, 2, \ldots$, from (6.25) into (6.24) and (6.23) and using known tools from Legandre polynomial theory, we have:

$$\begin{aligned} \phi_{11}^R(\mathbf{r}) \cong V_s &\left[\frac{r_s}{r_1} + R_{1s}\frac{r_s}{r'}\left(\frac{r_s}{r_2} - \frac{r_s}{r}\right) + \right. \\ &\left. + \frac{R_{1s}T_{1s}}{2}\left(\frac{r_s}{r'}\right)\left(\frac{r_s}{r}\right)\ln\frac{r - r''\cos\theta + r_2}{2r}\right], r \geq r_s, \end{aligned} \tag{6.27}$$

$$\begin{aligned} \phi_{s1}^R(\mathbf{r}) \cong V_s &\left[T_{1s}\frac{r_s}{r_1} - R_{1s}\frac{r_s}{r'} + \right. \\ &\left. + \frac{R_{1s}T_{1s}}{2}\left(\frac{r_s}{r'}\right)\ln\frac{r' - r\cos\theta + r_1}{2r'}\right], r \leq r_s. \end{aligned} \tag{6.28}$$

The ESP solution when $r' \leq r_s$ is obtained in a similar way. After obtaining the unknown constants, satisfying the condition (6.5) and approximately satisfying the condition (6.6), we have:

$$\phi_{1s}^{R}(\mathbf{r}) \cong V_s\left[T_{1s}\frac{r_s}{r_1} - R_{1s}\frac{r_s}{r} + \frac{R_{1s}T_{1s}}{2}\left(\frac{r_s}{r}\right)\ln\frac{r - r'\cos\theta + r_1}{2r}\right], r \geq r_s, \tag{6.29}$$

$$\begin{aligned}\phi_{ss}^{R}(\mathbf{r}) \cong V_s\Bigg[\frac{T_{1s}}{T_{s1}}\frac{r_s}{r_1} - R_{1s}\frac{T_{1s}}{T_{s1}}\frac{r_s}{r'}\frac{r_s}{r_2} - R_{1s}+\\ + \frac{R_{1s}T_{1s}}{2}\ln\frac{r'' - r\cos\theta + r_2}{2r''}\Bigg], r \leq r_s.\end{aligned} \tag{6.30}$$

6.2.5 Analysis of the Presented ESP Solutions

- In contrast to the exact solution [19], both approximate ones have a closed form.
- Obtained approximate solutions (6.27) and (6.28) are very similar to (6.19) and (6.20). The only difference is between (6.19) and (6.27) in factor r_s/r that factors the Ln-function. The same respectively goes for solutions (6.29) and (6.30) when compared to approximate ones (6.29) and (6.30).
- Solutions (6.19) and (6.20) satisfy boundary conditions (6.5) and (6.6), but do not have a general solution for $r \geq r_s$ that follows from the solution for the Poisson, i.e. Laplace equation [19]. The same goes for (6.21) and (6.22). A solution to one technical problem of this kind, solved applying this "V" model, is given in [3, 4].
- Solutions (6.27) and (6.28) satisfy the general solution for the Poisson, i.e. Laplace partial differential equation, satisfy boundary condition (6.5), and approximately satisfy boundary condition (6.6). The same goes for solutions given by (6.29) and (6.30).
- Expressions (6.27)–(6.30) can also be obtained starting from accurate ones (6.9)–(6.12) under a condition $n_{1s} \ll 1$. In that case, the addend $T_{1s}/2$ in the denominator under the sum symbol is $|\ T_{1s}/2\ |=|\ \bar{n}_{1s}^2/(1+\bar{n}_{1s}^2)\ |= 1$, so, it can be neglected in relation to the sum index $n \geq 1$. Since, for example the sum term in (6.9) is then approximately $-\sum_{n=1}^{\infty}\frac{1}{n}(\frac{r''}{r})^{n+1}P_n(\cos\theta) = \frac{r''}{r}\ln\frac{r-r''\cos\theta+r_2}{2r}$, then, the expression (6.9) becomes identical to (6.27). Similarly, we obtain remaining expressions (6.28)–(6.30). Accordance of the results obtained applying the approximate model "R" and the exact one "S" is better for all values of the refraction coefficient $n_{1s} < 1$, then for the case of $n_{1s} > 1$. This can be easily concluded analysing the given expressions "S" and "R". This conclusion is also confirmed by numerical experiments.

6.2.6 Error Estimation Using the Approximate Expressions for the ESP

All the ESP expressions, accurate ones (6.9)–(6.12) according to [19], approximate ones (6.19)–(6.22) according to [20, 21] and approximate ones according to R-model proposed in this paper, evolved towards the same form so they could be directly compared. Firstly, the terms that associate to spherical mirror imaging are singled out, and they correspond to images with weight coefficients multiplied by quasi-stationary reflection R_{1s}, or transmission T_{1s}, coefficients. Remaining part of the solution is an infinite sum in the case of the exact solution, and in the case of approximate ones, a closed form expressed by Ln-functions.

Error estimation of the ESP calculation is done according to the general expression

$$\delta_Q = 100 \mid \frac{\varphi_{ij}^s(\mathbf{r}) - \varphi_{ij}^Q(\mathbf{r})}{\varphi_{ij}^s(\mathbf{r})} \mid, \text{ in } [\%], \tag{6.31}$$

where $i, j = 1, s$ and $Q = V, R$.

Relative error estimation of satisfying boundary condition (6.6) can be evaluated according to the following expression:

$$\delta_Q = 100 \mid \frac{e\{J_{11r}^{tot}\}}{-\bar{\sigma}_1 \partial \varphi_{11}^s / \partial r} \mid_{r=r_s}, \text{ in } [\%]. \tag{6.32}$$

6.3 Numerical Results

Based on the presented ESP expressions a number of numerical experiments were performed in order to establish the validity of the proposed approximate solutions compared to the exact ESP calculations using expressions (6.9)–(6.12) according to [19].

The results presented graphically in the figures that follow will be denoted as:

- S-model, Eqs. (6.9)–(6.12), [19];
- V-model, Eqs. (6.19)–(6.22), [20, 21]; and
- R-model, Eqs. (6.27)–(6.30), i.e. the ESP model proposed in this paper.

The first group of numerical results deals with the electrostatic problem of the point charge (PCh) in the presence of the spherical dielectric inhomogeneity. In this case $\varepsilon_i = \varepsilon_0 \varepsilon_{ri}$, $i = 1, s$, should replace $\bar{\sigma}_1$ in all the expressions, having $V_s = Q/(4\pi\varepsilon_1 r_s)$ and $p_{1s} = \varepsilon_1/\varepsilon_s$. Normalized ESP (R-model, solid line) versus radial distance r for different values of angle $\theta = 0°, 5°, 45°$, and $90°$ and different values of relative permittivity $\varepsilon_{rs} = 1, 1.5, 2, 3, 5, 10, 20, 36, 80$ and 1000 as parameters, are given in the left column of the Fig. 6.3 (the PCh inside the sphere: $r' = 0.7 r_s$).

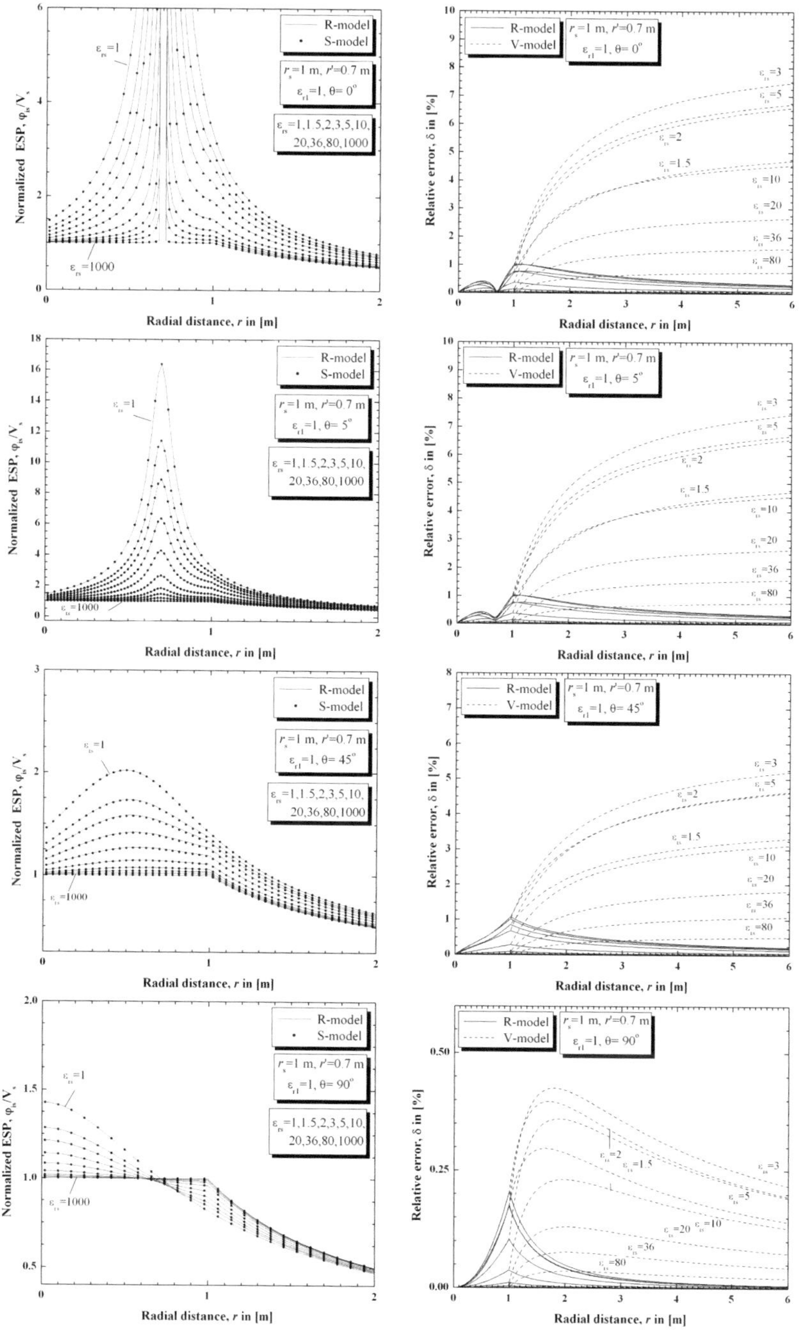

Fig. 6.3 Point charge inside dielectric sphere. Normalized ESP and corresponding relative error versus radial distance r for different values of angle θ and relative permittivity ε_{rs} taken as parameters

For the sake of comparing, the exact normalized ESP values according to S-model (solid circle) are presented in the same figures. Corresponding relative error, δ in [%], calculated for both R- and V- models is presented in the right column of Fig. 6.3. Normalized ESP (R- and S-models) and corresponding relative errors (R- and V-models) for the case of $r' = 1.5r_s$ are presented in Fig. 6.4 (the PCh outside the sphere).

The second group of numerical experiments consider the quasi-stationary field, i.e. the case of the semi-conducting spherical inhomogeneity and the PGE fed by VLF current, $f = 50$ [Hz]. The ESP is calculated as a function of r, and angle $\theta = 0°, 45°$ and ratio $p_{1s} = \sigma_1/\sigma_s = 0.1, 10$ are taken as parameters. The rest of system parameters are given in figures. For the sake of comparing, the ESP values obtained applying the R-, S- and V-models are presented in the same figures. For each example, relative errors δ in [%], done using the approximate models are also calculated. The results for the case of the PGE placed inside the sphere, $r' = 0.9r_s$, are presented in Fig. 6.5, and in Fig. 6.6 the results for the case of the PGE placed outside the sphere, $r' = 1.1r_s$.

Based on graphically illustrated results one can conclude that the relative error for the R-model is always $\delta < 1$ % when the refraction index is $n_{1s} < 1$. For the other case, $n_{1s} > 1$, the maximal error is $\delta < 15$ %, but only for the worst case, i.e. when the field point P is on the sphere surface, $r = r_s$. This conclusion does not apply on the V-model, i.e. the error δ is for certain parameters in a wide range of radial distance r greater than 30 % (see Figs. 6.5 and 6.6).

6.4 Technical Application

Considering the fact that the author uses a method of numerical solving of integral equations when dealing with technical problems of modelling a design of EM field structure of wire structures in the presence of inhomogeneous media, it is very important to have simple expressions for Green function for that purpose.

One of those models refers to problems of modeling and design of groundings in the presence of different ground inhomogeneities. Thus, direct application of the models deployed in this paper combined with quasi-stationary image theory is on modeling and design of grounding electrical characteristics in the presence of a spherical and semi-spherical ground inhomogeneity. Direct technical application on real technical problems are schematically illustrated in the Fig. 6.7 labeled as a–d, [1, 2, 5–9].

6.5 Conclusion

A new approximate solution for the Green function of the ESP that originates from PGE current in the presence of a spherical ground inhomogeneity, when the PGE is fed by a VLF current through a thin isolated ground conductor, was proposed in this

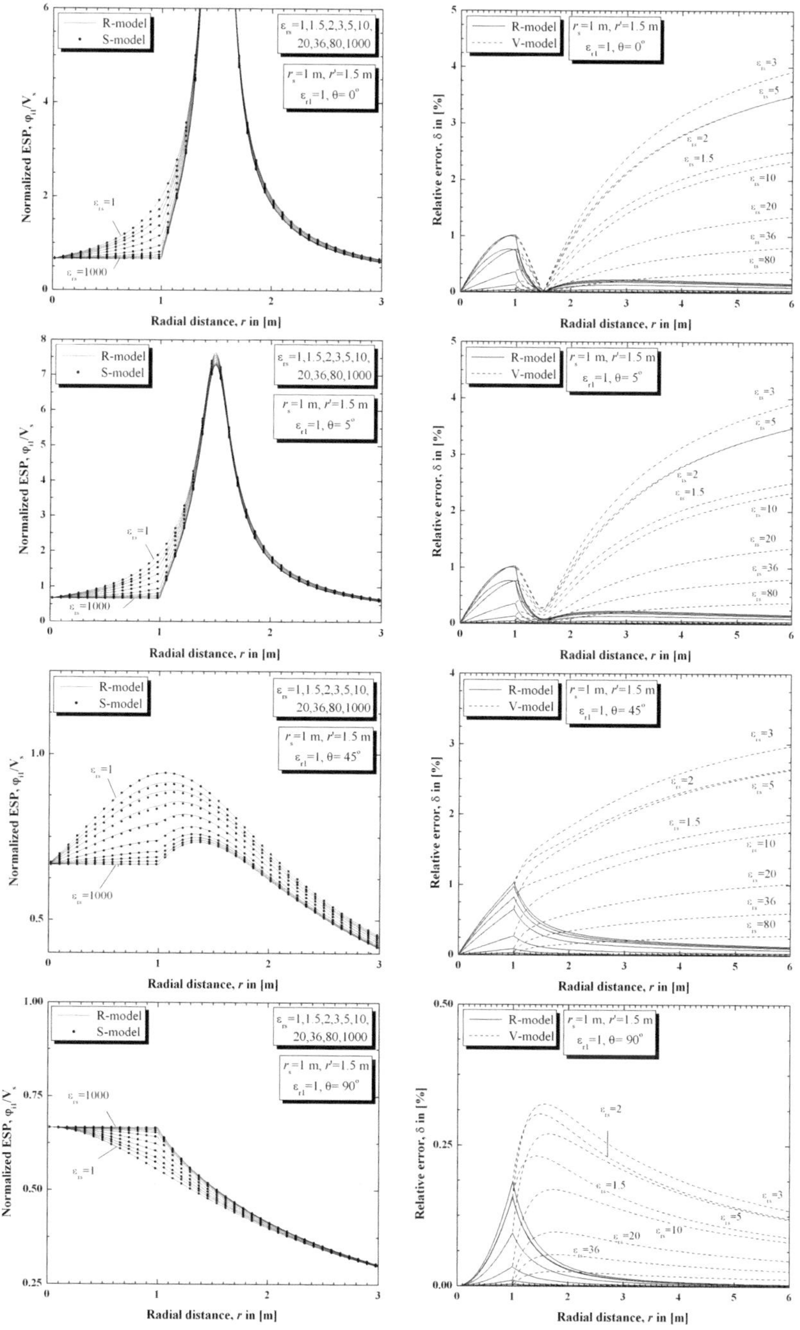

Fig. 6.4 Point charge outside dielectric sphere. Normalized ESP and corresponding relative error versus radial distance r for different values of angle θ and relative permittivity ε_{rs} taken as parameters

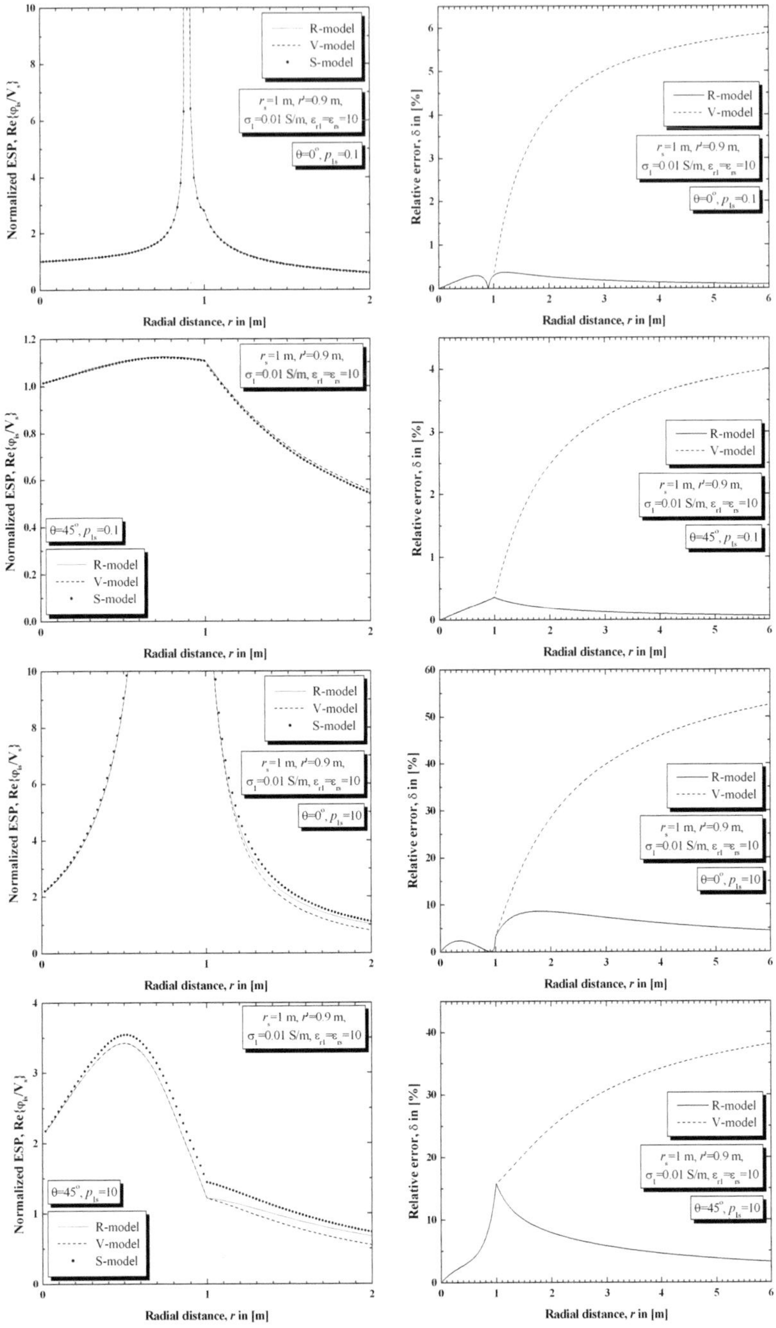

Fig. 6.5 The PGE inside the sphere. Normalized ESP, φ_{is}/V_s, $i = 1, s$, and corresponding relative errors versus radial distance r, for different values of geometry parameters and relation $p_{1s} = \sigma_1/\sigma_s$ as parameters

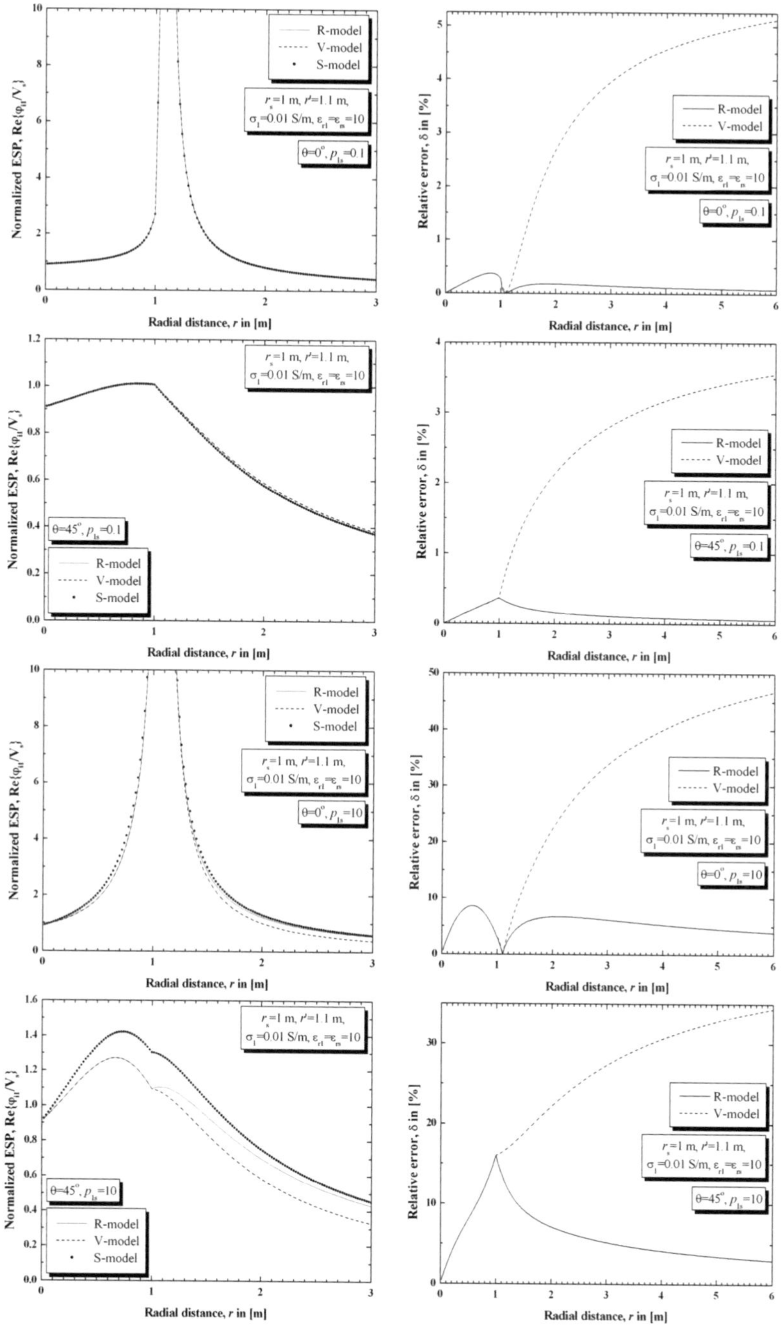

Fig. 6.6 The PGE outside the sphere. Normalized ESP, $\varphi_{i1}/V_s, i = 1, s$, and corresponding relative errors versus radial distance r, for different values of geometry parameters and relation $p_{1s} = \sigma_1/\sigma_s$ as parameters

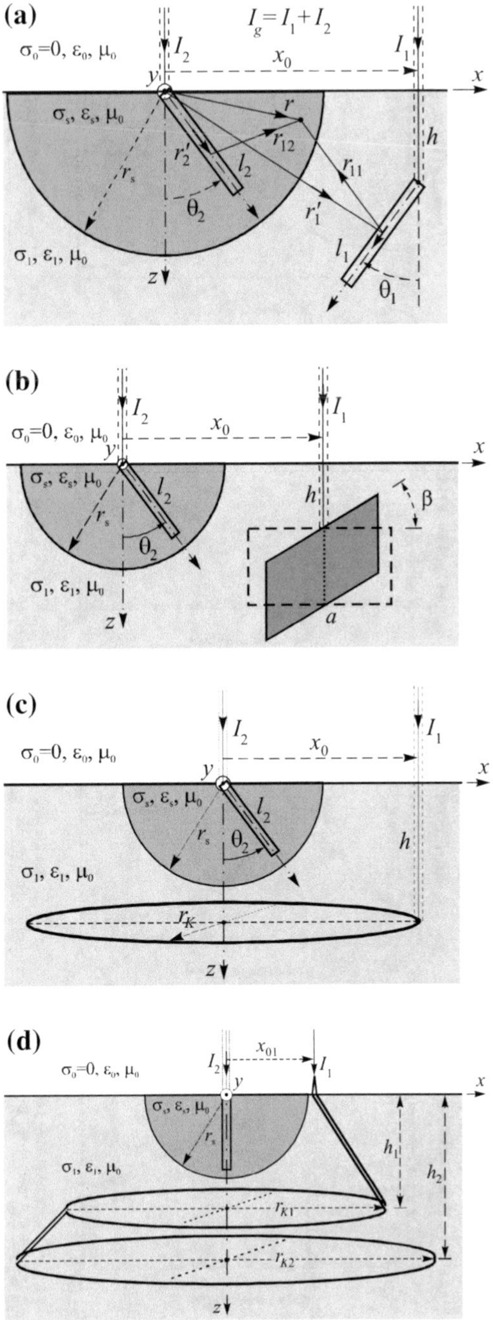

Fig. 6.7 Direct technical application to real technical problems

paper. The obtained solution is compared to the exact one from [19, pp. 201–205] and also, according to author's opinion, to the approximate solution from [20, pp. 97–98] and [21].

This conclusion (regarding the V-model) is theoretically explained and numerically verified in this paper. Both approximate solutions are in a closed form, which is not the case for the exact one according to [19].

Based on the numerical experiments, one can conclude that using the proposed approximate solution, smaller error in ESP evaluation is done, then when the approximate solution from [20, 21] is used, where the error is estimated in relation to the exact solution from [19]. This is also evident analysing the presented ESP expressions. The error is almost negligible in special cases, e.g. when the refraction coefficient is $n_{1s} < 1$.

Based on everything that was presented, one can conclude that the proposed solution can be successfully used for modelling grounding characteristics in the presence of a spherical and also semi-spherical ground inhomogeneity, but also for other problems of this kind.

The proposed approximate R-model can be also applied to derivation of expressions for the Green function of electrical dipole in the presence of a spherical material inhomogeneity and also on other problems of this kind.

References

1. Cvetkovic, N.N.: Modelling of hemispherical ground inhomogeneities. In: Proceedings of 10th International Conference on Telecommunications in Modern Satellite. Cable and Broadcasting Services - TELSIKS 2011, pp. 436–439. Niš, Serbia (2011)
2. Cvetkovic, N.N., Rancic P.D.: The point ground electrode in vicinity of the semi-spherical inhomogenity. In: CD Proceedings of 7th International Conference on Applied Electromagnetics - PES 2005, Niš, Serbia (2005). (Abstract. In: Proceedings of extended abstracts, pp. 139–140)
3. Cvetkovic, N.N., Rancic P.D.: Influence of the semi-spherical semi-conducting ground inhomogenity on the grounding characteristics. In: Proceedings of VII International Symposium on Electromagnetic Compatibility - EMC BARCELONA '06, pp. 918–923, Barcelona, Spain (2006). (Abstract. In: Book of abstracts, p. 29)
4. Cvetkovic, N.N., Rancic, P.D.: The influence of semi-spherical inohomogenity on the linear grounding system characteristics. FACTA UNIVERSITATIS. Ser. Electr. Energy **20**(2), 147–161 (2007)
5. Cvetkovic, N.N., Rancic P.D.: Conductive semi-sphere and linear ground electrode as pillar foundation grounding system. In: CD Proceedings of 8th International Conference on Applied Electromagnetics - PES 2007, Paper O3-6, Niš, Serbia (2007). (Abstract. In: Proceedings of extended abstracts, pp. 43–44)
6. Cvetkovic, N.N., Rancic, P.D.: A simple model for a numerical determination of electrical characteristics of a pillar foundation grounding system. Eng. Anal. Bound. Elem. **33**, 555–560 (2009)
7. Cvetkovic, N.N., Rancic, P.D.: Influence of foundation on pillar grounding system's characteristics. COMPEL: Int. J. Comput. Math. Electr. Electron. Eng. **28**(2), 471–492 (2009)
8. Cvetkovic, N.N., Rancic P.D.: Conductive semi-sphere and two ring ground electrodes as pillar foundation grounding system. In: CD Proceedings of 9th International Conference on Applied

Electromagnetics - PES 2009, Paper O3-2, Niš, Serbia (2009). (Abstract. In: Proceedings of extended abstracts, pp. 43–44)
9. Cvetkovic, N.N., Rancic P.D.: Conductive semi-sphere and two ring ground electrodes as pillar foundation grounding system. Elektrotechnica and Elektronica E+E, The Union of Electronics, Electrical Engineering and Telecommunications. vol. 45, no. 1–2, pp. 8–11 (2010)
10. Hannakam, L.: Allgemeine lösung des randwert problems für eine kugel durch integration des unggestörten erregenden feldes. Archiv für Elektrotechnik. **54**, 187–199 (1971). (in German)
11. Hannakam, L., Sakaji, N.: Disturbance of the potential distribution in the d-c feeded earth due to an ore deposit. Archiv für Elektrotechnik. **68**, 57–62 (1985). (in German)
12. Lavrov, G.A., Knyazev, A.S.: Prizemnie i Podzemnie Antenni: Teoriya i Praktika Antenn. Razmeshchennikh Vblizi Poverkhnosti Zemli, Izdatelstvo Sovetskoe Radio, Moskva, Russia (1965). (in Russian)
13. Lindell, I.V.: Electrostatic image theory for the dielectric sphere. Radio Sci. **27**(1), 1–8 (1992)
14. Lindell, I.V., Sten, J.C.-E.: Low-frequency image theory for the dielectric sphere. J. EM Waves Appl. **8**(3), 295–313 (1994)
15. Rancic, P.D., Kitanovic, M.I.: A new model for analysis of vertical asymmetrical linear antenna above a lossy half-space. Int. J. Electron. Commun. AEÜ **51**(3), 155–162 (1997)
16. Rancic, M.P., Rancic, P.D.: Vertical linear antennas in the presence of a lossy half-space: an improved approximate model. Int. J. Electron. Commun. AEÜ **60**(5), 376–386 (2006)
17. Reiß, K.: Deformation of the potential of a point charge by a spherical inhomogenity of material. Archiv für Elektrotechnik **74**, 135–144 (1990). (in German)
18. Sten, J.C.-E., Lindell, I.V.: Electrostatic image theory for the dielectric sphere with an internal source. McW Opt. T.L. **5**(11), 597–602 (1992)
19. Stratton, J.A.: Electromagnetic Theory. Mc Grow-Hill Book Company, New York (1941)
20. Uhlmann, H., Velickovic, D.M., Brandisky, K., Stancheva, R.D., Brauer, H.: Fundamentals of Modern Electromagnetics for Engineering-Textbook for Graduate Students, Part I: Static and Stationary EM Field. TU Ilmenau, Ilmenau, Germany (2005)
21. Velickovic, D.M.: Green's function of spherical body. In: Proceedings of EUROEM '94, THp-09-04, Bordeaux, France (1994)

Chapter 7
The Electromagnetic–Thermal Dosimetry Model of the Human Brain

Mario Cvetković and Dragan Poljak

Abstract The electromagnetic–thermal dosimetry model for the human brain exposed to EM radiation is developed. The electromagnetic (EM) model based on the surface integral equation (SIE) formulation is derived using the equivalence theorem for the case of a lossy homogeneous dielectric body. The thermal dosimetry model of the brain is based on the form of Pennes' equation of heat transfer in biological tissue. The numerical solution of the EM model is carried using the Method of Moments (MoM) while the bioheat equation is solved using the finite element method. Developed electromagnetic thermal model has been applied in internal dosimetry of the human brain to assess the absorbed electromagnetic energy and consequent temperature rise due to exposure of 900 MHz plane wave.

Keywords Electromagnetic-thermal model · Human brain · Numerical dosimetry · Surface integral equation approach

7.1 Introduction

The exposure of a modern man to artificially generated EM fields has raised some controversies as well as unanswered questions regarding the potentially harmful effects on the human health. This is, in particular, the case for the human head and brain exposed to radiation of nowadays ubiquitous cellular phones and base station antennas. Due to this fact the set of techniques for measuring and for calculation of the absorbed EM radiation in the human body referred to as the electromagnetic dosimetry have been developed.

M. Cvetković (✉)
Department of Power Engineering, University of Split, FESB,
R. Boskovica 32, 21000 Split, Croatia
e-mail: mcvetkov@fesb.hr

D. Poljak
Department of Electronics, University of Split, FESB,
R. Boskovica 32, 21000 Split, Croatia
e-mail: dpoljak@fesb.hr

S. Silvestrov and M. Rančić (eds.), *Engineering Mathematics I*,
Springer Proceedings in Mathematics & Statistics 178,
DOI 10.1007/978-3-319-42082-0_7

It is a well established fact that the principal biological effect of high frequency EM radiation is predominantly thermal in nature [1, 10, 12]. If the body absorbs high enough dose of EM power, it could lead to the harmful effects due to a breakdown of the protective thermoregulatory mechanisms. These can be quantified by the analysis of the thermal response of the particular body organ [17].

A direct experimental measurement of the brain thermal response in humans is not possible, and the indirect methods such as magnetic resonance imaging cannot record fine variations in temperature, hence lacking necessary resolution. On the other hand, animal studies are questionable due to a difference in interspecies size and tissue parameters. Consequently, the computational modeling provides the powerful alternative.

This paper describes an electromagnetic–thermal dosimetry model of the human brain. In the first part the electromagnetic model based on the SIE formulation is derived by using the equivalence theorem and the appropriate boundary conditions for the case of lossy dielectric object of an arbitrary shape. The second part outlines the thermal dosimetry model of the human brain based on the form of Pennes' equation of heat transfer in biological tissue. The obtained numerical results for the electric and magnetic fields, respectively, on the brain surface are presented, as well as the distribution of specific absorption rate (SAR) and the related temperature increase.

7.2 Electromagnetic Dosimetry Model

The human brain exposed to incident EM radiation is treated as a classical scattering problem.

The human brain, represented by an arbitrary shape S of a complex parameters (ε_2, μ_2) is placed in a free space with given properties (ε_1, μ_1), as shown in Fig. 7.1a. The complex permittivity of the brain is given by

$$\varepsilon_2 = \varepsilon_0 \varepsilon_r - j\frac{\sigma}{\omega}, \tag{7.1}$$

where ε_0 is permittivity of the free space, ε_r is relative permittivity, σ is electrical conductivity of the brain, and $\omega = 2\pi f$ is the operating frequency. The value for the permeability of the brain is that of free space, i.e. $\mu_0 = 4\pi \times 10^{-7}$ Vs/Am, due to the fact that biological tissues do not posses magnetic properties.

The lossy homogeneous object representing the human brain is exposed to the electromagnetic field $(\mathbf{E}^{inc}, \mathbf{H}^{inc})$. This incident field is present regardless of the scattering object. Due to the scattering object, a scattered field denoted by $(\mathbf{E}^{sca}, \mathbf{H}^{sca})$ is also present. The electric and magnetic fields exterior and interior to the surface S are, $(\mathbf{E}_1, \mathbf{H}_1)$ and $(\mathbf{E}_2, \mathbf{H}_2)$, respectively.

Applying the equivalence theorem, the equivalent problems for both regions 1 and 2 are formulated in terms of the equivalent electric and magnetic current densities **J**

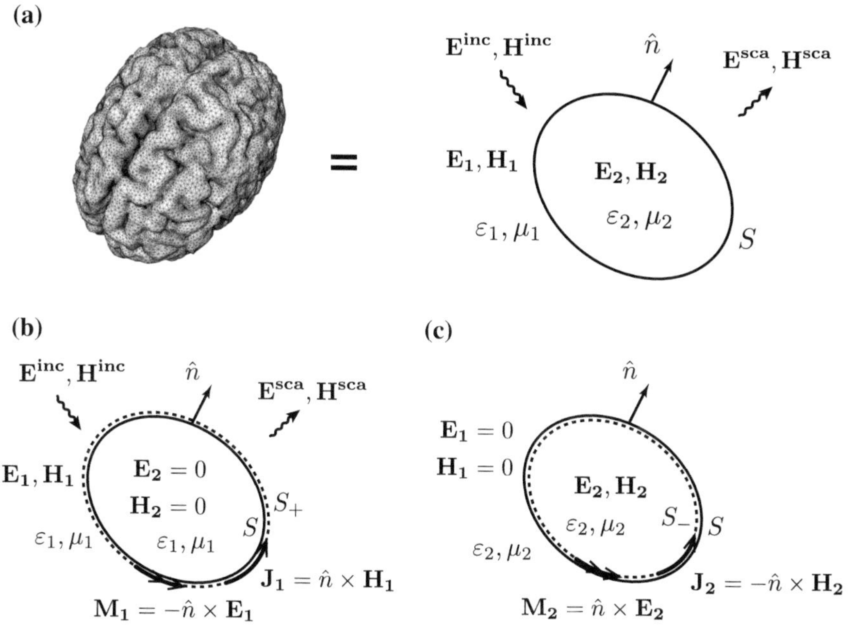

Fig. 7.1 Scattering from arbitrarily shaped lossy homogeneous dielectric (human brain) placed in the incident field ($\mathbf{E}^{inc}, \mathbf{H}^{inc}$). **a** Original problem, **b** Region 1 equivalent problem, **c** Region 2 equivalent problem

and **M** placed on the scatterer surface S [3, 9, 16, 22]. Two equivalent problems are shown on Fig. 7.1b and c, for the external and internal region, respectively.

In case of region 1 equivalent problem, shown in Fig. 7.1b, the field inside is assumed zero, ($\mathbf{E}_2 = 0, \mathbf{H}_2 = 0$), allowing one to arbitrarily choose material properties for this region. Selecting the properties of the exterior region, a homogeneous domain of (ε_1, μ_1) is obtained, enabling the use of the free space Green's function. The boundary conditions on the surface S are satisfied by introducing the equivalent surface currents $\mathbf{J}_1$ and $\mathbf{M}_1$ at the surface S. Applying the same procedure for the region 2, it follows another homogeneous domain of (ε_2, μ_2). Here as well, the equivalent surface currents $\mathbf{J}_2 = -\mathbf{J}_1$ and $\mathbf{M}_2 = -\mathbf{M}_1$, as shown in Fig. 7.1c, are introduced at the surface S.

Since both equivalent problems represent the equivalent current densities radiating in a homogeneous medium, following expressions for the scattered fields due to these sources can be used:

$$\mathbf{E}_n^{sca}(\mathbf{J}, \mathbf{M}) = -j\omega \mathbf{A}_n - \nabla \varphi_n - \frac{1}{\varepsilon_n} \nabla \times \mathbf{F}_n, \tag{7.2}$$

$$\mathbf{H}_n^{sca}(\mathbf{J}, \mathbf{M}) = -j\omega \mathbf{F}_n - \nabla \psi_n + \frac{1}{\mu_n} \nabla \times \mathbf{A}_n, \tag{7.3}$$

where $n = 1, 2$ is index of the medium where equivalent surface currents radiate, and φ, $\mathbf{F}$, ψ i $\mathbf{A}$ are electric and magnetic, scalar and vector potentials, respectively. These potentials are given in terms of integrals over the sources, i.e.

$$\mathbf{A}_n(\mathbf{r}) = \mu_n \int_S \mathbf{J}(\mathbf{r}')G_n(\mathbf{r}, \mathbf{r}')\, \mathrm{d}S', \tag{7.4}$$

$$\mathbf{F}_n(\mathbf{r}) = \varepsilon_n \int_S \mathbf{M}(\mathbf{r}')G_n(\mathbf{r}, \mathbf{r}')\, \mathrm{d}S', \tag{7.5}$$

$$\varphi_n(\mathbf{r}) = \frac{j}{\omega\varepsilon_n} \int_S \nabla'_S \cdot \mathbf{J}(\mathbf{r}')G_n(\mathbf{r}, \mathbf{r}')\, \mathrm{d}S', \tag{7.6}$$

$$\psi_n(\mathbf{r}) = \frac{j}{\omega\mu_n} \int_S \nabla'_S \cdot \mathbf{M}(\mathbf{r}')G_n(\mathbf{r}, \mathbf{r}')\, \mathrm{d}S', \tag{7.7}$$

where the electric and magnetic charge from (7.6) and (7.7) is replaced with the divergence of the electric and magnetic current, respectively, featuring the use of a continuity equation. $G_n(\mathbf{r}, \mathbf{r}')$ is homogeneous medium Green's function given by

$$G_n(\mathbf{r}, \mathbf{r}') = \frac{e^{-jk_n R}}{4\pi R}, \quad R = |\mathbf{r} - \mathbf{r}'|, \tag{7.8}$$

where R is the distance from the observation point $\mathbf{r}$ to the source point $\mathbf{r}'$, and k_n is the wave number in medium n.

Applying the boundary conditions for the electric field at the interface of the two equivalent problems, i.e. the surface S, the following is obtained

$$\left[-\mathbf{E}_n^{sca}(\mathbf{J}, \mathbf{M})\right]_{tan} = \begin{cases} \left[\mathbf{E}^{inc}\right]_{tan} & , \mathrm{n} = 1, \\ 0 & , \mathrm{n} = 2. \end{cases} \tag{7.9}$$

Equation (7.9) represents the electric field integral equation (EFIE) formulation in the frequency domain for the lossy homogeneous object, i.e. the human brain. The incident field $\mathbf{E}^{inc}$ is known, while $\mathbf{J}$ and $\mathbf{M}$ represent unknown surface currents, to be solved for.

Substituting (7.4)–(7.7) into (7.2) and (7.3), and the resulting expressions into (7.9), we arrive at the coupled set of integral equations

$$\begin{aligned} &j\omega\mu_n \int_S \mathbf{J}(\mathbf{r}')G_n(\mathbf{r}, \mathbf{r}')\, \mathrm{d}S' - \\ -&\frac{j}{\omega\varepsilon_n}\nabla \int_S \nabla'_S \cdot \mathbf{J}(\mathbf{r}')G_n(\mathbf{r}, \mathbf{r}')\, \mathrm{d}S' + \\ +&\nabla \times \int_S \mathbf{M}(\mathbf{r}')G_n(\mathbf{r}, \mathbf{r}')\, \mathrm{d}S' \end{aligned} = \begin{cases} \left[\mathbf{E}^{inc}\right]_{tan} & , \mathrm{n} = 1, \\ 0 & , \mathrm{n} = 2. \end{cases} \tag{7.10}$$

Following some mathematical manipulations on the second and third integral of (7.10), the nabla operator can be transferred to the Green's function leading to

$$\left.\begin{aligned} & j\omega\mu_n \int_S \mathbf{J}(\mathbf{r}')G_n(\mathbf{r},\mathbf{r}')\,\mathrm{d}S' - \\ & -\frac{j}{\omega\varepsilon_n}\int_S \nabla'_S\cdot\mathbf{J}(\mathbf{r}')\nabla G_n(\mathbf{r},\mathbf{r}')\,\mathrm{d}S' + \\ & +\int_S \mathbf{M}(\mathbf{r}')\times\nabla' G_n(\mathbf{r},\mathbf{r}')\,\mathrm{d}S' \end{aligned}\right. = \begin{cases} \left[\mathbf{E}^{inc}\right]_{tan} & ,\ \mathrm{n}=1, \\ 0 & ,\ \mathrm{n}=2, \end{cases} \tag{7.11}$$

where the property for the Green's function gradient, $\nabla G_n(\mathbf{r},\mathbf{r}') = -\nabla' G_n(\mathbf{r},\mathbf{r}')$, is used in (7.11).

7.2.1 Numerical Solution

For complex geometry of surface S, such as the human brain, the coupled integral equations set (7.11) cannot be solved analytically, hence the numerical approach is necessary. The corresponding numerical solution is carried out via the method of moments (MoM). It is a technique for finding an approximate solution to the system of a linear operator equations. Inserting the approximated function back into the operator equation, while multiplying it by a set of a known test functions, leads to a system of a linear equations. Solving the matrix system, one obtain the unknown coefficients from which equivalent surface currents are determined.

This work features an efficient MoM scheme in which the equivalent electric and magnetic currents $\mathbf{J}$ and $\mathbf{M}$ in (7.11), are first expanded by a linear combination of basis functions $\mathbf{f}_n$ and $\mathbf{g}_n$, respectively [5]

$$\mathbf{J}(\mathbf{r}) = \sum_{n=1}^{N} J_n \mathbf{f}_n(\mathbf{r}), \tag{7.12}$$

$$\mathbf{M}(\mathbf{r}) = \sum_{n=1}^{N} M_n \mathbf{g}_n(\mathbf{r}), \tag{7.13}$$

where J_n and M_n are unknown coefficients, and N is the number of elements used to discretize the surface S.

The brain surface S is discretized using the triangular elements or patches enabling one to use the Rao-Wilton-Glisson (RWG) basis functions [18] specially developed for triangular patches.

RWG function $\mathbf{f}_n$ is defined on T_n^+ and T_n^- pair of triangles that share a common edge (hence, sometimes the name edge-element is used), while on the rest of the surface S function vanishes.

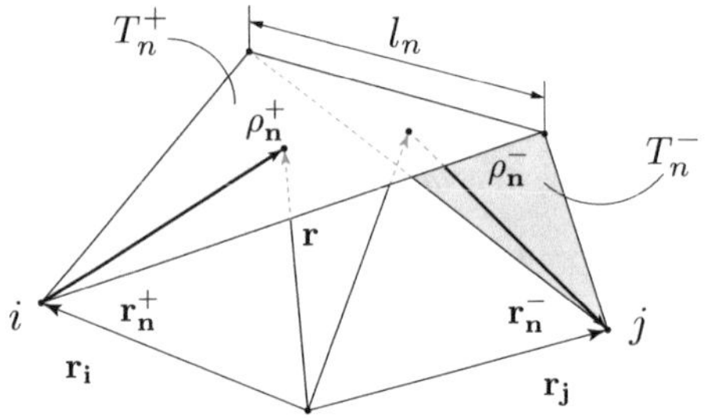

Fig. 7.2 RWG basis function $\mathbf{f}_n(\mathbf{r})$ defined on a pair of triangles in $\mathbf{R}^3$ [18]

Namely, the function is given by

$$\mathbf{f}_n^{\pm}(\mathbf{r}) = \begin{cases} \dfrac{l_n}{2A_n^{\pm}}\boldsymbol{\rho}_n^{\pm} & ,\mathbf{r} \in T_n^{\pm}, \\ 0 & ,\mathbf{r} \notin T_n^{\pm}, \end{cases} \tag{7.14}$$

where l_n is the edge length at the interface of triangles T_n^+ and T_n^-, while A_n^+ and A_n^- are the surface areas of those triangles. The vector $\boldsymbol{\rho}_n^+ = \mathbf{r} - \mathbf{r}_n^+$ is directed from the free vertex of T_n^+ and $\boldsymbol{\rho}_n^- = \mathbf{r}_n^- - \mathbf{r}$ is directed to the free vertex of T_n^-, as shown on Fig. 7.2.

While the surface electric current $\mathbf{J}$ is approximated by the RWG function $\mathbf{f}_n$, the surface magnetic current $\mathbf{M}$ is approximated by $\mathbf{g}_n = \hat{n} \times \mathbf{f}_n$, i.e. the function point wise orthogonal to the RWG function. The unknown equivalent currents $\mathbf{J}(\mathbf{r}')$ and $\mathbf{M}(\mathbf{r}')$ from (7.11) are substituted by (7.12) and (7.13). Equation (7.11) is next multiplied by the set of a test functions $\mathbf{f}_m$, where $\mathbf{f}_m = \mathbf{f}_n$, and integrated over the surface S. After some mathematical manipulations, it follows

$$\begin{aligned} & j\omega\mu_i \sum_{n=1}^{N} J_n \int_S \mathbf{f}_m(\mathbf{r}) \cdot \int_{S'} \mathbf{f}_n(\mathbf{r}')G_i \, dS' \, dS + \\ & + \frac{j}{\omega\varepsilon_i} \sum_{n=1}^{N} J_n \int_S \nabla_S \cdot \mathbf{f}_m(\mathbf{r}) \int_{S'} \nabla_S' \cdot \mathbf{f}_n(\mathbf{r}')G_i \, dS' \, dS \pm \\ & \pm \sum_{n=1}^{N} M_n \int_S \mathbf{f}_m(\mathbf{r}) \cdot [\hat{n} \times \mathbf{g}_n(\mathbf{r}')] \, dS + \\ & + \sum_{n=1}^{N} M_n \int_S \mathbf{f}_m(\mathbf{r}) \cdot \int_{S'} \mathbf{g}_n(\mathbf{r}') \times \nabla' G_i \, dS' \, dS \end{aligned} = \begin{cases} \int_S \mathbf{f}_m(\mathbf{r}) \cdot \mathbf{E}^{inc} \, dS & ,i = 1, \\ 0 & ,i = 2, \end{cases} \tag{7.15}$$

where subscript i is now the index of the medium. The third and the fourth integrals on the left hand side of (7.15) represent the residual term and the Cauchy principal value, respectively, of the last integral from (7.11). The residual term is calculated in the limiting case when $\mathbf{r} \rightarrow \mathbf{r}'$.

After extracting the two sums, (7.15) can be written in the form of the following linear equations system

$$\sum_{n=1}^{N}\left(j\omega\mu_i A_{mn,i} + \frac{j}{\omega\varepsilon_i}B_{mn,i}\right)J_n + \sum_{n=1}^{N}\left(C_{mn,i} + D_{mn,i}\right)M_n = \begin{cases} V_m & ,i=1, \\ 0 & ,i=2, \end{cases} \tag{7.16}$$

or in the matrix form as

$$[\mathbf{Z}]\cdot\{\mathbf{I}\} = \{\mathbf{V}\}, \tag{7.17}$$

where **Z** and **V** represents the system matrix, and the source vector, respectively, while $A_{mn,i}$, $B_{mn,i}$, $C_{mn,i}$ and $D_{mn,i}$ represent the surface integrals calculated for each $m - n$ combination of basis and testing functions, respectively.

Solution to the (7.17) is a vector **I** containing the unknown coefficients J_n and M_n. From these coefficients, the equivalent surface electric and magnetic currents **J** and **M**, respectively, placed on the surface S of the dielectric object, i.e. the human brain, can be determined from (7.12) and (7.13), respectively. Knowing these currents, the electric field can be determined at an arbitrary point in space, i.e. the electric field inside the human brain represented by parameters (ε_2, μ_2), can be calculated from the following integral expression:

$$\begin{aligned}\mathbf{E}_2(\mathbf{r}) = & -j\omega\mu_2\int_S \mathbf{J}(\mathbf{r}')G_2(\mathbf{r},\mathbf{r}')\,dS' - \\ & -\frac{j}{\omega\varepsilon_2}\int_S \nabla'_S\cdot\mathbf{J}(\mathbf{r}')G_2(\mathbf{r},\mathbf{r}')\,dS' - \\ & -\int_S \mathbf{M}(\mathbf{r}')\times\nabla G_2(\mathbf{r},\mathbf{r}')\,dS'.\end{aligned} \tag{7.18}$$

Once obtained the electric field distribution inside the brain, the distribution of the SAR can be readily found using the following relation

$$SAR = \frac{\sigma}{2\rho}|\mathbf{E}|^2, \tag{7.19}$$

where σ and ρ are the electric conductivity and the brain tissue density, respectively. The SAR distribution can be latter used as the input information to the thermal part of the brain model.

7.3 Thermal Dosimetry Model

It is well known that two most important factors for sustaining biological system are the metabolism and the blood flow [14]. The complex network of blood vessels significantly complicates mathematical modeling of heat transfer in biological tissues, unless a distributed heat source or sink is assumed.

The most commonly used model taking the flow of blood in this manner is the Pennes bioheat transfer equation [15]

$$\nabla \cdot (\lambda \nabla T) + w\rho_b c_b (T_a - T) + Q_m + Q_{ext} = \rho C \frac{\partial T}{\partial t}. \quad (7.20)$$

According to (7.20) the temperature rise in the given volume of tissue is based on the energy balance between the conductive heat transfer, the heat generated due to metabolic processes Q_m, the heat loss (generation) due to blood perfusion, and the influence of external heat sources Q_{ext}. The volumetric perfusion rate is given by ω, ρ_b and c_b are the density and the specific heat capacity of blood, respectively, λ is the thermal conductivity of the tissue, while T_a is the temperature of the arterial blood.

The analytical solutions of the bioheat transfer equation (7.20) are limited to cases of relatively high degree of symmetry [21], thus making numerical approach necessary for problems with complex geometry of the domain arising for realistic exposure scenarios. In this work the problem of determining the temperature distribution in the human brain is addressed using the finite element method (FEM) [7].

The steady-state temperature distribution in the brain, exposed to an incident time harmonic EM field, is governed by the stationary form of the bioheat equation (7.20)

$$\nabla \cdot (\lambda \nabla T) + W_b c_b (T_a - T) + Q_m + Q_{ext} = 0 \quad (7.21)$$

extended with Q_{ext}. This term represents the amount of heat generated per unit time per unit volume due to absorption of EM energy in the biological tissue [4, 6, 7]:

$$Q_{ext} = \rho \cdot SAR, \quad (7.22)$$

where SAR is defined by (7.19).

The bioheat equation (7.21) is supplemented by the corresponding boundary conditions, as shown in Fig. 7.3.

This work features the use of Neumann or the natural boundary conditions given by

$$-\lambda \frac{\partial T}{\partial \hat{n}} = h_s (T - T_{amb}), \quad (7.23)$$

where, λ is the thermal conductivity of the brain, and h_s is the convection coefficient between the surface and the surroundings, T and T_{amb} are the surface and the ambient

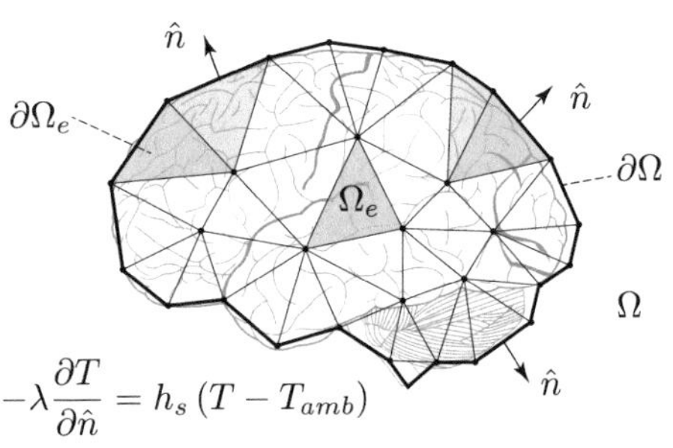

Fig. 7.3 Illustration of the finite element mesh with boundary conditions on the brain surface

temperature, respectively. Unit normal $\hat{n}$ is directed from the surface S, as shown on Fig. 7.3.

Note that the heat loss due to radiation, and the forced convection are neglected. Neverthless, (7.23) satisfactorily describes the heat exchange between the surface of the brain and the environment.

Since the human brain is separated from the scalp by various other tissues, when using the homogeneous brain model, it is necessary to account for the heat exchange through all of them. This is ensured by using the effective thermal convection coefficient h_{eff} [23] between the brain and the surroundings.

The widely adopted value for the effective thermal convection coefficient, typical for the human brain, is $h_{eff} = 1.2 \times 10^{-3}$ W/cm^2°C [20]. This value is used in our homogeneous thermal model of the brain, as well.

7.3.1 Finite Element Solution

The finite element formulation of (7.21) is based on the weighted residual approach. The approximate solution of (7.21) is expanded in terms of the known basis functions N_i and the unknown coefficients α_i

$$T(x, y, z) = \sum_{i=1}^{m} N_i(x, y, z)\alpha_i, \tag{7.24}$$

where i is the node index, m is the number of nodes per finite element, and N_i is the basis function given by

$$N_i(x, y, z) = \frac{1}{D}(V_i + a_i x + b_i y + c_i z), \quad i = 1, 2, 3, 4, \tag{7.25}$$

where expressions for the coefficients a_i, b_i, c_i, V_i and D can be found in [19].

Multiplying (7.21) by a set of weighting functions W_j and integrating over the domain $\Omega = V$, yields

$$\int_{\Omega} [\nabla \cdot (\lambda \nabla T) + W_b c_b (T_a - T) + Q_m + Q_{ext}] W_j \, d\Omega = 0. \tag{7.26}$$

Applying the same procedure on (7.23), it follows

$$-\lambda \int_{\partial\Omega} \frac{\partial T}{\partial \hat{n}} W_j \, dS = \int_{\partial\Omega} h_s T W_j \, dS - \int_{\partial\Omega} h_s T_{amb} W_j \, dS. \tag{7.27}$$

Taking the integration by parts in the first term of (7.26), the Gauss' divergence theorem is applied, resulting in

$$\int_{\Omega} \nabla \cdot \left[(\lambda \nabla T)\, W_j \right] \mathrm{d}\Omega = \lambda \int_{\partial\Omega} \frac{\partial T}{\partial \hat{n}} W_j \,\mathrm{d}S. \tag{7.28}$$

Inserting (7.28) into (7.26), and after some rearranging, a suitable expression for the FEM implementation is obtained [7]

$$\int_{\Omega} \lambda \nabla T \cdot \nabla W_j \,\mathrm{d}\Omega + \int_{\Omega} W_b c_b T W_j \,\mathrm{d}\Omega =$$
$$= \int_{\partial\Omega} \lambda \frac{\partial T}{\partial \hat{n}} W_j \,\mathrm{d}S + \int_{\Omega} (W_b c_b T_a + Q_m + Q_{ext}) W_j \,\mathrm{d}\Omega. \tag{7.29}$$

Having discretized the brain surface by triangular elements, performed in the electromagnetic part of the model [5], the interior of the brain depicted as Ω in Fig. 7.3 was discretized by the tetrahedral elements.

Implementing the Galerkin-Bubnov procedure, followed by the standard finite element discretization of (7.29), the weak formulation for the finite element domain Ω_e can be written in the matrix form

$$[K]^e \{T\}^e = \{M\}^e + \{P\}^e, \tag{7.30}$$

where $[K]^e$, $\{M\}^e$ and $\{P\}^e$ are the finite element matrix

$$[K]^e_{ji} = \int_{\Omega_e} \lambda^e \nabla W_i \cdot \nabla W_j \,\mathrm{d}\Omega_e + \int_{\Omega_e} W^e_b c^e_b W_i W_j \,\mathrm{d}\Omega_e, \tag{7.31}$$

the flux vector on the boundary $\partial\Omega_e$ of the finite element

$$\{M\}^e_j = \int_{\partial\Omega_e} \lambda^e \frac{\partial T}{\partial \hat{n}} W_j \,\mathrm{d}S_e, \tag{7.32}$$

and the finite element source vector

$$\{P\}^e_j = \int_{\Omega_e} (W^e_b c^e_b T_a + Q^e_m + Q^e_{ext}) W_j \,\mathrm{d}\Omega_e, \tag{7.33}$$

respectively.

Solving (7.31)–(7.33) for each N elements, the global matrix is assembled from the contribution of the local finite element matrices, while the global flux and the source vectors are assembled from the local flux and the local source vectors, respectively:

$$[K]\{T\} = \{M\} + \{P\}. \tag{7.34}$$

The solution of the matrix system (7.34) is the vector $\{T\}$ whose elements represent the values of temperature in the tetrahedra nodes.

7.4 Computational Example

The numerical results for our homogeneous three-dimensional brain model are presented in this section. The dimensions of the average adult human brain are used (length 167 mm, width 140 mm, height 93 mm, volume of 1400 cm^3) [2]. The surface of the brain is disretized using the $T = 696$ triangular elements and $N = 1044$ edge-elements, while the interior of the brain is discretized using 1871 tetrahedral elements. The frequency dependent parameters of the human brain are taken from [8]. The value for the relative permittivity and the electrical conductivity of the brain are $\varepsilon_r = 45.805$ and $\sigma = 0.766$ S/m, respectively, taken as the average values between white and gray matter at 900 MHz. Value for the density of the brain tissue is $\rho = 1046$ kg/m^3.

The incident plane wave of power density of $P = 5$ mW/cm^2 is directed perpendicular to the right side of the brain (positive x coordinate), the polarization of the wave is in the horizontal (y coordinate) direction, while the operating frequency is 900 MHz.

Using our EM model based on the SIE formulation [5], the distribution of the electric and magnetic fields on the brain surface, shown on Fig. 7.4, are determined first.

From the electric field values in the brain interior, SAR can be calculated using (7.19). The obtained peak and average SAR values are 0.856 W/kg and 0.174 W/kg, respectively. The calculated results show that the peak SAR value in the human brain does not exceed the limit set by ICNIRP [11] as a basic restriction for localized SAR (in the head and the trunk), for the occupational exposure (10 W/kg). Figure 7.5 shows the distribution of the SAR obtained for the brain model.

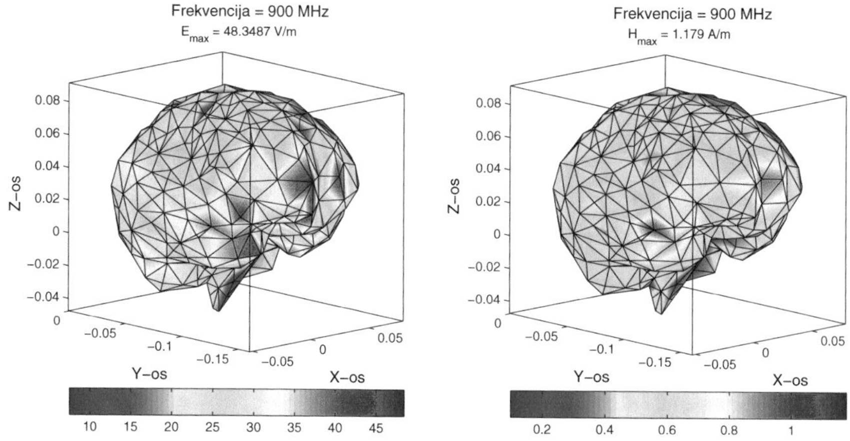

Fig. 7.4 Distribution of electric and magnetic fields on the brain surface. Horizontally polarized plane wave of frequency 900 MHz

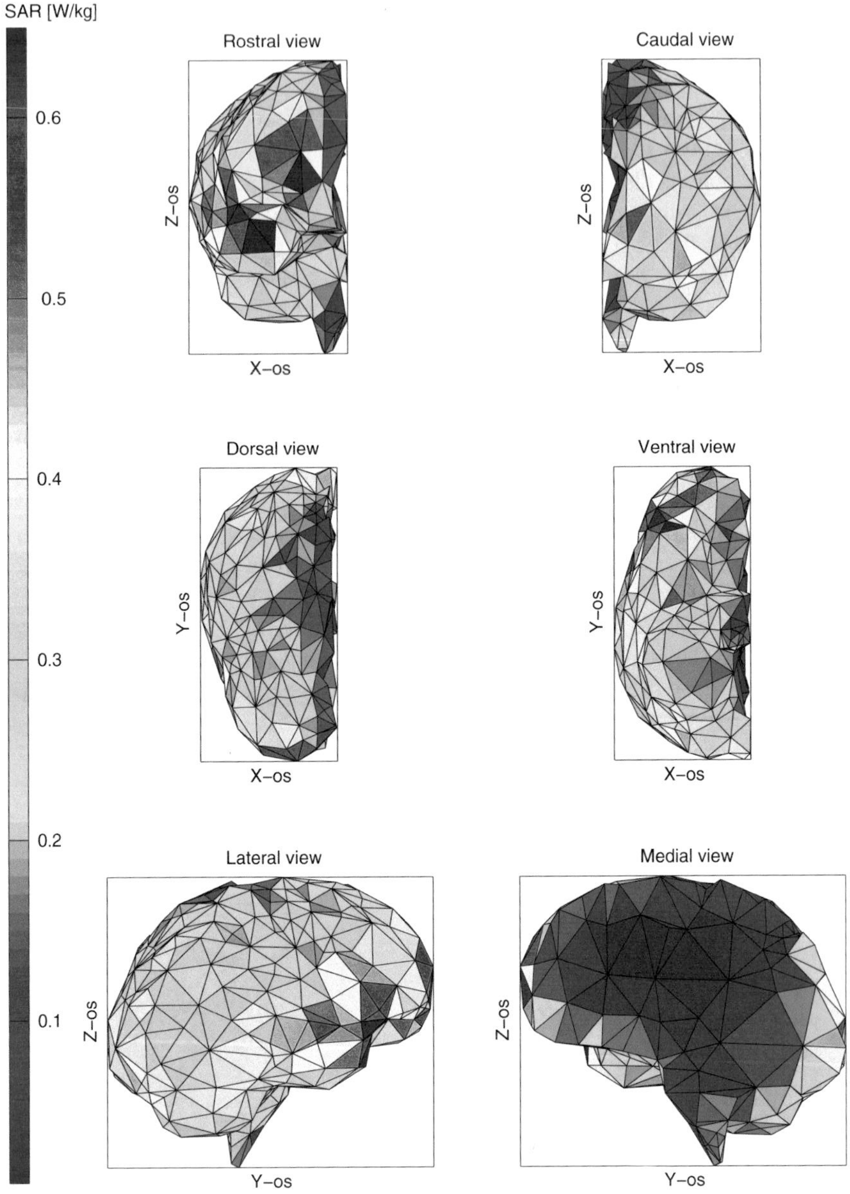

Fig. 7.5 Distribution of SAR for the case of horizontally polarized plane wave of frequency 900 MHz, power density $P = 5$ mW/cm^2

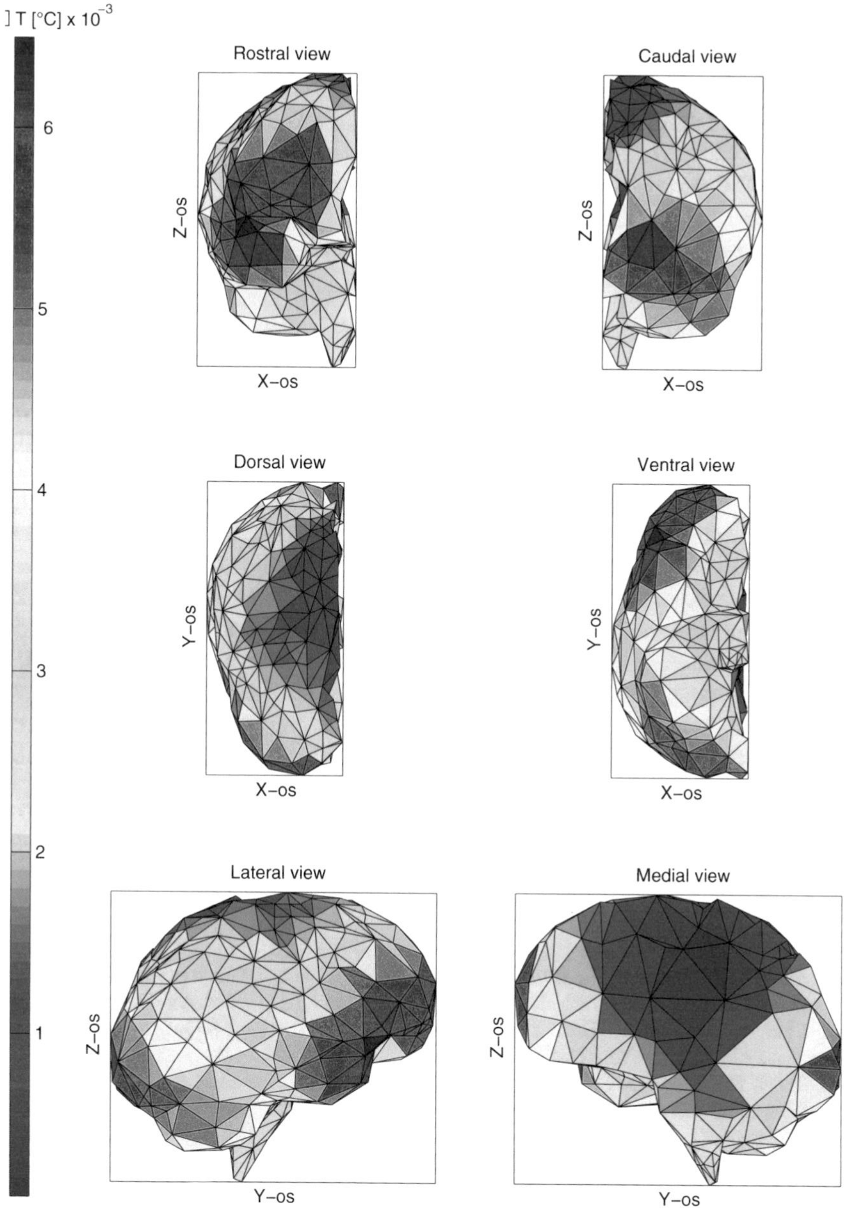

Fig. 7.6 Temperature rise in the human brain model due to incident 900 MHz horizontally poralized plane wave, power density $P = 5\,\text{mW/cm}^2$

The human brain parameters used in the thermal dosimetry model are taken from [13]: the heat conductivity $\lambda = 0.513$ W/m °C, the volumetric perfusion rate of blood $W_b = 33297$ kg/m^3, the specific heat capacity of blood $c_b = 1$ J/kg °C, the heat generated due to metabolism $Q_m = 6385$ W/m^3, and the arterial blood temperature $T_{art} = 37$ °C.

Figure 7.6 shows the results for the temperature rise in the human brain. The maximum temperature rise is $\Delta T = 7.11 \times 10^{-3}$ °C, which is rather negligible compared to the values proven to cause adverse health effects.

7.5 Conclusion

This work deals with the electromagnetic–thermal dosimetry model for the human brain exposed to EM radiation. The electromagnetic model based on the surface integral equation formulation is first derived from the equivalence theorem and using the boundary conditions for the electric field. The human brain is represented by an arbitrarily shaped lossy homogeneous dielectric. The thermal model of the brain is based on the extended form of the Pennes' bioheat equation supplemented by the natural boundary condition on the brain surface. The numerical results for the electric and magnetic fields are presented for the brain exposed to a radiation of 900 MHz horizontally polarized plane wave. The calculated peak SAR value in the human brain does not exceed the basic restriction for the occupational exposure set by the ICNIRP. Also, the resulted temperature rise in the human brain is rather negligible compared to established health based threshold.

References

1. Adair, E.R., Petersen, R.: Biological effects of radiofrequency/microwave radiation. IEEE Trans. Microw. Theory Tech. **50**(3), 953–962 (2002). doi:10.1109/22.989978
2. Blinkov, S.M., Glezer, I.I.: The Human Brain in Figures and Tables: A Quantitative Handbook. Plenum Press, New York (1968)
3. Chew, W.C., Tong, M.S., Hu, B.: Integral Equation Methods for Electromagnetic and Elastic Waves. Morgan & Claypol Publishers, California (2009)
4. Cvetković, M., Čavka, D., Poljak, D., Peratta, A.: 3D FEM temperature distribution analysis of the human eye exposed to laser radiation. Adv. Comput. Methods Exp. Heat Transf. XI. WIT Trans. Eng. Sci. **68**, 303–312 (2009)
5. Cvetković, M., Poljak, D.: An efficient integral equation based dosimetry model of the human brain. In: Proceedings of the International Symposium on Electromagnetic Compatibility (EMC EUROPE) 2014, Gothenburg, Sweden, 1–4 September, pp. 375–380 (2014)
6. Cvetković, M., Poljak, D., Peratta, A.: Thermal modelling of the human eye exposed to laser radiation. In: Proceedings of 2008 International Conference on Software, Telecommunications and Computer Networks, Split, Croatia, 25–26 September, pp. 16–20 (2008)
7. Cvetković, M., Poljak, D., Peratta, A.: FETD computation of the temperature distribution induced into a human eye by a pulsed laser. Prog. Electromagn. Res. PIER **120**, 403–421 (2011)

8. Gabriel, C.: Compilation of the dielectric properties of body tissues at RF and microwave frequencies. Technical report, Brooks Air Force Base, TX. Report: AL/OE-TR-1996-0037 (1996)
9. Harrington, R.: Boundary integral formulations for homogeneous material bodies. J. Electromagn. Waves Appl. **3**(1), 1–15 (1989)
10. Hirata, A., Shiozawa, T.: Correlation of maximum temperature increase and peak SAR in the human head due to handset antennas. IEEE Trans. Microw. Theory Tech. **51**(7), 1834–1841 (2003). doi:10.1109/TMTT.2003.814314
11. International Commission on Non-Ionizing Radiation Protection: ICNIRP): Guidelines forlimiting exposure to time-varying electric, magnetic and electromagnetic fields (up to 300GHz). Health Physics **74**(4), 494–522 (1998)
12. International Commission on Non-Ionizing Radiation Protection: ICNIRP): Guidelines forlimiting exposure to time-varying electric and magnetic fields (1 Hz to 100 kHz). Health Physics **99**(6), 818–836 (2010). doi:10.1097/HP.0b013e3181f06c86
13. McIntosh, R.L., Anderson, V.: A comprehensive tissue properties database provided for the thermal assessment of a human at rest. Biophys. Rev. Lett. **5**(3), 129–151 (2010)
14. Minkowycz, W.J., Sparrow, E.M., Murthy, J.Y.: Handbook of Numerical Heat Transfer, 2nd edn. Wiley, New York (2006)
15. Pennes, H.H.: Analysis of tissue and arterial blood temperatures in the resting human forearm. 1948. J. Appl. Physiol. **85**(1), 5–34 (1998)
16. Poggio, A.J., Miller, E.K.: Integral equation solutions of three–dimensional scattering problems. In: Mittra, R. (ed.) Computer Techniques for Electromagnetics, Second Edition, Chap. 4, pp. 159–264. Hemisphere Publishing Corporation, Washington (1987)
17. Poljak, D., Peratta, A., Brebbia, C.A.: The boundary element electromagnetic-thermal analysis of human exposure to base station antennas radiation. Eng. Anal. Bound. Elem. **28**(7), 763–770 (2004). doi:10.1016/j.enganabound.2004.02.004
18. Rao, S., Wilton, D.R., Glisson, A.: Electromagnetic scattering by surfaces of arbitrary shape. IEEE Trans. Antennas Propag. **30**(3), 409–418 (1982)
19. Silvester, P.P., Ferrari, R.L.: Finite Elements for Electrical Engineers, 3rd edn. Cambridge University Press, Cambridge (1996)
20. Sukstanskii, A., Yablonskiy, D.: Theoretical model of temperature regulation in the brain during changes in functional activity. Proc. Natl. Acad. Sci. **103**(32), 12144–12149 (2006). doi:10.1073/pnas.0604376103
21. Sukstanskii, A.L., Yablonskiy, D.A.: An analytical model of temperature regulation in human head. J. Therm. Biol. **29**(7), 583–587 (2004). doi:10.1016/j.jtherbio.2004.08.028
22. Umashankar, K., Taflove, A., Rao, S.: Electromagnetic scattering by arbitrary shaped three-dimensional homogeneous lossy dielectric objects. IEEE Trans. Antennas Propag. **34**(6), 758–766 (1986)
23. Zhu, M., Ackerman, J.J.H., Sukstanskii, A.L., Yablonskiy, D.A.: How the body controls brain temperature: the temperature shielding effect of cerebral blood flow. J. Appl. Physiol. **101**(5), 1481–1488 (2006). doi:10.1152/japplphysiol.00319.2006

Chapter 8
Quasi-TEM Analysis of Multilayered Shielded Microstrip Lines Using Hybrid Boundary Element Method

Mirjana Perić, Saša Ilić and Slavoljub Aleksić

Abstract In this paper multilayered shielded structures have been analyzed using the hybrid boundary element method. The method is based on the equivalent electrodes method, on the point-matching method for the potential of the perfect electric conductor electrodes and for the normal component of electric field at boundary surface between any two dielectric layers. The quasi-static TEM analysis is applied. The characteristic parameters (characteristic impedance and effective relative permittivity) of shielded multilayered microstrip lines are determined. The method can be use to analyze microstrip transmission lines with arbitrary configurations, arbitrary number of conductors and dielectric layers, infinitesimally thin or finite metallization thickness and finite width of substrate. It is a simple and an accurate procedure comparing to the other numerical and semi-numerical methods. In order to verify the obtained results, they have been compared with the finite element method and results that have already been reported in the literature. A very good results agreement with available data can be noticed.

Keywords Characteristic impedance · Hybrid boundary element method · Multilayered structures · Shielded microstrip lines

8.1 Introduction

Analysis of microwave transmission lines is the main subject of research in the world for more than six decades. From the first days of the stripline origin, back in 1949, and its modifications that followed in the coming years, an "army" of scientists trying

M. Perić (✉) · S. Ilić · S. Aleksić
Faculty of Electronic Engineering, University of Niš, Niš, Serbia
e-mail: mirjana.peric@elfak.ni.ac.rs

S. Ilić
e-mail: sasa.ilic@elfak.ni.ac.rs

S. Aleksić
e-mail: slavoljub.aleksic@elfak.ni.ac.rs

S. Silvestrov and M. Rančić (eds.), *Engineering Mathematics I*,
Springer Proceedings in Mathematics & Statistics 178,
DOI 10.1007/978-3-319-42082-0_8

to analyze it simpler and to design the structures, which, due to their characteristics, have found wide application in microwave integrated circuits, for microwave filters and antennas design, delay lines, directional couplers, etc.

Microstrip lines with multilayered media have been investigated during the years using various numerical and analytical techniques such as variational method [1, 5], boundary element method (BEM)/method of moments [2, 4, 7, 11], conformal mapping, so-called moving perfect electric wall (MPEW) method [22, 23], finite element method (FEM) [10, 13, 15], finite difference method (FDM) [3], etc. Those methods evaluate, in different manners, the capacitance per unit length of the microstrip line, from which the characteristic impedance can be calculated. The application of some of those methods is limited by the number of dielectric layers, conductor's thickness or shape. The multiple image method can be used for deriving Green's function for the microstrip line, but there is difficult to extend it to the case of multilayered and shielded stripline. The equivalent electrodes method (EEM) [25] was successfully applied for analysing transmission and microstrip lines in [26]. Generally, the application of the EEM depends on the Green's function for the observed problem. The method is based on the combination of analytical derivation of the Green's function in the closed form and the numerical procedure for solving simplified problems.

Combining the EEM with the boundary element method, in order to solve problems of arbitrarily shaped multilayered structures, where finding the Green's function is very difficult or even impossible, an improvement of the EEM has been done. This method, called in [18] the hybrid boundary element method (HBEM), is developed at our Department. It is based on the EEM, on the point-matching method (PMM) for the potential of the perfect electrode conductor (PEC) electrodes and for the normal component of electric field at the boundary surface between any two dielectric layers. Until now, it is successfully applied to solve large scale of electromagnetic problems [8, 9, 16, 19, 20, 27]. The method is capable to analyze microstrip transmission lines with arbitrary configurations, arbitrary number of conductors and dielectric layers, infinitesimally thin or finite metallization thickness and finite width of substrate. It is valid if a conductor touches a dielectric interface, straddles a dielectric interface or is a totally within one dielectric media.

In this paper, as an illustration of the HBEM application, numerical solutions for shielded microstrip lines with multilayered media are presented. A quasi-TEM analysis is applied, which is often adequate for microwave frequencies. Quasi-static methods are based on the assumption that the dominant mode of the wave, which propagates along the transmission line, can be approximated (with good accuracy) by the transversal electromagnetic (TEM) wave. This assumption is valid on low microwave frequencies (typically by 5–10 GHz). If frequency increases, then the value of longitudinal components of electromagnetic field rises, and hence, it cannot be neglected.

The characteristic impedance and effective relative permittivity of several shielded microstrip lines will be determined. With the aim to test the accuracy of the method, the results will be compared to those obtained by the FEMM software [12] and the results already reported in the literature.

8.2 Theoretical Background

In order to explain the application of the hybrid boundary element method, the multilayered shielded microstrip line with an arbitrary cross-section and inhomogeneous dielectric layer is considered, Fig. 8.1. The following procedure can also be applied for analysis of open microstrip line structures. The each subregion of the layer is isotropic, linear, homogeneous dielectric with different permittivities ε_i $(i = 1, \ldots, N)$. The line is uniform along z-axis. The quasi-TEM analysis assumes that the dominant mode propagating along the line is the TEM mode.

During the HBEM application, each arbitrary shaped surface of the PEC electrode as well as an arbitrary shaped boundary surface between any two dielectric layers is divided into a large number of segments. Each of those segments on PEC electrode is replaced by equivalent electrodes (EEs) placed at their centres. The equivalent electrodes can be: toroidal electrodes, in the case of 2D problems with axial symmetry, cylindrical electrodes (line charges) for planparallel problems and spherical electrodes for 3D systems. The potential of equivalent electrodes, obtained in this manner, is the same as the potential of PECs themselves:

$$\varphi = \varphi_k,\ k = 1, \ldots, N. \tag{8.1}$$

where φ_k is the potential of k-th electrode.

The segments at any boundary surfaces between two layers are replaced by discrete equivalent total charges placed in the air. The Green's function for the electric scalar potential of the charges is used. However, the problem occurs during the determination of polarized charges at the boundary surfaces of any two layers, because it must be taken into consideration a dielectric influence on the electric potential and electric field distribution. In the electrostatic field theory it is well

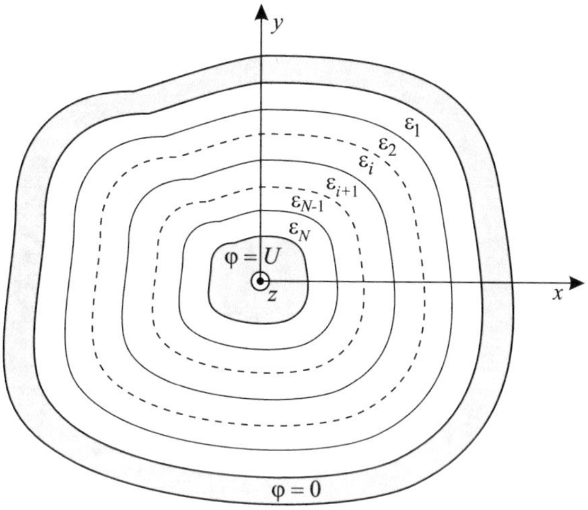

Fig. 8.1 Cross-section of a shielded microwave transmission line

known that the influence of dielectric can be replaced with polarized (surface and volume) charges, placed in the air. In that way, the electric field strength and electric potential will remain unchanged at all system points. Inside an isotropic, linear and homogeneous dielectric, the polarized volume charges, ρ_v, does not exist,

$$\rho_{\mathrm{v}} = -\mathrm{div}\boldsymbol{P} = 0, \tag{8.2}$$

where $\boldsymbol{P}$ is the polarization vector. Thus, only the polarized surface charges, η_v, exist at the dielectric surfaces.

Total surface charges at boundary surface of two dielectric layers are equal to the polarized surface charges, because the free charges at this surface do not exist. The free charges exist only at the PEC surfaces.

The boundary surface between any two dielectric layers is divided into a large number of segments. Those segments are replaced with equivalent electrodes placed at the centers of the segments. The electrodes are placed in the air and represent the polarized charges.

At the boundary surface of two dielectric, the boundary condition for the normal components of polarization vector is satisfied, and

$$\eta_{\mathrm{v}} = P_{2\mathrm{n}} - P_{1\mathrm{n}}. \tag{8.3}$$

Using described procedure, the equivalent HBEM system is formed, Fig. 8.2.

At the boundary surface between the PEC and the dielectric, the free and polarized charges exist. Their sum gives the total charges. But, the satisfying results are obtain using an approximation that polarized charges at that boundary surface can be neglected and only free charges taken into account. Those charges are placed in the corresponding dielectric layer. At the boundary surface between the PEC and the air, only free surface charges exist, placed in the air. The polarized charges exist at

Fig. 8.2 HBEM model

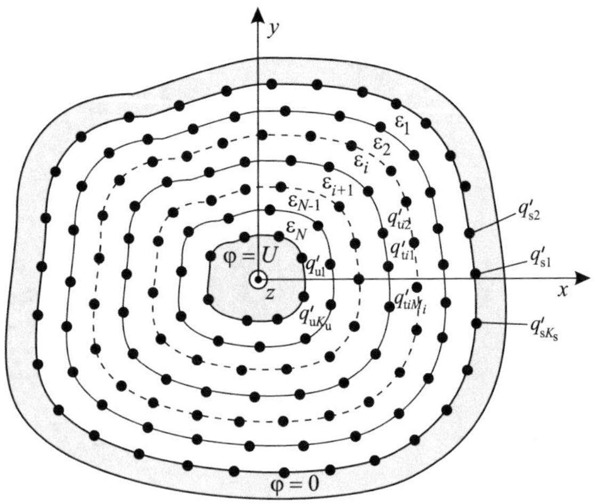

the boundary surface between two dielectrics or dielectric and the air. Those EEs are placed in the air. The electric potential for the system from Fig. 8.2 is:

$$\varphi = \varphi_0 - \sum_{k=1}^{K_\mathrm{u}} \frac{q'_{\mathrm{u}k}}{2\pi\varepsilon_N} \ln\sqrt{(x - x_{\mathrm{u}k})^2 + (y - y_{\mathrm{u}k})^2} - \\ - \sum_{k=1}^{K_\mathrm{s}} \frac{q'_{\mathrm{s}k}}{2\pi\varepsilon_1} \ln\sqrt{(x - x_{\mathrm{s}k})^2 + (y - y_{\mathrm{s}k})^2} - \\ - \sum_{i=1}^{N-1}\sum_{m=1}^{M_i} \frac{q'_{\mathrm{t}im}}{2\pi\varepsilon_0} \ln\sqrt{(x - x_{\mathrm{t}im})^2 + (y - y_{\mathrm{t}im})^2},\ N \geq 2, \tag{8.4}$$

where

- K_u is the number of equivalent electrodes at the inner conductor, with free line charges $q'_{\mathrm{u}k}$ $(k = 1, \ldots, K_\mathrm{u})$;
- K_s is the number of equivalent electrodes at the outer conductor, with free line charges $q'_{\mathrm{s}k}$ $(k = 1, \ldots, K_\mathrm{s})$;
- M_i is the number of EEs on the i-th boundary surface between any two dielectric layers, with polarized line charges $q'_{\mathrm{t}im}$ $(m = 1, \ldots, M_i,\ i = 1, \ldots, N-1)$;
- $N - 1$ is the number of boundary surfaces (between two dielectric layers);
- $(x_{\mathrm{u}k}, y_{\mathrm{u}k})$, $(x_{\mathrm{s}k}, y_{\mathrm{s}k})$ and $(x_{\mathrm{t}im}, y_{\mathrm{t}im})$ are EE coordinates;
- ε_n $(n = 1, \ldots, N)$ is the relative permittivity of n-th dielectric layer;
- φ_0 is an additive constant which value depends on the position of zero potential point.

$$N_\mathrm{tot} = K_\mathrm{u} + K_\mathrm{s} + \sum_{i=1}^{N-1} M_i + 1 \tag{8.5}$$

is total number of unknowns.

The electric field strength is

$$\boldsymbol{E} = -\mathrm{grad}(\varphi) = E_x\hat{x} + E_y\hat{y}, \tag{8.6}$$

and the corresponding components:

$$E_x = -\frac{\partial\varphi}{\partial x} = \sum_{k=1}^{K_\mathrm{u}} \frac{q'_{\mathrm{u}k}}{2\pi\varepsilon_N} \frac{x - x_{\mathrm{u}k}}{(x - x_{\mathrm{u}k})^2 + (y - y_{\mathrm{u}k})^2} + \\ + \sum_{k=1}^{K_\mathrm{s}} \frac{q'_{\mathrm{s}k}}{2\pi\varepsilon_1} \frac{x - x_{\mathrm{s}k}}{(x - x_{\mathrm{s}k})^2 + (y - y_{\mathrm{s}k})^2} + \sum_{i=1}^{N-1}\sum_{m=1}^{M_i} \frac{q'_{\mathrm{t}im}}{2\pi\varepsilon_0} \frac{x - x_{\mathrm{t}im}}{(x - x_{\mathrm{t}im})^2 + (y - y_{\mathrm{t}im})^2}, \tag{8.7}$$

$$E_y = -\frac{\partial \varphi}{\partial y} = \sum_{k=1}^{K_u} \frac{q'_{uk}}{2\pi\varepsilon_N} \frac{y - y_{uk}}{(x - x_{uk})^2 + (y - y_{uk})^2} +$$

$$+ \sum_{k=1}^{K_s} \frac{q'_{sk}}{2\pi\varepsilon_1} \frac{y - y_{sk}}{(x - x_{sk})^2 + (y - y_{sk})^2} + \sum_{i=1}^{N-1} \sum_{m=1}^{M_i} \frac{q'_{tim}}{2\pi\varepsilon_0} \frac{y - y_{tim}}{(x - x_{tim})^2 + (y - y_{tim})^2}. \quad (8.8)$$

The relation between the normal component of electric field strength and total surface charges, η_t, is

$$\hat{\boldsymbol{n}}_{im} \cdot \boldsymbol{E}_{im}^{(0+)} = \frac{-\varepsilon_{i+1}}{\varepsilon_0(\varepsilon_i \varepsilon_{i+1})} \eta_{tim}, \ \eta_{tim} = \frac{q'_{tim}}{\Delta l_{im}}, \ m = 1, \ldots, M_i, \ i = 1, \ldots, N-1, \quad (8.9)$$

where $\hat{\mathbf{n}}_{im}$ is the unit normal vector oriented from the layer ε_{i+1} towards the layer ε_i and Δl_{im} is the segment width.

Positions of the matching points for the potential of the inner and the outer PECs are: $x_{un} = x_{uk} + \delta_{nk} a_{euk} \hat{\boldsymbol{n}}_{uk} \cdot \hat{x}$, $y_{un} = y_{uk} + \delta_{nk} a_{euk} \hat{\boldsymbol{n}}_{uk} \cdot \hat{y}$, $x_{sn} = x_{sk} + \delta_{nk} a_{esk} \hat{\boldsymbol{n}}_{sk} \cdot \hat{x}$, $y_{sn} = y_{sk} + \delta_{nk} a_{esk} \hat{\boldsymbol{n}}_{sk} \cdot \hat{y}$ and $a_{euk} = \Delta l_{uk}/4$, $a_{esk} = \Delta l_{sk}/4$, where δ_{nk} is the Kronecker's delta function,

$$\delta_{nk} = \begin{cases} 1, \ n = k; \\ 0, \ n \neq k. \end{cases} \quad (8.10)$$

while a_{euk} and a_{esk} are corresponding EEs radii.

The boundary surface matching points' for the normal component of the electric field on the i-th boundary surface are: $x_{tin} = x_{tim} + \delta_{nm} a_{etim} \hat{\boldsymbol{n}}_{tim} \cdot \hat{x}$ and $y_{tin} = y_{tim} + \delta_{nm} a_{etim} \hat{\boldsymbol{n}}_{tim} \cdot \hat{y}$, where $a_{etim} = \Delta l_{tim}/\pi$ are the EEs radii.

It is necessary to add only one equation to the system of linear equations for the electrical neutrality of the whole observed microwave transmission line,

$$\sum_{k=1}^{K_u} q'_{uk} + \sum_{k=1}^{K_s} q'_{sk} = 0. \quad (8.11)$$

The aim is to obtain the quadratic system of linear equations with unknown free charges of PECs, total charges per unit length at boundary surfaces between dielectric layers, and unknown additive constant φ_0 that depends on the chosen referent point for the electric scalar potential. Using the PMM for the potential of the inner and the outer conductor given by (8.4), the PMM for the normal component of the electric field (8.9), and the electrical neutrality condition (8.11), it is possible to determine unknown free charges per unit length on conductors, total charges per unit length on the boundary surfaces between layers and the unknown constant φ_0.

Increasing the number of the EEs the distances between them becomes smaller. In order to keep stability of the formed system of equations it is necessary that the distances between EEs be larger than their radius. The quadratic system of linear equations, formed in that way, is well-conditioned. The system matrix always has the greatest values at the main diagonal.

After solving the system of linear equations it is possible to calculate capacitance per unit length of the microwave transmission line given by

$$C' = \sum_{k=1}^{K_u} \frac{q'_{uk}}{U}. \tag{8.12}$$

Characteristic impedance of the transmission line is calculated as

$$Z_c = \frac{Z_{c0}}{\sqrt{\varepsilon_r^{eff}}}, \tag{8.13}$$

where $\varepsilon_r^{eff} = C'/C'_0$ is the effective (dielectric) permittivity, and Z_{c0} is the characteristic impedance of the transmission line without dielectrics (free space).

The expressions given in (8.12) and (8.13) are for the single line, but the same procedure can be applied to find the capacitance and characteristic impedance of coupled microwave transmission lines for even and odd modes.

8.3 Numerical Results

On the basis of described procedure, the computer codes have been written to obtain numerical solutions for several multilayered structures.

A shielded, multilayered microstrip line with finite metallization thickness is considered, Fig. 8.3.

The structure of this type has been recently used as a part of integrated microwave circuits, [24]. In some of the papers which deal with this structures, it is considered that strip has zero thickness.

Convergence of the effective relative permittivity and the characteristic impedance as well as the computation time are shown in Table 8.1. Microstrip line parameters are: $\varepsilon_{r1} = \varepsilon_{r3} = 1, \varepsilon_{r2} = 9.35, a/w = 4.0, b/w = 2.0, h_1/w = 0.8, h_2/w = 0.4, h_3/w =$

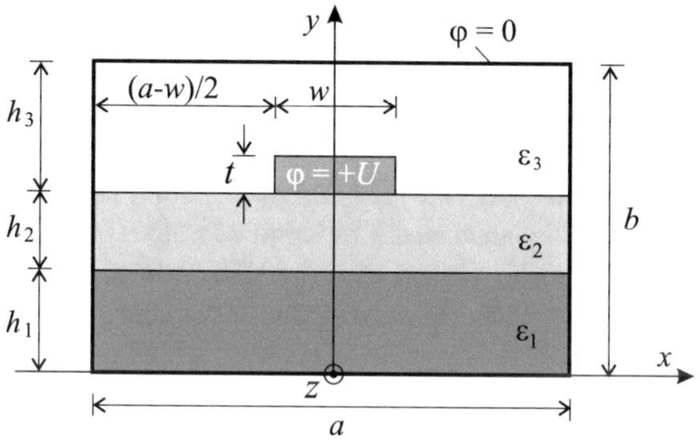

Fig. 8.3 Shielded multilayered microstrip line

Table 8.1 Convergence of the results and computation time

N	N_{tot}	ε_r^{eff}	Z_c [Ω]	t(s)
50	189	1.5903	45.519	1.25
100	375	1.6707	44.982	4.7
150	556	1.6971	44.808	10.4
200	735	1.7106	44.723	18.1
250	919	1.7186	44.680	28.5
300	1100	1.7246	44.642	41.0
350	1281	1.7289	44.614	56.5
400	1465	1.7317	44.602	74.9
450	1644	1.7341	44.587	94.6
500	1825	1.7362	44.573	116.5
550	2011	1.7378	44.566	138.8

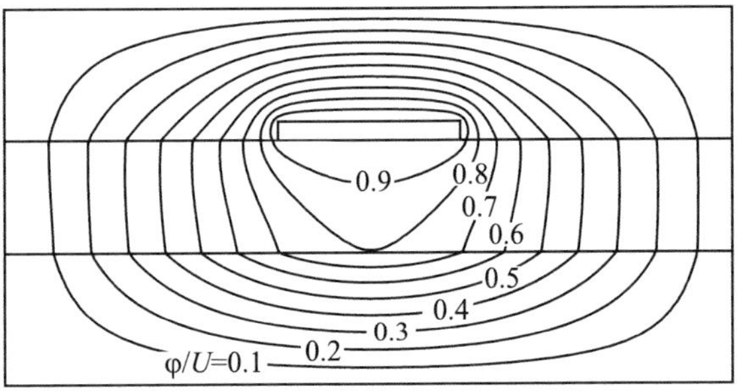

Fig. 8.4 Equipotential contours

0.8 and $t/w = 0.4$. N is the initial number and N_{tot} denotes the total number of unknowns. A good convergence of the results is achieved in a short computation time.

The "computation time" is a term which describes the time spent for determining the number of unknowns using initial number, their positioning, forming a matrix elements, solving the system of linear equations, the characteristic parameters determination. Most of the calculation time goes to the matrix fill.

Equipotential contours are shown in Fig. 8.4 for: $\varepsilon_{r1} = \varepsilon_{r3} = 1, \varepsilon_{r2} = 9.35, a/w = 4.0, b/w = 2.0, h_1/w = 0.7, h_2/w = 0.6, h_3/w = 0.7$ and $t/w = 0.1$. The influence of dielectric layers is evident.

In order to verify the obtained HBEM values, a comparison of HBEM results with FEMM software [12] and results from [15, 28] is given in Fig. 8.5. The characteristic impedance as a function of ratio a/b and t/b obtained using the FEMM is denoted with dashed line, the HBEM results are shown with solid line, and the results from [15, 28] with square and circle points, respectively. In [15] the finite element method is applied. The Green's function method is used in [28]. A very good results agreement can be noticed. The influence of the side walls on the characteristic impedance values can be neglected if the walls are sufficiently far away from the strip

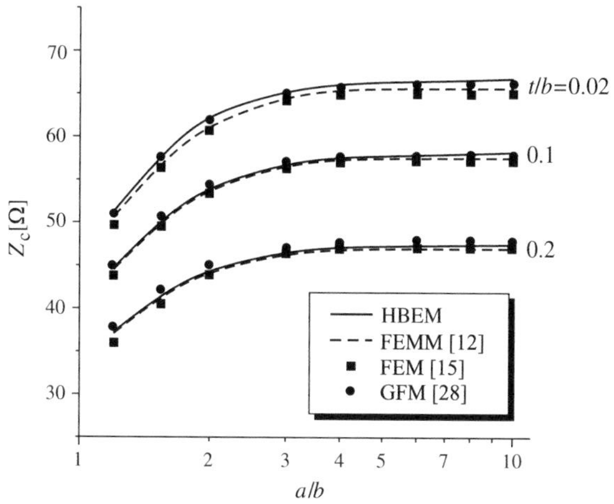

Fig. 8.5 Characteristic impedance distribution for: $\varepsilon_{r1} = \varepsilon_{r3} = 1$, $\varepsilon_{r2} = 9.35$, $b/w = 2.0$, $h_1/w = 0.8$, $h_2/w = 0.4$, and $h_3/w = 0.8$

$(a/b > 5)$. Increasing the strip thickness, the characteristic impedance decreases. Shielded microstrip line parameters are: $\varepsilon_{r1} = \varepsilon_{r3} = 1$, $\varepsilon_{r2} = 9.35$, $b/w = 2.0$, $h_1/w = 0.8$, $h_2/w = 0.4$, and $h_3/w = 0.8$. During the HBEM application, the total number of unknowns was about 1500. All calculations in this paper were performed on computer with dual core INTEL processor 2.8 GHz and 4 GB of RAM.

An influence of relative permittivity ε_{r3} on characteristic impedance distribution for different values of parameter h_2/w is given in Fig. 8.6. The parameters of the shielded microstrip line from Fig. 8.3 are: $\varepsilon_{r1} = 1$, $\varepsilon_{r2} = 9.35$, $a/w = 4.0$, $b/w = 2.0$, $h_1/w = h_3/w = (b - h_2)/w$ and $t/w = 0.2$.

From Fig. 8.6 is evident that increasing the relative permittivity of third layer, the characteristic impedance decreases. Also, increasing the height of second layer, the characteristic impedance decreases. The HBEM results have also been compared with the FEMM [12] results. An excellent results agreement is obtained.

A special case of the structure presented in Fig. 8.3 is obtained for $\varepsilon_1 = \varepsilon_2 = \varepsilon$ and $\varepsilon_3 = \varepsilon_0$, Fig. 8.7.

The characteristic impedance distribution versus h/w and ε_r, obtained for: $a/w = 4.0$, $b/w = 2.0$ and $t/w = 0.2$ is shown in Fig. 8.8. Also, the HBEM and FEMM results comparison is given in this figure. The results deviation is less than 1 %.

Increasing the substrate height the characteristic impedance increases first, then decreases. The reason for this variation is due to the influence of shield upper side.

Using the HBEM coupled structures can also be analyzed. The shielded microstrip line with partial dielectric support is shown in Fig. 8.9 [17].

The convergence of the results for the characteristic impedance and effective relative permittivity is shown in Table 8.2. Both modes ("even" and "odd") are taken into account. The parameters of the microstrip line are: $\varepsilon_{r1} = 1$, $\varepsilon_{r2} = 2.35$, $a/w = b/w = 2.5$, $d/w = 0.5$ and $t/w = 0.01$. The good results convergence is obtained.

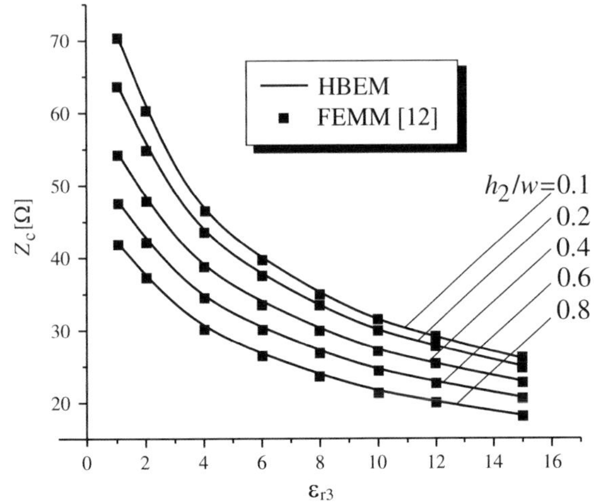

Fig. 8.6 Characteristic impedance distribution versus ε_{r3} and h_2/w for: $\varepsilon_{r1} = 1$, $\varepsilon_{r2} = 9.35$, $a/w = 4.0$, $b/w = 2.0$, $h_1/w = h_3/w = (b - h_2)/w$ and $t/w = 0.2$

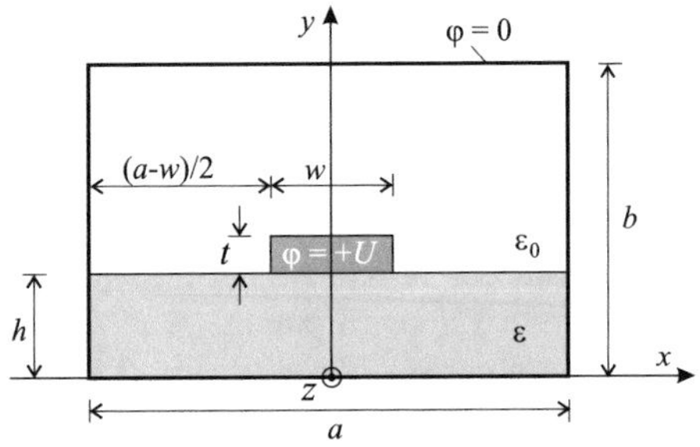

Fig. 8.7 Shielded microstrip line as a special case of Fig. 8.3

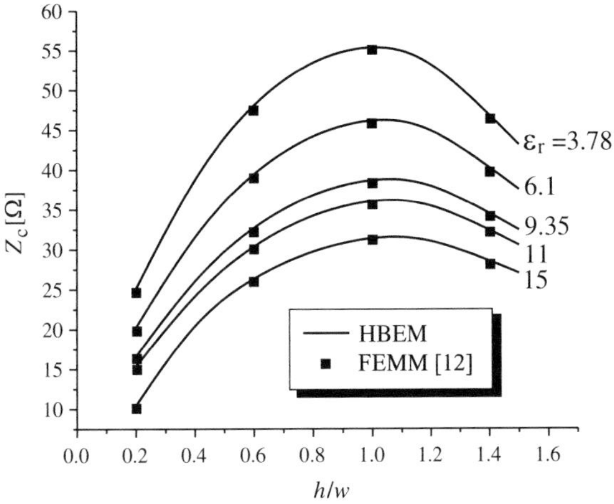

Fig. 8.8 Characteristic impedance distribution versus h/w and ε_r for: $a/w = 4.0$, $b/w = 2.0$ and $t/w = 0.2$

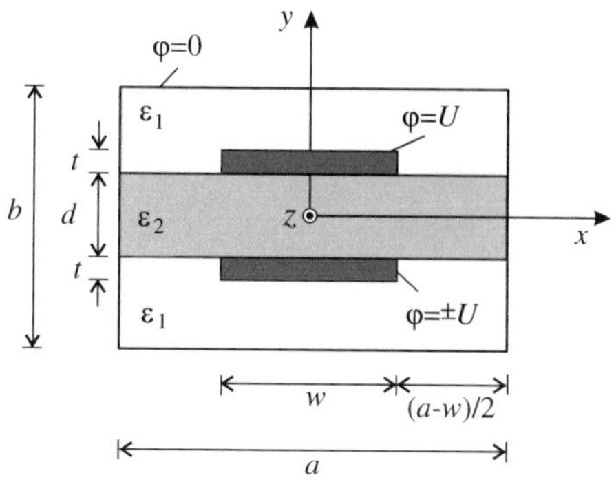

Fig. 8.9 Shielded coupled microstrip line with partial dielectric support

Table 8.2 Convergence of the results and computation time

"even" mode			"odd" mode		
N_{tot}	ε_r^{eff}	Z_c [Ω]	ε_r^{eff}	Z_c [Ω]	t(s)
784	1.1058	66.702	2.0388	36.869	20.0
872	1.1394	65.732	2.0157	37.096	25.3
960	1.1621	65.101	2.0005	37.252	29.2
1044	1.1790	64.646	1.9901	37.362	35.6
1128	1.1920	64.306	1.9825	37.444	41.6
1216	1.2022	64.043	1.9768	37.509	47.4
1300	1.2104	63.835	1.9722	37.561	54.7
1384	1.2173	63.661	1.9686	37.603	61.5
1468	1.2231	63.516	1.9656	37.639	72.3
1565	1.2282	63.392	1.9632	37.669	81.3
1640	1.2325	63.288	1.9610	37.696	90.9

Equipotential curves for "even" and "odd" modes are shown in Figs. 8.10 and 8.11, respectively. The microstrip parameters are: $\varepsilon_{r1} = 1$, $\varepsilon_{r2} = 2.35$, $a/w = b/w = 2.5$, $d/w = 0.5$ and $t/w = 0.1$.

Table 8.3, taken from [17], shows the comparison of the obtained HBEM results for the characteristic impedance (for "even" mode) with the ones from [6, 14, 21], for parameters: $\varepsilon_{r1} = 1$, $\varepsilon_{r2} = 2.35$, $a/w = b/w = 2.5$, $d/w = 0.5$ and $t/w = 0$, adopted from [21]. Good results agreement, within 3 %, is obtained. Some disagreement is the result of different values of parameter t/w in the case when the FEMM and HBEM are applied. In those two cases the conductors are of finite thickness, while in the other ones are infinitesimally thin.

Influences of dielectric substrate height and relative permittivity ε_{r2} are given in Figs. 8.12 and 8.13 for "even" and "odd" modes, respectively. Some of typical materials which can be used as microstrip substrate are mentioned in [29]: $\varepsilon_{r3} = 3.78$ (quartz), $\varepsilon_{r3} = 6.1$ (99 % berylia), $\varepsilon_{r3} = 9.35$ (99.5 % alumina) and $\varepsilon_{r3} = 11$ (sapphire).

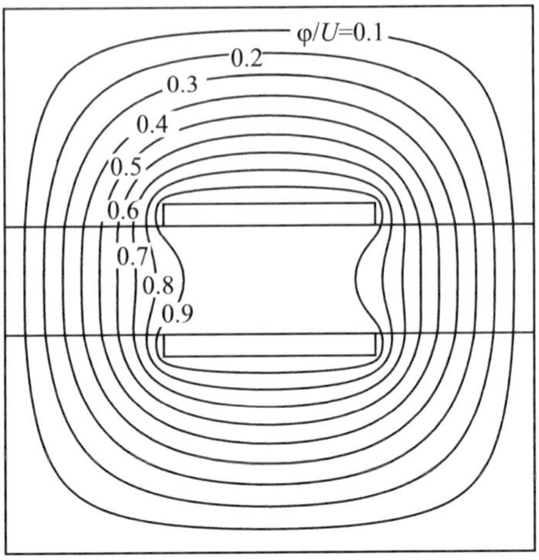

Fig. 8.10 Equipotential contours for "even" mode

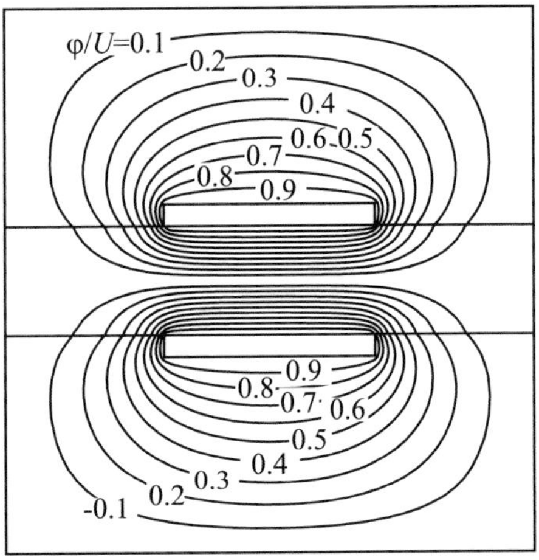

Fig. 8.11 Equipotential contours for "odd" mode

Table 8.3 Comparison of results ("even" mode)

Method/Reference	Z_c [Ω]
Gish and Graham [6]	62.50
FEMM [12], $t/w = 0.01$	61.63
Naiheng and Harrington [14]	65.02
FDM [17]	64.67
FDM [21]	61.53
HBEM, $t/w = 0.01$	63.29

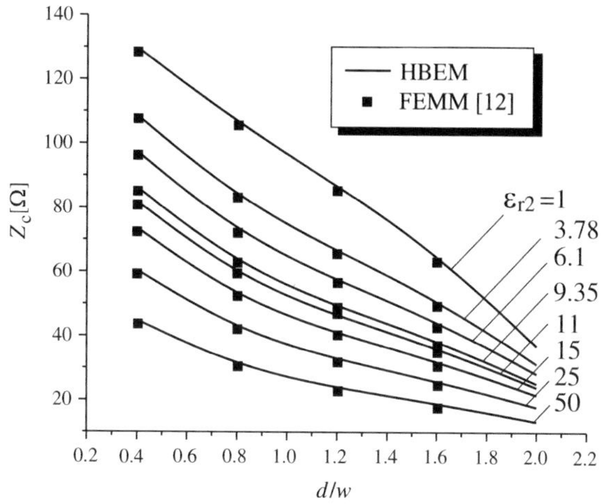

Fig. 8.12 Characteristic impedance distribution versus ε_{r2} and d/w for: $\varepsilon_{r1} = 1$, $a/w = b/w = 2.5$ and $t/w = 0$ ("even" mode)

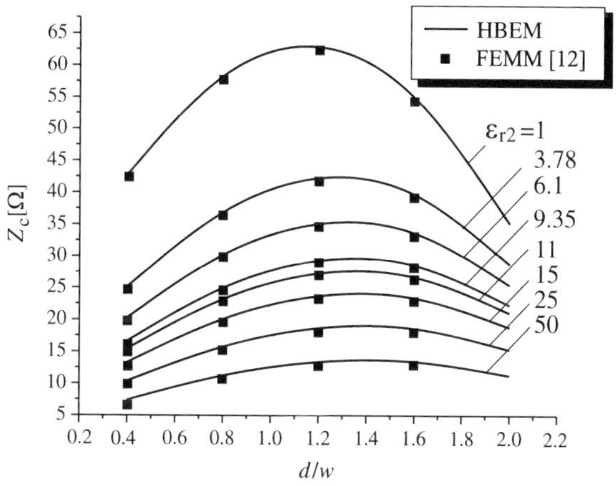

Fig. 8.13 Characteristic impedance distribution versus ε_{r2} and d/w for: $\varepsilon_{r1} = 1$, $a/w = b/w = 2.5$ and $t/w = 0$ ("odd" mode)

In "even" mode, increasing the substrate height, the characteristic impedance decreases, Fig. 8.12. Increasing the substrate height in the "odd" mode, the characteristic impedance increases first, then decreases as the conductor approaches to the shield upper side, Fig. 8.13. The dielectric permittivity of substrate also has the influence on the characteristic impedance value. Increasing the substrate permittivity, the characteristic impedance values decrease in both modes.

In Figs. 8.12 and 8.13 the HBEM results have been compared with the FEMM results. A very good results agreement can be noticed.

8.4 Conclusion

This paper presents the hybrid boundary element method application for shielded microstrip lines analysis. The method is capable to solve arbitrary shaped, multilayered configuration of microstrip lines, with finite strip thickness, without any numerical integration. The quasi-TEM analysis is applied. The convergence of the results is very good and the computation time is very short. Analysis of shielded microstrip lines with two and three layers was performed for different values of microstrip parameters and different configurations. The results comparison with those reported in the literature and obtained using software package gives a very good agreement.

Although this paper describes only the HBEM application on shielded microstrip structures, the method can be applied without any restriction for analysis of all types of microwave transmission lines. In the following research, the method will be extended to the structures with bi-isotropic and anisotropic layers.

Acknowledgements This research was partially supported by funding from the Serbian Ministry of Education and Science in the frame of the project TR 33008.

References

1. Bazdar, M.B., Djordjevic, A.R., Harrington, R.F., Sarkar, T.K.: Evaluation of quasi-static matrix parameters for multiconductor transmission lines using Garlekin's method. IEEE Trans. Microw. Theory Techn. **42**(7), 1223–1228 (1994)
2. Bryant, T.G., Weiss, J.A.: Parameters of microstrip transmission lines and of coupled pairs of microstrip lines. IEEE Trans. Microw. Theory Techn. **16**, 1021–1027 (1968)
3. Cantaragiu, S.: Analysis of shielded microstrip lines by finite-difference method. In: Proceedings of ICECS'99, pp. 565–567. Pafos, Cyprus (1999)
4. Farrar, A., Adams, A.T.: Characteristic impedance of microstrip by the method of moments. IEEE Trans. Microw. Theory Techn. **18**, 65–66 (1970)
5. Fukuda, T., Sugie, T., Wakino, K., Lin, Y.-D., Kitazawa, T.: Variational method of coupled strip lines with an inclined dielectric substrate. In: Asia Pacific Microwave Conference - APMC 2009, pp. 866–869. Suntec City, Singapore (2009)
6. Gish, D.L., Graham, O.: Characteristic impedance and phase velocity of a dielectric-supported air strip transmission lines with side walls. IEEE Trans. Microw. Theory Techn. **18**, 131–148 (1970)
7. Harrington, R.F.: Field Computation by Moment Methods. Macmillan, New York (1968)
8. Ilic, S.S., Aleksic, S.R., Raicevic, N.B.: TEM analysis of multilayered transmission lines using new hybrid boundary element method. Proceedings of 10th International Conference on Telecommunications in Modern Satellite. Cable and Broadcasting Services - TELSIKS 2011, vol. 2, pp. 428–431. Nis, Serbia (2011)
9. Ilic, S., Peric, M., Aleksic, S., Raicevic, N.: Hybrid boundary element method and quasi-TEM analysis of two-dimensional transmission lines - Generalization. Electromagnetics. **33**(4), 292–310 (2013)
10. Kakria, P., Marwaha, A., Manna, M.S.: Optimized design of shielded microstrip lines using adaptive finite element method. In: CD-Procreedings of COMSOL Users Conference, Bangalore, India (2011)

11. Li, K., Fujii, Y.: Indirect boundary element method of applied to generalized microstrip line analysis with applications to side-proximity effect in MMICs. IEEE Trans. Microw. Theory Techn. **44**, 237–244 (1992)
12. Meeker, D.: FEMM 4.2. http://www.femm.info/wiki/Download (2009)
13. Musa, S., Sadiku, M.: Quasi-static analysis of shielded microstrip lines. In: CD-Procreedings of COMSOL Users Conference, Las Vegas, USA (2006)
14. Naiheng, Y., Harrington, R.F.: Characteristic impedance of transmission lines with arbitrary dielectrics under the TEM approximation. IEEE Trans. Microw. Theory Techn. **34**(4), 472–475 (1986)
15. Pantic, Z., Mittra, R.: Quasi-TEM analysis of microwave transmission lines by the finite-element method. IEEE Trans. Microw. Theory Techn. **34**(11), 1096–1103 (1986)
16. Peric, M., Ilic, S., Aleksic, S., Raicevic, N.: Application of hybrid boundary element method to 2D microstrip lines analysis. Int. J. Appl. Electromagn. Mech. **42**(2), 179–190 (2013)
17. Peric, M., Ilic, S., Aleksic, S., Raicevic, N., Monsefi F., Rancic, M., Silvestrov, S.: Analysis of shielded coupled microstrip line with partial dielectric support. In: Procreedings of 18th International Symposium on Electrical Apparatus and Technologies SIELA 2014. pp. 165–168, Burgas, Bugaria (2014)
18. Raicevic, N., Aleksic, S.: One method for electric field determination in the vicinity of infinitely thin electrode shells. Eng. Anal. Bound. Elem. **34**, 97–104 (2010)
19. Raicevic, N.B., Aleksic, S.R., Ilic, S.S.: A hybrid boundary element method for multilayer electrostatic and magnetostatic problems. Electromagnetics. **30**(6), 507–524 (2010)
20. Raicevic, N.B., Ilic S.S., Aleksic, S.R.: Application of new hybrid boundary element method on the cable terminations. In: Proceedings of 14th International IGTE'10 Symposium. pp. 56–61, Graz, Austria (2010)
21. Sadiku M.N.O.: Numerical Techniques in Electromagnetics. CRC Press (2001)
22. Svacina, J.: Analytical models of width-limited microstrip lines. Microw. Opt. Techn. Let. **36**, 63–65 (2003)
23. Svacina, J.: New method for analysis of microstrip with finite-width ground plane. Microw. Opt. Techn. Let. **48**(2), 396–399 (2006)
24. Tatsuguchi, I., Aslaksen, E.W.: Integrated 4-GHz balanced mixer assembly. IEEE J. Solid-State Circuits **SC–3**, 21–26 (1968)
25. Velickovic, D.M.: Equivalent electrodes method. Sci. Rev. **21–22**, 207–248 (1996)
26. Velickovic, D.M.: TEM analysis of transmission lines using equivalent electrodes method. Proceedings of 3rd Internatinal Conference on Telecommunications in Modern Satellite. Cable and Broadcasting Services - TELSIKS 97, pp. 13–22. Nis, Serbia (1997)
27. Vuckovic, A.N., Raicevic, N.B., Peric, M.T., Aleksic, S.R.: Magnetic force calculation of permanent magnet systems using hybrid boundary element method. In: CD Proceedings of the Sixteenth Biennial IEEE Conference on Electromagnetic Field Computation CEFC 2014, 113/PB3:17. France, Grenoble (2014)
28. Yamashita, E., Mittra, R.: Variational method for the analysis of microstrip lines. IEEE Trans. Microw. Theory Techn. **16**, 251–256 (1968)
29. Yamashita, E., Atsuki, K.: Strip line with rectangular outer conductor and three dielectric layers. IEEE Trans. Microw. Theory Techn. **18**(5), 238–244 (1970)

Chapter 9
Modified Transmission Line Models of Lightning Strokes Using New Current Functions and Attenuation Factors

Vesna Javor

Abstract New engineering modified transmission line models of lightning strokes are presented in this paper. Their computational results for lightning electromagnetic field (LEMF) at various distances from lightning discharges are in good agreement with experimental results that are usually used for validating electromagnetic, engineering and distributed-circuit models. Electromagnetic theory relations, thin-wire antenna approximation of a lightning channel without tortuosity and branching, so as the assumption of perfectly conducting ground, are used for electric and magnetic field computation. An analytically extended function (AEF), suitable for approximating channel-base currents in these models, may also represent typical lightning stroke currents as given in IEC 62305-1 Standard, as well as the IEC 61000-4-2 Standard electrostatic discharge current.

Keywords Lightning electromagnetic field · Modified transmission line model · Return stroke

9.1 Introduction

Modeling of lightning strokes and computation of lightning electromagnetic field (LEMF), based on these models, are important for electromagnetic compatibility applications such as estimation of lightning effects and lightning protection of power systems, electrical equipment and other objects in such a field. Experimental results are given in literature for lightning discharge currents at striking points, so as electric and magnetic field values and waveshapes at some distances from lightning channels, usually at the ground surface or nearby above. Based on the channel-base current, speed of the propagating front and channel luminosity, an engineering model is an attempt to achieve agreement between calculated and measured LEMF results at

V. Javor (✉)
Faculty of Electronic Engineering, Department of Power Engineering,
University of Niš, Niš, Serbia
e-mail: vesna.javor@elfak.ni.ac.rs

S. Silvestrov and M. Rančić (eds.), *Engineering Mathematics I*,
Springer Proceedings in Mathematics & Statistics 178,
DOI 10.1007/978-3-319-42082-0_9

distances from tens of meters to hundreds of kilometers. In engineering models, either current distribution or charge density distribution along the channel is considered, and afterwards, LEMF is calculated based on it. Although lightning events differ very much from each other, there are some noticed features in experimental LEMF results of typical lightning strokes. Engineering models from literature provide a few of these features, but neither one provides all of them [17, 19]. Calculated LEMF values and waveshapes may differ from experimental results at a certain distance more than at other, which depends on the applied model. Although channel-base current and return stroke speed are more studied in literature than current attenuation factors and channel-heights, the latter two have great influence on LEMF results. Channel-base current, as the most important for direct lightning discharges, is defined in IEC 62305 Standard [8] for typical lightning strokes. However, current attenuation factors are usually taken in one of just a few forms given in literature [19]. Return stroke speed is often taken as constant in all models, although it is well known that it varies along the channel. It is based on optical measurements. The influence of lightning channel-height on LEMF results is rarely addressed to [14].

Review and evaluation of lightning stroke models from literature, so as their computational LEMF results are given in detail in papers [17, 19]. Most of comparisons usually refer to experimental results given in [1, 2, 16, 20] for natural lightning, and other references for triggered lightning. Although double-exponential function has been widely used for approximation of channel-base currents, results are often given in literature for Heidler's function [7], the sum of a few of its terms, the sum of it and other functions [17], or other pulse functions from literature. For the first stroke lightning currents having more discontinuities in waveshapes than other typical lightning strokes, the sum of seven Heidler's function terms is needed for their representation [3]. More terms mean more parameters to adjust and often non-analytical solutions of current integrals and derivatives which are necessary for LEMF calculations.

An analytically extended function (AEF), proposed as the sum of same terms with different parameters values [15], results in the variety of waveshapes [11, 12] that may represent typical lightning stroke currents [9] given in IEC 62305-1 Standard, measured lightning stroke currents, but also the IEC 61000-4-2 Standard [5] electrostatic discharge current, as proposed in [13].

New engineering models and their current attenuation factors are presented in Sect. 9.2 of this paper. AEF and results of using this function for representing different lightning stroke currents is presented in Sect. 9.3. LEMF computation results of new models are given in Sect. 9.4 and compared to experimental and other models, results.

9.2 Current Attenuation Factors in Engineering Models of Lightning Strokes

In an engineering model, current distribution $i(z', t)$ at time t and height z' along the lightning channel (Fig. 9.1) is assumed as a product of the channel-base current $i(0, t)$, height- and time-dependent attenuation factor $P(z', t)$, and Heaviside

Fig. 9.1 Thin-wire antenna representation of a lightning channel

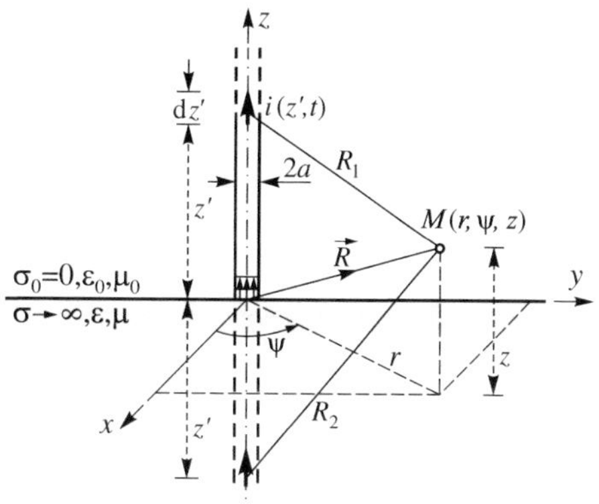

function equal to unity for $t > z'/v_f$ and to zero for $t < z'/v_f$, thus showing if the current pulse front has either reached that point along the channel, or not. $P(z',\ t)$ introduces uneven current weakening along the channel. Longitudinal channel-current is given with the following expression

$$i(z',\ t) = h(t - z'/v_f)\, i(0,\ t - z'/v)\ P(z',\ t), \tag{9.1}$$

where v_f is the return stroke speed (upward-propagating front speed) and v is the current-wave propagation speed, $i(0,\ t - z'/v)$ is the channel-base current function, delayed in time for z'/v due to the current-wave propagation. σ_0, μ_0, ε_0 are electrical parameters of the air and σ, μ, ε electrical parameters of the ground. The diameter of the channel is $2a << H$, and H is the total channel-height. Position of the field point M having cylindrical coordinates r, ψ, z is defined with $\mathbf{R}$.

The current attenuation factor in (9.1) was introduced as height-dependent in [18]. Based on different values of current-wave propagation speeds, attenuation factors and other parameters, engineering models are grouped [19] into transmission line models (TL), modified transmission line models with linear decay of the current with height (MTLL), modified transmission line models with exponential decay of the current with height (MTLE), modified transmission line models with distortion of the current (MTLD), Bruce-Golde model (BG) [6], travelling current source models (TCS) [7], and other, as Master-Uman-Lin-Standler (MULS), Diendorfer-Uman (DU), etc.

The difference between two major types of engineering models, TL and TCS, is that the direction of current wave propagation is upward in TL type of models and downward in TCS type of models, whereas channel-base current may be chosen the same. Although the direction of wave propagation is not taken the same in these types of models, the direction of the current is the same for modeling the transport of the same sign charge [19]. BG may be viewed as a special case of either TL or TCS types of models. Current wave propagates in BG model with infinite speed and the propagation direction is not defined, whereas the return stroke speed is of finite value as in other models given in Table 9.1.

Table 9.1 Attenuation factors and pulse propagation speed for some engineering models

Engineering model	Attenuation factor $P(z')$	Pulse propagation speed v
TL	1	v_f
MTLL	$1 - z'/H$	v_f
MTLE	$\exp(-z'/\lambda)$	v_f
MTLE	$\exp(-z'/H)$	v_f
MTLTCOS	$0.95 - 0.9z'/H + 0.05\cos(5\pi z'/H)$	v_f
MTLT	$[1 + (1 - 2z'/H)^3]/2$	v_f
MTLTS	$[1 + (1 - 2z'/H)^3]^2/4$	v_f
BG	1	∞
TCS	1	$-c$

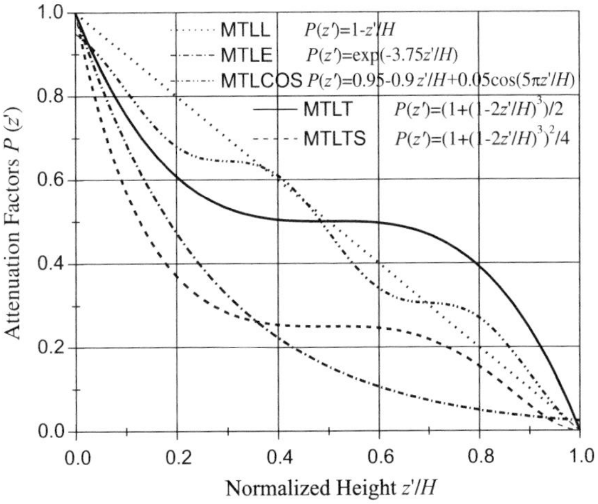

Fig. 9.2 Attenuation factors in MTLL, MTLE, MTLCOS, MTLT and MTLTS models

Current attenuation factors in the often used engineering models are also presented in Table 9.1. TL model is without any current decay along the channel, so its attenuation factor is $P(z') = 1$. In MTLL the attenuation factor is $P(z') = 1 - z'/H$, so that current decays to zero value at the channel top (Fig. 9.2). In MTLE, the current attenuation factor is exponential function $P(z') = \exp(-z'/\lambda)$, where the constant is often chosen to be $\lambda = 2000$ m, as in [17]. Its value may be discussed and specified otherwise according to LEMF results [14]. In MTLE model the total channel-height H may also be the parameter of its attenuation factor, as in fact the value at the channel top is defined by $z' = H$, so it has influence on LEMF results in both MTLE models given in Table 9.1. For MTLE model the attenuation factor is assumed in this paper as $P(z') = \exp(-z'/\lambda) = \exp(-7500z'/2000H) = \exp(-3.75z'/H)$, so to compare it with other models for the normalized height z'/H (Fig. 9.2). Thus, $H = 7500$ m is taken for the channel-height corresponding to $\lambda = 2000$ m as the constant in this model.

New models and their attenuation factors are presented in this paper and denoted with MTLT, MTLTS and MTLCOS (Fig. 9.2). MTLT denotes modified transmission line model with the third degree function of height z' in the attenuation factor, $P(z') = [1 + (1 - 2z'/H)^3]/2$, which results in zero value at the channel top, for $z' = H$. MTLTS denotes the square of the MTLT attenuation factor (sixth degree function of height z'), so that $P(z') = [1 + (1 - 2z'/H)^3]^2/4$, or simply $P_{\text{MTLTS}} = P^2_{\text{MTLT}}$. Obviously, it also results in zero value at $z' = H$, so as in factors of MTLT and MTLL models. Model denoted with MTLCOS has the current attenuation factor which gives results more similar to MTLL than MTLT and MTLTS. Its attenuation factor is $P(z') = 0.95 - 0.9z'/H + 0.05\cos(5\pi z'/H)$, also resulting in $P(0) = 1$ and $P(H) = 0$.

If MTLT is applied, the current peak near the channel-base and up to $z' = 0.2H$ decays faster with height than if MTLL is applied, later very slow up to $z' = 0.7H$, and afterwards faster than in all other models, up to the channel top. In MTLTS, due to its attenuation factor, current peak decays even faster than in other models up to $z' = 0.3H$, but very slow at about half height of the channel, and afterwards fast, near the assumed total channel-height. In the lower half of the channel this attenuation factor is more similar to MTLE, whereas in the upper half is more similar to MTLL attenuation factor (Fig. 9.2).

9.3 Channel-Base Current Functions

One-peaked pulse functions are usually used as channel-base currents in engineering models of lightning strokes. Double exponential (DEXP) function (at $z' = 0$) is given in [6] as

$$i(t) = I_m[e^{-\alpha t} - e^{-\beta t}], \tag{9.2}$$

where I_m is the current value, α and β are the constants. DEXP has non-realistic convex waveshape in the rising part and its first derivative is not equal to zero at $t = 0^+$, as it should be.

Heidler's function (at $z' = 0$) is given in [7] as

$$i(t) = \frac{I_0}{\eta} \frac{\left(\frac{t}{\tau_1}\right)^n}{1 + \left(\frac{t}{\tau_1}\right)^n} e^{-\frac{t}{\tau_2}}, \text{ for } \eta = e^{-\frac{\tau_1}{\tau_2}\left(\frac{n\tau_1}{\tau_2}\right)^{1/n}}, \tag{9.3}$$

where I_0 is the current value, η is the peak correction factor, τ_1 and τ_2 are time constants, and n is often chosen between 2 and 10. Heidler's function is also used for representation of the first and subsequent negative strokes and first positive strokes as defined in [8].

One-peaked AEF may approximate the IEC 62305 standard currents and other typical lightning stroke currents, as given in [9, 15], with the following expression

$$i(t) = \begin{cases} I_m (t/t_m)^a e^{a(1-t/t_m)}, & 0 \le t \le t_m, \\ I_m \sum_{i=1}^{n} c_i (t/t_m)^{b_i} e^{b_i(1-t/t_m)}, & t_m \le t \le \infty, \end{cases} \tag{9.4}$$

for t_m the rise time to the maximum current value I_m, whereas n is the number of terms in the decaying part, a and b_i are parameters, and c_i weighting coefficients, so that $\sum_1^n c_i = 1$. In the simplest case $n = 1$, $c_1 = 1$ and $b_1 = b$, so the function has four parameters I_m, t_m, a and b. The function normalized to the maximum value at $t_m = 1.9$ µs, for $a = 4$ and $b = 0.03$ is presented in Fig. 9.3. This waveshape may represent high-voltage pulse 1.2/50 µs. Its rising part is shown in more detail in Fig. 9.4, and both rising and decaying part in Fig. 9.3.

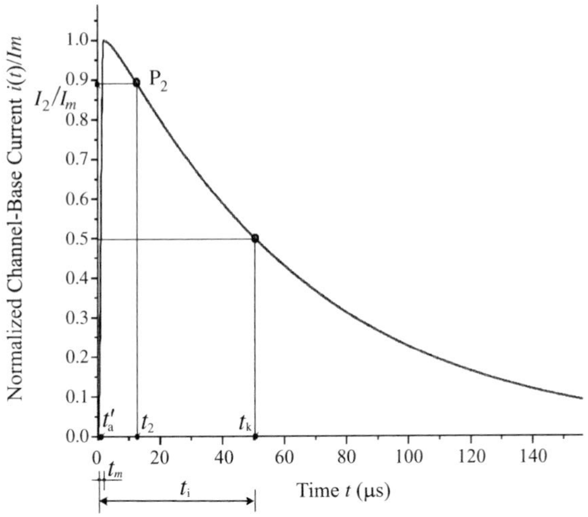

Fig. 9.3 Normalized channel-base current AEF

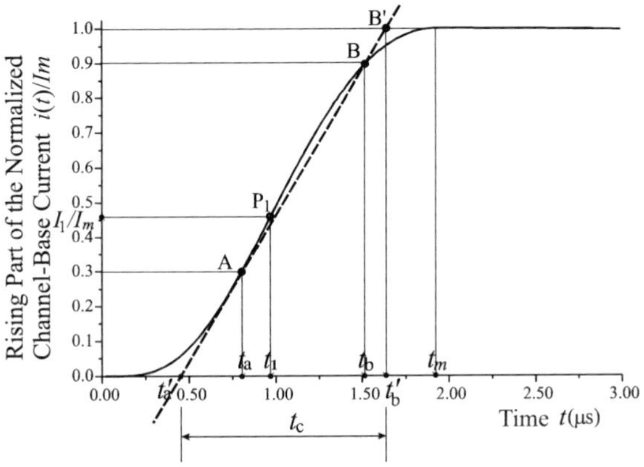

Fig. 9.4 Rising part of the normalized channel-base current AEF

Parameters of this function are calculated according to the IEC 62305 lightning currents of the first negative strokes, subsequent negative strokes and positive strokes and given in [15], and for other typical lightning strokes in [9].

First stroke current, as measured in experiments at Monte San Salvatore [1], is represented by the sum of seven Heidler's functions [3], given in Fig. 9.5 and denoted by MSS and the dash line. First stroke current measured in experiments at Morro do Cachimbo Station [20] is approximated also by the sum of seven Heidler's functions [3], given in Fig. 9.6 and denoted by MCS and the dash line. These two first stroke currents may be well represented with the double-peaked AEF (solid lines in Figs. 9.5 and 9.6) and parameters given in [11]. Double-peaked AEF has the following expression:

$$i(t) = \begin{cases} I_{m1} \sum_{i=1}^{m} d_i (t/t_{m1})^{a_i} e^{a_i(1-t/t_{m1})}, & 0 \le t \le t_{m1}, \\ I_{m1} + I_{m2} \sum_{i=1}^{l} f_i \left(\frac{t/t_{m1} - 1}{t_{m2}/t_{m1} - 1} \right)^{b_i} e^{b_i\left(1 - \frac{t/t_{m1}-1}{t_{m2}/t_{m1}-1}\right)}, & t_{m1} \le t \le t_{m2}, \\ (I_{m1} + I_{m2}) \sum_{i=1}^{n} g_i (t/t_{m2})^{c_i} e^{c_i(1-t/t_{m2})}, & t_{m2} \le t \le \infty, \end{cases} \quad (9.5)$$

with parameters a_i, b_i, c_i, and weighting coefficients d_i, f_i, g_i, so that $\sum_1^m d_i = \sum_1^l f_i = \sum_1^n g_i = 1$.

Three-peaked AEF, as given in [12], is used for the computation of LEMF results in this paper. Its peaks are I_{m1} at t_{m1}, $I_{m1} + I_{m2}$ at t_{m2}, $I_{m1} + I_{m2} + I_{m3}$ at t_{m3}, and it is given with the following expression:

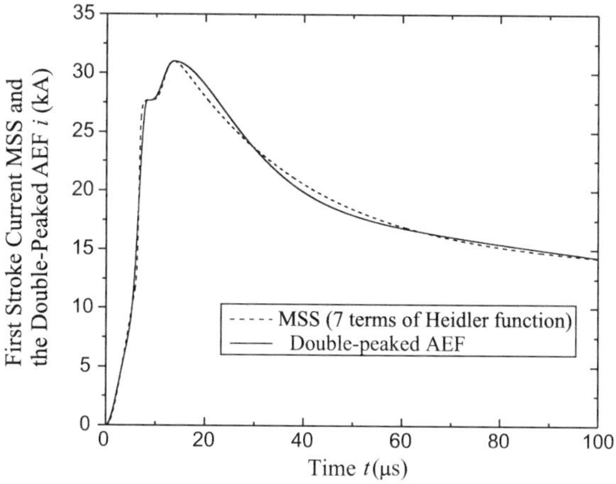

Fig. 9.5 Double-peaked AEF representing first stroke current MSS

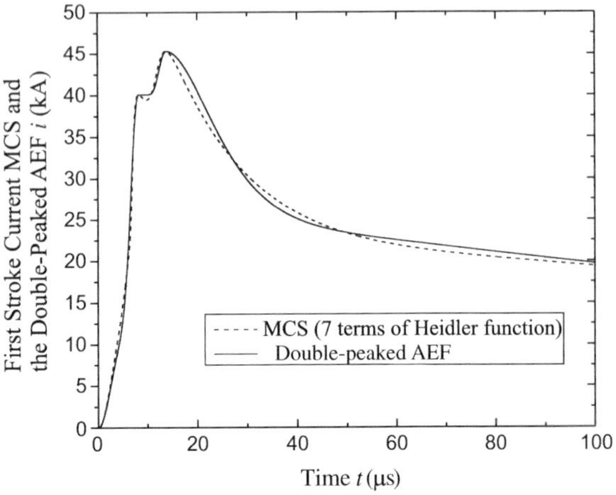

Fig. 9.6 Double-peaked AEF representing first stroke current MCS

$$i(t) = \begin{cases} I_{m1} \sum_{i=1}^{j} e_i (t/t_{m1})^{a_i} e^{a_i(1-t/t_{m1})}, & 0 \le t \le t_{m1}, \\ I_{m1} + I_{m2} \sum_{i=1}^{k} f_i \left(\frac{t/t_{m1} - 1}{t_{m2}/t_{m1} - 1} \right)^{b_i} e^{b_i\left(1 - \frac{t/t_{m1}-1}{t_{m2}/t_{m1}-1}\right)}, & t_{m1} \le t \le t_{m2}, \\ I_{m1} + I_{m2} + I_{m3} \sum_{i=1}^{l} g_i \left(\frac{t/t_{m2} - 1}{t_{m3}/t_{m2} - 1} \right)^{c_i} e^{c_i\left(1 - \frac{t/t_{m2}-1}{t_{m3}/t_{m2}-1}\right)}, & t_{m2} \le t \le t_{m3}, \\ (I_{m1} + I_{m2} + I_{m3}) \sum_{i=1}^{n} h_i (t/t_{m3})^{d_i} e^{d_i(1-t/t_{m3})}, & t_{m3} \le t \le \infty, \end{cases} \tag{9.6}$$

for parameters a_i, b_i, c_i, d_i, weighting coefficients e_i, f_i, g_i, h_i, and j, k, l, n the number of terms chosen for better approximation in the corresponding time interval. The expression reducing the number of unknown coefficients is

$$\sum_{1}^{j} e_i = \sum_{1}^{k} f_i = \sum_{1}^{l} g_i = \sum_{1}^{n} h_i = 1. \tag{9.7}$$

Three-peaked AEF (Fig. 9.7) approximating measurements results from [4] has the following parameters: the first peak $I_{m1} = 11$ kA at $t_{m1} = 2$ μs, the second peak $I_{m1} + I_{m2} = 8.3$ kA at $t_{m2} = 22$ μs, the third peak $I_{m1} + I_{m2} + I_{m3} = 4.4$ kA at $t_{m3} = 110$ μs, and other parameters: $a_1 = 2.2$, $a_2 = 0.5$, $e_1 = 0.37$, $e_2 = 1 - e_1$, $b_1 = 2$, $b_2 = 0.5$, $f_1 = 0.9$, $f_2 = 1 - f_1$, $c_1 = 2$, $g_1 = 1$, $d_1 = 5$, $d_2 = 0.55$, $h_1 = 0.6$, $h_2 = 1 - h_1$.

The advantage of AEF is analytically calculated first derivative and integral, both necessary for LEMF computation at the perfectly conducting ground. Fourier trans-

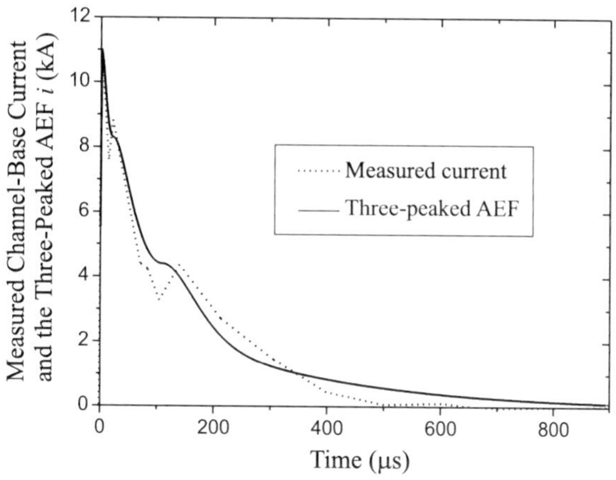

Fig. 9.7 Three-peaked AEF representing measured current [4]

form, needed for calculations above lossy ground, is analytically calculated for (9.4), (9.5) and (9.6). The integral of the square of the AEF, needed for calculating specific energy of lightning strokes, is also given in [10].

9.4 LEMF Computation

At the perfectly conducting ground surface, electric field has only vertical component and magnetic field its azimuthal component, whereas other components of electric and magnetic field do not exist. Vertical electric field at the ground surface points can be calculated from

$$E_z(\mathbf{R},t) = \frac{1}{4\pi\varepsilon_0}\int_{-H}^{H}\left[\frac{2(z-z')^2-r^2}{R^5}\int_{\tau=0}^{\tau=t} i(z',\tau-R/c)\mathrm{d}\tau + \frac{2(z-z')^2-r^2}{cR^4}i(z',t-R/c) - \frac{r^2}{c^2R^3}\frac{\partial i(z',t-R/c)}{\partial t}\right]\mathrm{d}z', \tag{9.8}$$

and azimuthal magnetic field from

$$H_\psi(\mathbf{R},t) = \frac{1}{4\pi}\int_{-H}^{H}\left[\frac{r}{R^3}i(z',t-R/c) + \frac{r}{cR^2}\frac{\partial i(z',t-R/c)}{\partial t}\right]\mathrm{d}z', \tag{9.9}$$

for c the speed of light and R the distance from the elementary current source to the field point $\mathrm{M}(r,\psi,z)$, as in Fig. 9.1. For all the results of engineering models applied in this paper, the return stroke speed is $v_f = 1.3\ 10^8$ m/s, and the maximum channel-base current value is $I_m = 11$ kA.

For AEF with one term in the rising part and two terms in the decaying part (9.4), which well approximates the channel-base current given in [17], the current along channel of the total height $H = 7000$ m is given in Fig. 9.8a in three time moments, and in Fig. 9.8b at three heights. If the decaying constant is $\lambda = 4500$ m in the applied MTLE model, the current along the channel at $t = 10$ μs, $t = 20$ μs, and $t = 30$ μs in Fig. 9.8a, and at $z' = 0$, $z' = 2$ km and $z' = 4$ km in Fig. 9.8b has greater values than if $\lambda = 2000$ m. Smaller λ means greater attenuation along the channel.

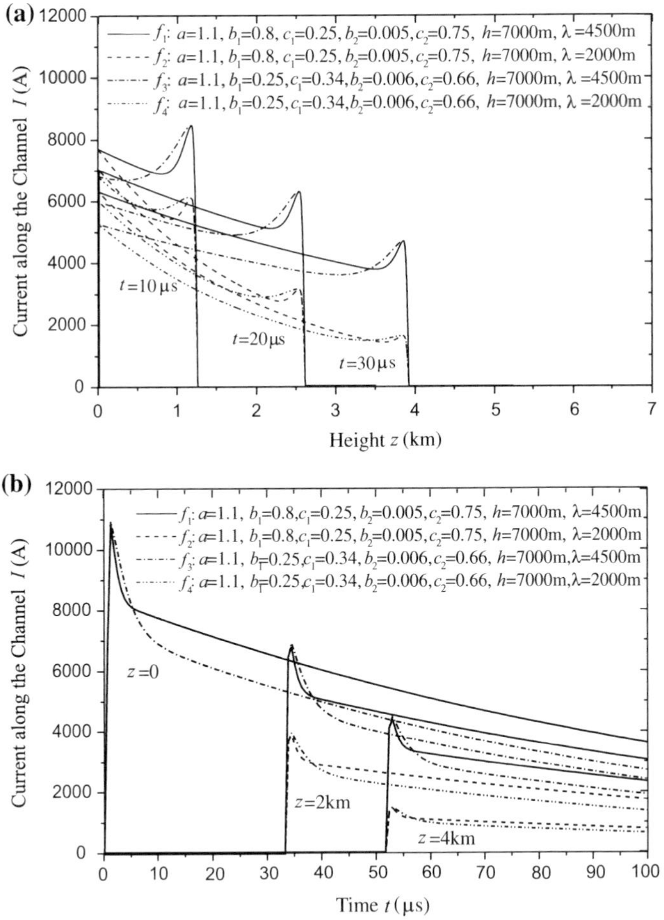

Fig. 9.8 Current along the channel for $H = 7000$ m, $\lambda = 4500$ m and $\lambda = 2000$ m, at $t = 10$ μs, $t = 20$ μs, and $t = 30$ μs (**a**), and at $z' = 0$, $z' = 2$ km and $z' = 4$ km (**b**)

For parameters of the three-peaked AEF (9.6) representing measured results from [4], given also in Fig. 9.7, the influence of the channel-height on electric field is great at $r = 50$ m, as can be noticed for MTLE and MTLL in Fig. 9.9a, but not on magnetic field results as presented in Fig. 9.9b for all the models. In Fig. 9.9, the results are given for the channel heights $H = 2600$ m and $H = 7500$ m, and in

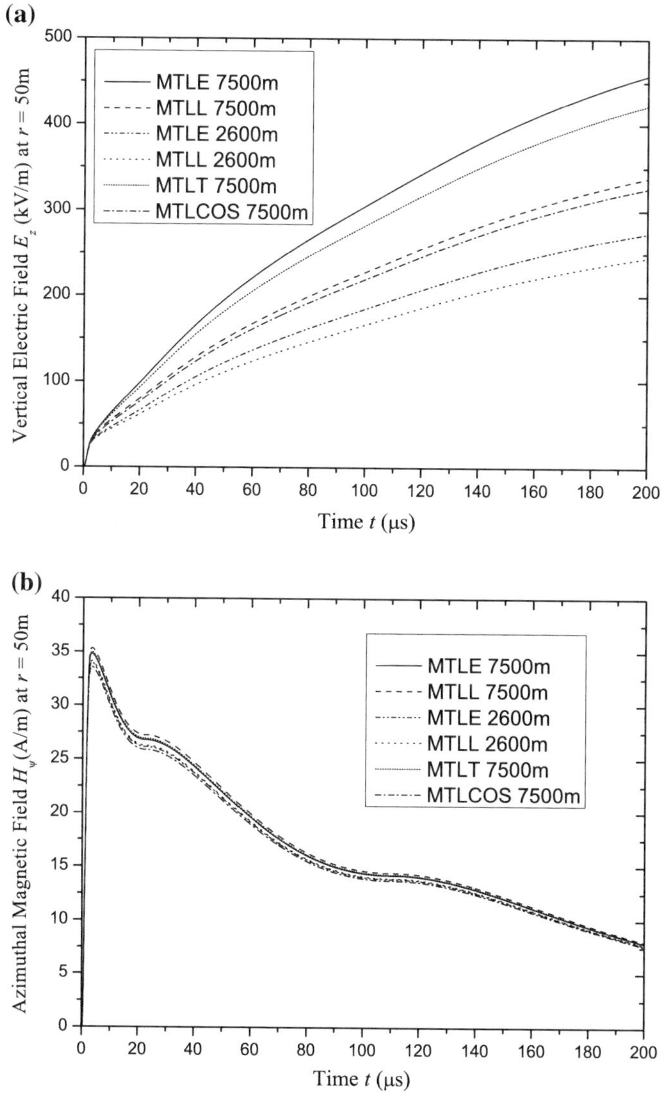

Fig. 9.9 Vertical electric (**a**) and azimuthal magnetic field (**b**) results of MTLL, MTLE, MTLCOS, and MTLT models at $r = 50$ m for channel heights $H = 2600$ m and $H = 7500$ m

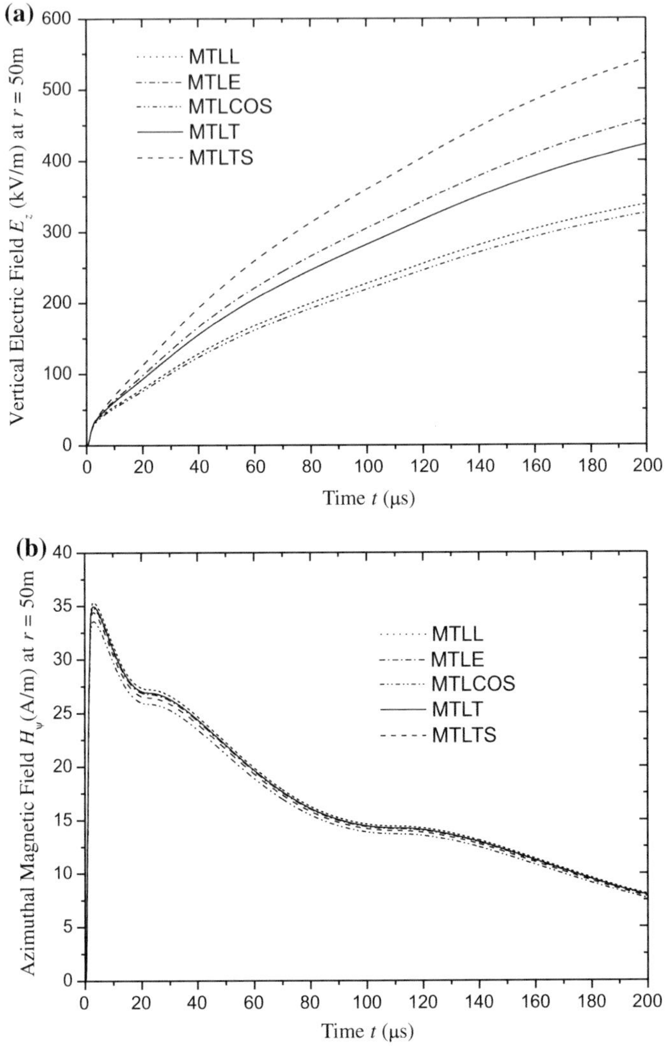

Fig. 9.10 Vertical electric field (**a**), and azimuthal magnetic field (**b**) results at $r = 50$ m from the channel-base for MTLL, MTLE, MTLCOS, MTLT and MTLTS models

Fig. 9.10 for MTLL, MTLE, MTLT, MTLTS and MTLCOS just for $H = 7500$ m, with the greatest electric field values obtained for MTLTS.

Results for electric and magnetic field at the distance of $r = 5$ km are presented in Fig. 9.11. They are in better agreement with the experimental results from [16], given in Fig. 9.12, compared to the results for other models (Fig. 9.13), given in [17] for Heidler's function and other parameters the same. LEMF results obtained with

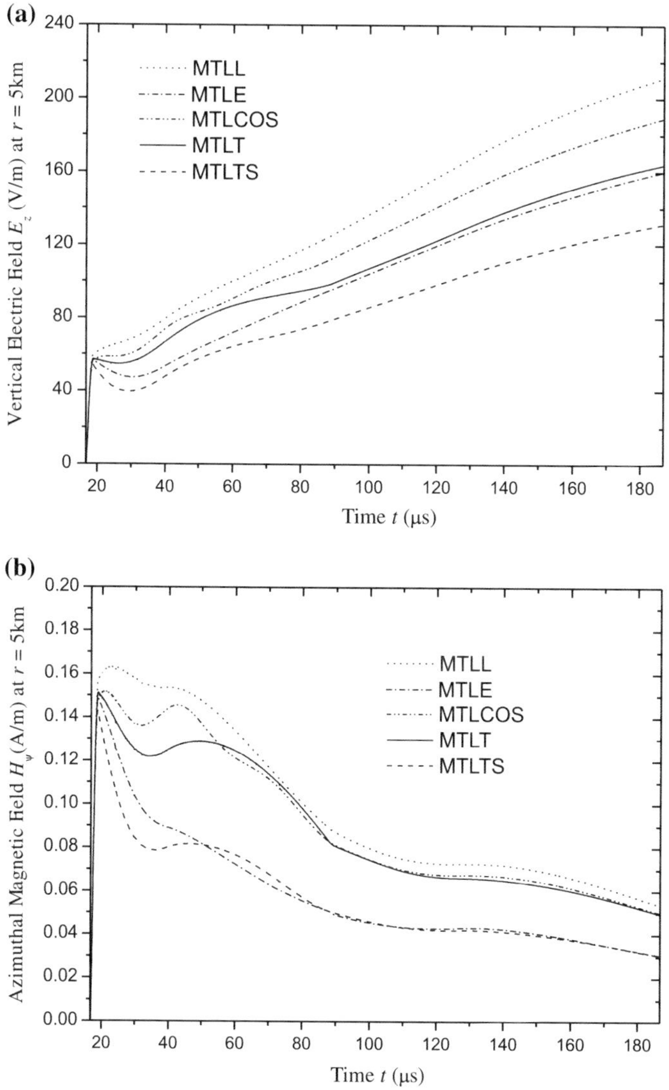

Fig. 9.11 Vertical electric field (**a**) and azimuthal magnetic field (**b**) at $r = 5$ km from the channel-base for MTLL, MTLE, MTLCOS, MTLT and MTLTS models

MTLE, MTLL, MTLT, MTLTS and MTLCOS at $r = 15$ km are presented in Fig. 9.14 and experimental results in Fig. 9.15. At such distances waveshapes for MTLT and MTLTS are approximately the same, whereas the intensities differ. Figure 9.16 shows calculated results for all these models at $r = 200$ km, whereas experimental results from [16] are given in Fig. 9.17. For other models and Heidler's function in [17],

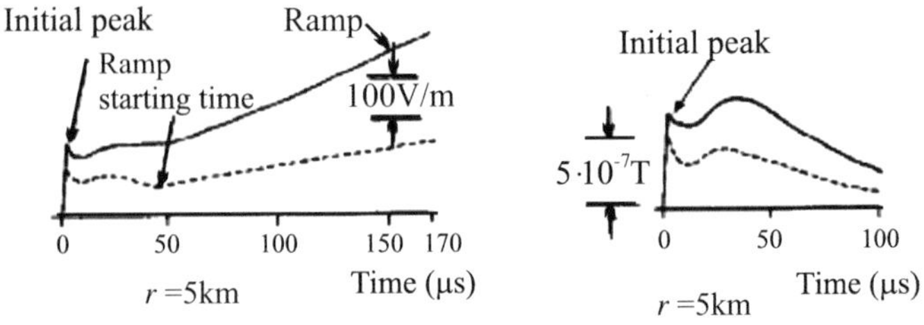

Fig. 9.12 Measurements results for electric field and magnetic flux at the distance $r = 5$ km, adopted from [16], *solid line* for first strokes and *dashed line* for subsequent strokes

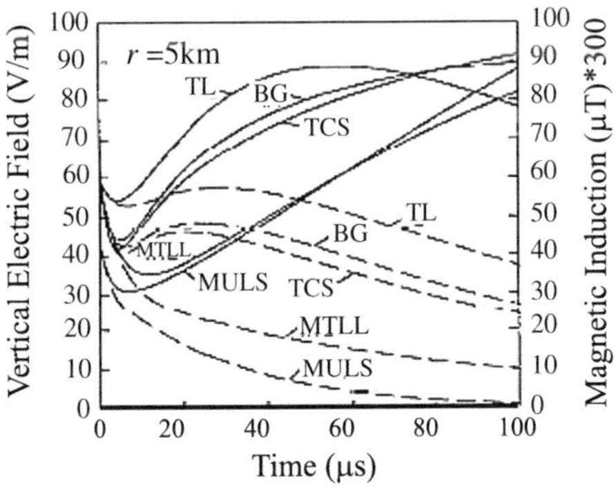

Fig. 9.13 LEMF results for TL, MTLL, MULS, BG and TCS models at $r = 5$ km, adopted from [17], *solid line* for vertical electric filed, *dashed line* for azimuthal magnetic flux density

LEMF results at $r = 100$ km are presented in Fig. 9.18. It should be noted that for 50 m the lightning electromagnetic pulse appears at $t_1 = 1/6$ μs after the current pulse starts propagating from the channel base (at $t = 0$). For 500 m this time is $t_2 = 5/3$ μs, for 5 km is $t_3 = 50/3$ μs, for 200 km is $t_4 = 2000/3\ \mu s$. All LEMF results are given in this paper for the first 170 μs of the pulse appearing at the corresponding distance, as results in [16] are given in such interval.

Some features of measured electric and magnetic fields are given in [17] as benchmark for the validation of models. The characteristics of typical lightning strokes are: (1) a sharp initial peak in both electric and magnetic fields, (2) a slow ramp in electric field waveshape within a few tens of kilometres, (3) a hump in magnetic field within a few tens of kilometers, and (4) a zero crossing in both electric and magnetic fields, at all the distances over 50 km. New models MTLT, MTLTS and MTLCOS provide all the features (Table 9.2) for the channel-base current approximated by three-peaked AEF.

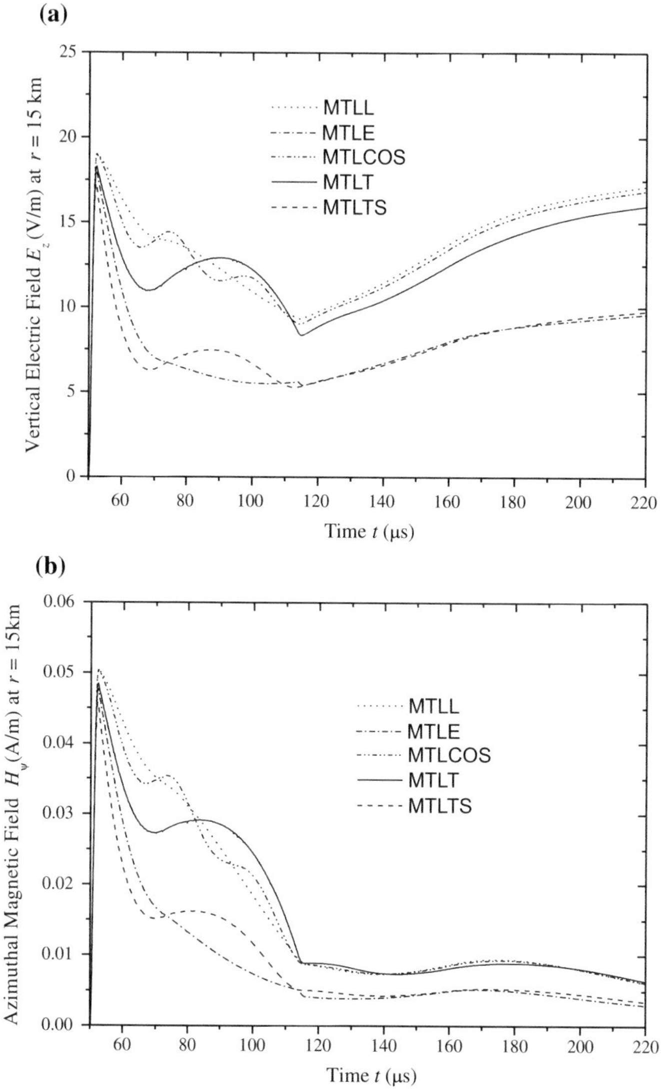

Fig. 9.14 Vertical electric field (**a**) and azimuthal magnetic field (**b**) at $r = 15$ km from the channel-base for MTLL, MTLE, MTLCOS, MTLT and MTLTS models

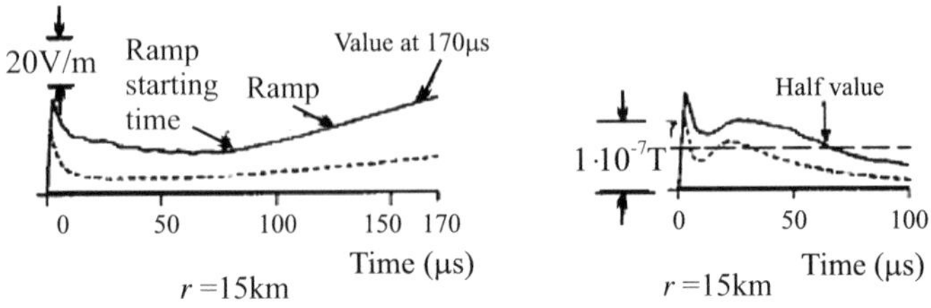

Fig. 9.15 Measurements results for electric field and magnetic flux at the distance $r = 15$ km, adopted from [16], *solid line* for first strokes and *dashed line* for subsequent strokes

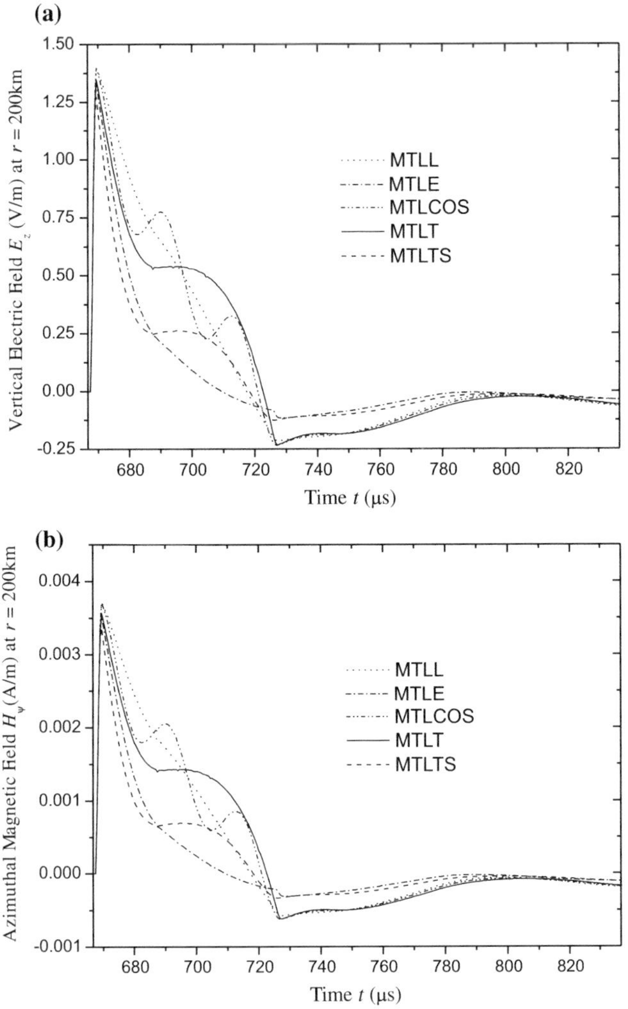

Fig. 9.16 Vertical electric field (**a**) and azimuthal magnetic field (**b**) at $r = 200$ km from the channel-base for MTLL, MTLE, MTLCOS, MTLT and MTLTS models

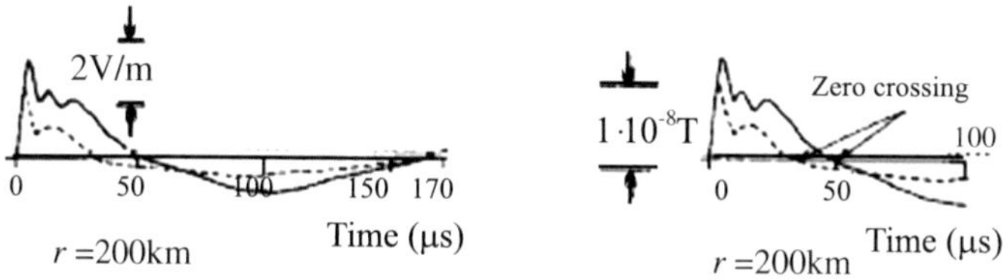

Fig. 9.17 Measurements results for electric field and magnetic flux at the distance $r = 200$ km, adopted from [16], *solid line* for first strokes and *dashed line* for subsequent strokes

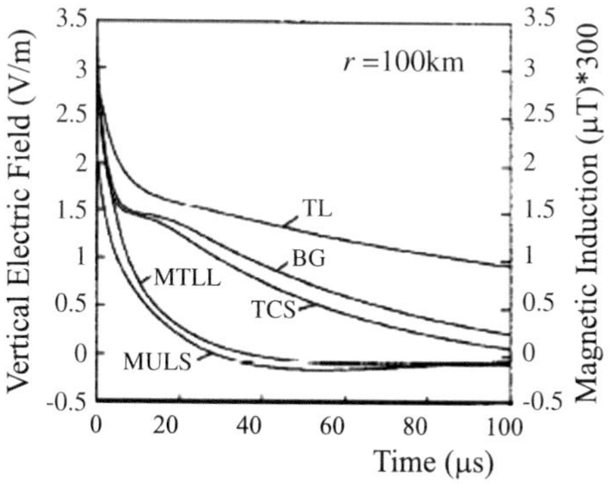

Fig. 9.18 Calculated LEMF results at $r = 100$ km for five models, adopted from [17], *solid line* for vertical electric field and *dashed line* for azimuthal magnetic field

Table 9.2 LEMF characteristic features of lightning stroke models

Characteristic feature of the model	(1) Sharp initial peak in E- and H-field	(2) E-field ramp within a few tens of km	(3) H-field hump within few tens of km, max at 10 to 40 μs	(4) E- and H-field zero crossing at about 50 to 200 km
TL	Yes	No	Yes	No
MTLL	Yes	Yes	No	Yes
MTLE	Yes	Yes	No	Yes
MTLD	Yes	Yes	No	Yes
TCS	Yes	Yes	Yes	No
DU	Yes	Yes	Yes	No
MTLT	Yes	Yes	Yes	Yes
MTLTS	Yes	Yes	Yes	Yes
MTLCOS	Yes	Yes	Yes	Yes

9.5 Conclusion

Modified transmission line models of lightning strokes with new attenuation factors and using AEF as the channel-base current function provide LEMF results in better agreement with measurements results. MTLT better models first strokes, whereas MTLTS better models subsequent strokes. MTLCOS gives ripples in electric field and magnetic field waveshapes in far field, similar to noticed in experimental results.

It should be noticed that if using three-peaked channel-base current all these models perform zero crossing in their waveshapes of vertical electric and azimuthal magnetic field at the distances over 50 km, a ramp in vertical electric field and a hump in azimuthal magnetic field within a few tens of kilometers. New lightning stroke models are based on the fact that electric and magnetic field waveshapes at the distances over 50 km approximately follow the waveshape of the channel-base current before their zero crossing.

Multi-peaked AEF is suitable for approximating experimental results for lightning currents, double-peaked AEF is suitable for the first stroke currents, and the simplest, one-peaked AEF, may well represent the Standard IEC 62305 lightning currents. AEF may be also used to approximate measured electrostatic discharge currents, so as the Standard IEC 61000-4-2 electrostatic discharge current [13].

References

1. Anderson, R.B., Eriksson, A.J.: Lightning parameters for engineering applications. Electra **69**, 65–102 (1980)
2. Berger, K., Anderson, R.B., Kroeninger, H.: Parameters of lightning flashes. Electra **80**, 23–37 (1975)
3. De Conti, A., Visacro, S.: Analytical representation of single- and double-peaked lightning current waveforms. IEEE Trans. Electromagn. Compat. **49**, 448–451 (2007)
4. Delfino, F., Procopio, R., Rossi, M., Rachidi, F.: Prony series representation for the lightning channel base current. IEEE Trans. Electromagn. Compat. **54**, 308–315 (2012)
5. EMC Part 4-2: Testing and Measurement Techniques Electrostatic Discharge Immunity Test. IEC International Standard 61000-4-2, Ed. 2 (2009)
6. Golde, R.H.: Lightning currents and related parameters. In: Golde, R.H. (ed.) Lightning. Physics of Lightning, vol. 1, pp. 309–350. Academic press, London (1977)
7. Heidler, F.: Travelling current source model for LEMP calculation. In: Proceedings of 6th International Zurich Symposium EMC, Zurich, Switzerland, pp. 157–162 (1985)
8. IEC International Standard 62305-SER Ed. 2.0 B: Protection against lightning (2013)
9. Javor, V.: New functions for representing IEC 62305 standard and other typical lightning stroke currents. J. Light. Res., Ref. BSP-JLR-2011-4 (2011)
10. Javor, V.: Parameters of lightning stroke currents defining their mechanical and thermal effects. In: Proceedings of 2012 IEEE Symposium on EMC, Pittsburgh, Pennsylvania, USA, vol. 1, pp. 334–339 (2012)
11. Javor, V.: Approximation of a double-peaked channel-base current. Int. J. Comput. Math. Electr. Electron. Eng. COMPEL **31**, 1007–1017 (2012)
12. Javor, V.: Multi-peaked functions for representation of lightning channel-base currents. In: Proceedings of 31st International Conference on Lightning Protection - ICLP 2012, Vienna, Austria, (2012). doi:10.1109/ICLP.2012.6344384

13. Javor, V.: New function for representing IEC 61000-4-2 standard electrostatic discharge current. Facta Universitatis Ser. Electron. Energ. **27**(4), 509–520 (2014)
14. Javor, V., Rancic, P.D.: The effect of lightning return stroke channel height on electric and magnetic field waveforms. In: CD Proceedings of 29th Internatioanl Conference on Lightning Protection - ICLP 2008, Paper3a-07, Uppsala, Sweden (2008)
15. Javor, V., Rancic, P.D.: A channel-base current function for lightning return-stroke modeling. IEEE Trans. Electromagn. Compat. **53**, 245–249 (2011)
16. Lin, Y.T., Uman, M.A., Tiller, J.A., Brantley, R.D., Beasley, W.H., Krider, E.P., Weidman, C.D.: Characterization of lightning return stroke electric and magnetic fields from simultaneous two-station measurements. J. Geophys. Res. **84**, 6307–6314 (1979)
17. Nucci, C.A., Diendorfer, G., Uman, M.A., Rachidi, F., Ianoz, M., Mazzetti, C.: Lightning return stroke current models with specified channel-base current: A review and comparison. J. Geophys. Res. **95**, 20395–20408 (1990)
18. Rakov, V.A., Dulzon, A.A.: A modified transmission line model for lightning return stroke field calculations. In: Proceedings of 9th International Zurich Symposium Electromagnetic Compatibility, Zurich, Switzerland, pp. 229–235 (1991)
19. Rakov, V.A., Uman, M.A.: Review and evaluation of lightning return stroke models including some aspects of their application. IEEE Trans. Electromagn. Compat. **40**, 403–426 (1998)
20. Visacro, S., Soares Jr., A., Schroeder, M.A.O., Cherchiglia, L.C.L., Souza, V.J.: Statistical analysis of lightning current parameters: measurements at Morro do Cachimbo station. J. Geophys. Res. **109**, D01105 (2004). doi:10.1029/2003JD003662

Chapter 10
On Some Properties of the Multi-peaked Analytically Extended Function for Approximation of Lightning Discharge Currents

Karl Lundengård, Milica Rančić, Vesna Javor and Sergei Silvestrov

Abstract According to experimental results for lightning discharge currents, they are classified in the IEC 62305 Standard into waveshapes representing the first positive, first and subsequent negative strokes, and long-strokes. These waveshapes, especially shot-term pulses, are approximated with a few mathematical functions in literature, in order to be used in lightning discharge models for calculations of electromagnetic field and lightning induced effects. An analytically extended function (AEF) is presented in this paper and used for lightning currents modeling. The basic properties of this function with a finite number of peaks are examined. A general framework for estimating the parameters of the AEF using the Marquardt least-squares method (MLSM) for a waveform with an arbitrary (finite) number of peaks as well as for the given charge transfer and specific energy is described. This framework is used to find parameters for some common single-peak waveforms and some advantages and disadvantages of the approach are also discussed.

Keywords Analytically extended function · Lightning current function · Lightning strike · Marquardt least-squares method · Electromagnetic compatibility

---

K. Lundengård (✉) · M. Rančić · S. Silvestrov
Division of Applied Mathematics, The School of Education, Culture and Communication, Mälardalen University, Box 883, 721 23 Västerås, Sweden
e-mail: karl.lundengard@mdh.se

M. Rančić
e-mail: milica.rancic@mdh.se

S. Silvestrov
e-mail: sergei.silvestrov@mdh.se

V. Javor
Department of Power Engineering, Faculty of Electronic Engineering, University of Niš, Niš, Serbia
e-mail: vesna.javor@elfak.ni.ac.rs

S. Silvestrov and M. Rančić (eds.), *Engineering Mathematics I*,
Springer Proceedings in Mathematics & Statistics 178,
DOI 10.1007/978-3-319-42082-0_10

10.1 Introduction

Many different types of systems, objects and equipment are susceptible to damage from lightning discharges. Lightning effects are usually analysed using lightning discharge models. Most of the engineering and electromagnetic models imply channel-base current functions. Various single and multi-peaked functions are proposed in the literature for modelling lightning channel-base currents, examples include Heidler, Heidler and Cvetic [3], Javor and Rancic [7], Javor [5, 6]. For engineering and electromagnetic models, a general function that would be able to reproduce desired waveshapes is needed, such that analytical solutions for its derivatives, integrals, and integral transformations, exist. A multi-peaked channel-base current function has been proposed in Javor [5] as a generalization of the so-called TRF (two-rise front) function from Javor [6], which possesses such properties.

In this paper we analyse a modification of such a multi-peaked function, a so-called p-peak analytically extended function (AEF). Possibility of application of the AEF to modelling of various multi-peaked waveshapes is investigated. Estimation of its parameters has been performed using the Marquardt least-squares method (MLSM), an efficient method for the estimation of non-linear function parameters, Marquardt [14]. It has been applied in many fields, including lightning research for optimizing parameters of the Heidler function in Lovric et al. [10], or the Pulse function in Lundengård et al. [11, 12].

Some numerical results are presented, including those for the Standard IEC 62305 [4] current of the first-positive strokes, and an example of a fast-decaying lightning current waveform.

10.2 The p-Peak Analytically Extended Function

The p-peaked AEF is constructed using the function

$$x(\beta; t) = \left(t e^{1-t}\right)^{\beta}, \; 0 \leq t, \tag{10.1}$$

which we will refer to as the power exponential function. The power exponential function is qualitatively similar to the desired waveforms in the sense that it has a steeply rising initial part followed by a more slowly decaying part. The steepness of both the rising and decaying part is determined by the β-parameter. This is illustrated in Fig. 10.1.

This function is in some ways similar to the Heidler function [2] that is commonly used [4]. One feature of the Heidler function that the power exponential function does not share is that a Heidler function with a very steep rise and slow decay can be easily constructed. To construct the AEF so that it can imitate this feature we define it as a piecewise linear combinations of scaled and translated power exponential functions, the concept is illustrated in Fig. 10.2.

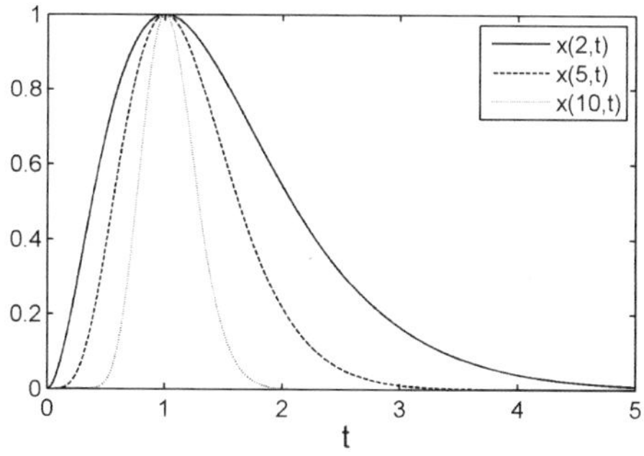

Fig. 10.1 An illustration of how the steepness of the power exponential function varies with β

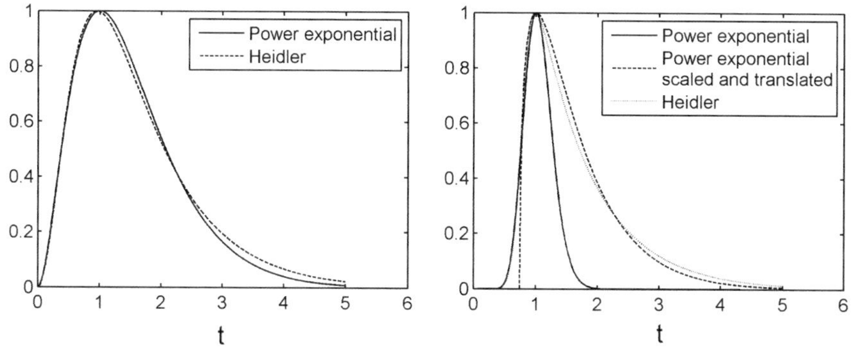

Fig. 10.2 An illustration of how the steepness of the power exponential function varies with β

In order to get a function with multiple peaks and where the steepness of the rise between each peak as well as the slope of the decaying part is not dependent on each other, we define the analytically extended function (AEF) as a function that consist of piecewise linear combinations of the power exponential function that has been scaled and translated so that the resulting function is continuous. Given the difference in height between each pair of peaks $I_{m_1}, I_{m_2}, \ldots, I_{m_p}$, the corresponding times t_{m_1}, $t_{m_2}, \ldots, t_{m_p}$, integers $n_q > 0$, real values $\beta_{q,k}, \eta_{q,k}$, $1 \le q \le p+1$, $1 \le k \le n_q$ such that the sum over k of $\eta_{q,k}$ is equal to one, the p-peaked AEF $i(t)$ is given by (10.2).

Definition 10.1 Given $I_{m_q} \in \mathbb{R}, t_{m_q} \in \mathbb{R}, q = 1, 2, \ldots, p$ such that $t_{m_0} = 0 < t_{m_1} \le t_{m_2} \le \ldots \le t_{m_p}$ along with $\eta_{q,k}$, $\beta_{q,k} \in \mathbb{R}$ and $0 < n_q \in \mathbb{Z}$ for $q = 1, 2, \ldots, p+1$, $k = 1, 2, \ldots, n_q$ such that $\sum_{k=1}^{n_q} \eta_{q,k} = 1$.

The *analytically extended function* (AEF), $i(t)$, with p peaks is defined as

$$i(t) = \begin{cases} \left(\sum_{k=1}^{q-1} I_{m_k}\right) + I_{m_q} \sum_{k=1}^{n_q} \eta_{q,k} x_q(t)^{\beta_{q,k}^2+1}, & t_{m_{q-1}} \leq t \leq t_{m_q},\ 1 \leq q \leq p, \\ \left(\sum_{k=1}^{p} I_{m_k}\right) \sum_{k=1}^{n_{p+1}} \eta_{p+1,k} x_{p+1}(t)^{\beta_{p+1,k}^2}, & t_{m_p} \leq t, \end{cases} \tag{10.2}$$

where

$$x_q(t) = \begin{cases} \frac{t - t_{m_{q-1}}}{\Delta t_{m_q}} \exp\left(\frac{t_{m_q} - t}{\Delta t_{m_q}}\right), & 1 \leq q \leq p, \\ \frac{t}{t_{m_q}} \exp\left(1 - \frac{t}{t_{m_q}}\right), & q = p + 1, \end{cases}$$

and $\Delta t_{m_q} = t_{m_q} - t_{m_{q-1}}$.

Sometimes the notation $i(t; \boldsymbol{\beta}, \boldsymbol{\eta})$ with

$$\boldsymbol{\beta} = \left[\beta_{1,1}\ \beta_{1,2}\ \ldots\ \beta_{q,k}\ \ldots\ \beta_{p+1,n_{p+1}}\right],\ \boldsymbol{\eta} = \left[\eta_{1,1}\ \eta_{1,2}\ \ldots\ \eta_{q,k}\ \ldots\ \eta_{p+1,n_{p+1}}\right]$$

will be used to clarify what the particular parameters for a certain AEF are.

Remark 10.1 The p-peak AEF can be written more compactly if we introduce the vectors

$$\boldsymbol{\eta}_q = [\eta_{q,1}\ \eta_{q,2}\ \ldots\ \eta_{q,n_q}]^\top, \tag{10.3}$$

$$\mathbf{x}_q(t) = \begin{cases} \left[x_q(t)^{\beta_{q,1}^2+1}\ x_q(t)^{\beta_{q,2}^2+1}\ \ldots\ x_q(t)^{\beta_{q,n_q}^2+1}\right]^\top, & 1 \leq q \leq p, \\ \left[x_q(t)^{\beta_{q,1}^2}\ x_q(t)^{\beta_{q,2}^2}\ \ldots\ x_q(t)^{\beta_{q,n_q}^2}\right]^\top, & q = p + 1. \end{cases} \tag{10.4}$$

The more compact form is

$$i(t) = \begin{cases} \left(\sum_{k=1}^{q-1} I_{m_k}\right) + I_{m_q} \cdot \boldsymbol{\eta}_q^\top \mathbf{x}_q(t), & t_{m_{q-1}} \leq t \leq t_{m_q},\ 1 \leq q \leq p, \\ \left(\sum_{k=1}^{q} I_{m_k}\right) \cdot \boldsymbol{\eta}_q^\top \mathbf{x}_q(t), & t_{m_q} \leq t,\ q = p + 1. \end{cases} \tag{10.5}$$

If the AEF is used to model an electrical current, than the derivative of the AEF determines the induced electrical voltage in conductive loops in the lightning field. For this reason it is desirable to guarantee that the first derivative of the AEF is continuous.

Since the AEF is a linear function of elementary functions its derivative can be found using standard methods.

Theorem 10.1 *The derivative of the p-peak AEF is*

$$\frac{\mathrm{d}i(t)}{\mathrm{d}t} = \begin{cases} I_{m_q} \dfrac{t_{m_q} - t}{t - t_{m_{q-1}}} \dfrac{x_q(t)}{\Delta t_{m_q}} \boldsymbol{\eta}_q^\top \mathbf{B}_q \mathbf{x}_q(t), & t_{m_{q-1}} \leq t \leq t_{m_q},\ 1 \leq q \leq p, \\ I_{m_q} \dfrac{x_q(t)}{t} \dfrac{t_{m_q} - t}{t_{m_q}} \boldsymbol{\eta}_q^\top \mathbf{B}_q \mathbf{x}_q(t), & t_{m_q} \leq t,\ q = p+1, \end{cases} \tag{10.6}$$

where

$$\mathbf{B}_{p+1} = \begin{bmatrix} \beta_{p+1,1}^2 & 0 & \dots & 0 \\ 0 & \beta_{p+1,2}^2 & \dots & 0 \\ \vdots & \vdots & \ddots & \vdots \\ 0 & 0 & \dots & \beta_{p+1,n_{p+1}}^2 \end{bmatrix}, \quad \mathbf{B}_q = \begin{bmatrix} \beta_{q,1}^2 + 1 & 0 & \dots & 0 \\ 0 & \beta_{q,2}^2 + 1 & \dots & 0 \\ \vdots & \vdots & \ddots & \vdots \\ 0 & 0 & \dots & \beta_{q,n_q}^2 + 1 \end{bmatrix},$$

for $1 \leq q \leq p$.

Proof From the definition of the AEF (see (10.2)) and the derivative of the power exponential function (10.1) given by

$$\frac{\mathrm{d}}{\mathrm{d}t} x(\beta; t) = \beta(1 - t)t^{\beta-1} e^{\beta(1-t)},$$

expression (10.6) can easily be derived since differentiation is a linear operation and the result can be rewritten in the compact form analogously to (10.5).

Illustration of the AEF function and its derivative for various values of $\beta_{q,k}$-parameters is shown in Fig. 10.3.

Lemma 10.1 *The AEF is continuous and at each t_{m_q} the derivative is equal to zero.*

Proof Within each interval $t_{m_{q-1}} \leq t \leq t_{m_q}$ the AEF is a linear combination of continuous functions and at each t_{m_q} the function will approach the same value from both directions unless all $\eta_{q,k} \leq 0$, but if $\eta_{q,k} \leq 0$ then $\sum_{k=1}^{n_q} \eta_{q,k} \neq 1$.

Noting that for any diagonal matrix $\mathbf{B}$ the expression

$$\boldsymbol{\eta}_q^\top \mathbf{B}\, \mathbf{x}_q(t) = \sum_{k=1}^{n_q} \eta_{q,k} \mathbf{B}_{kk} x_q(t)^{\beta_{q,k}^2+1}, \ 1 \leq q \leq p,$$

is well-defined and that the equivalent statement holds for $q = p$ it is easy to see from (10.6) that the factor $(t_{m_q} - t)$ in the derivative ensures that the derivative is zero every time $t = t_{m_q}$.

When interpolating a waveform with p peaks it is natural to require that there will not appear new peaks between the chosen peaks. This corresponds to requiring monotonicity in each interval. One way to achieve this is given in Lemma 10.2.

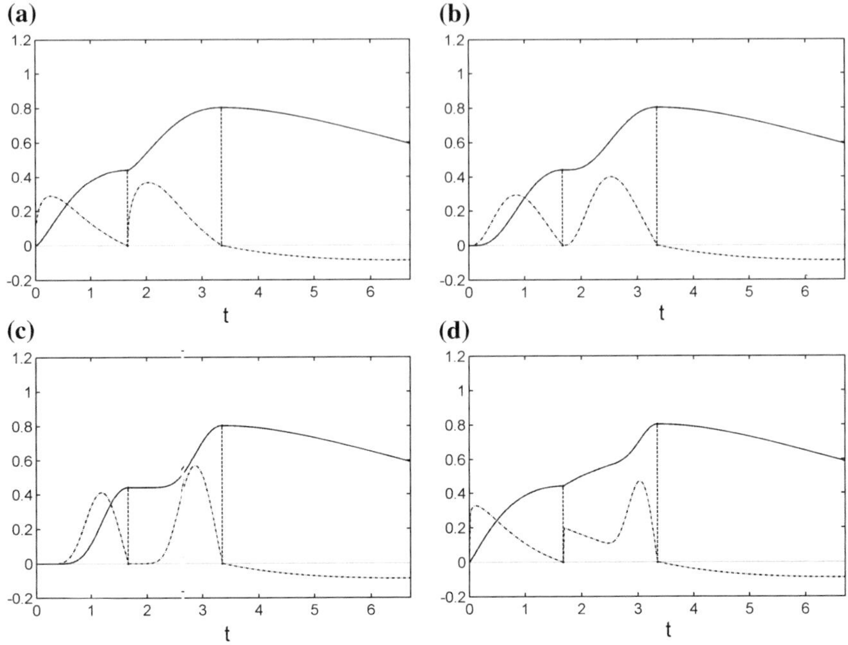

Fig. 10.3 Illustration of the AEF (*solid line*) and its derivative (*dashed line*) with the same I_{m_q} and t_{m_q} but different $\beta_{q,k}$-parameters. **a** $0 < \beta_{q,k} < 1$, **b** $4 < \beta_{q,k} < 5$, **c** $12 < \beta_{q,k} < 13$, **d** a mixture of large and small $\beta_{q,k}$-parameters

Lemma 10.2 *If* $\eta_{q,k} \geq 0$, $k = 1, \ldots, n_q$ *the AEF,* $i(t)$, *is strictly monotonic on the interval* $t_{m_{q-1}} < t < t_{m_q}$.

Proof The AEF will be strictly monotonic in an interval if the derivative has the same sign everywhere in the interval. That this is the case follows from (10.6) since every term in $\boldsymbol{\eta}_q^\top \mathbf{B}_q \, \mathbf{x}_q(t)$ is non-negative if $\eta_{q,k} \geq 0$, $k = 1, \ldots, n_q$, so the sign of the derivative it determined by I_{m_q}.

If we allow some of the $\eta_{q,k}$-parameters to be negative, the derivative can change sign the function might get an extra peak between two other peaks, see Fig. 10.4.

The integral of the electrical current represents the charge transfer. Unlike the Heidler function the integral of the AEF is relatively straightforward to find. How to do this is detailed in Lemmas 10.3, 10.4, Theorems 10.2, and 10.3.

Lemma 10.3 *For any* $t_{m_{q-1}} \leq t_0 \leq t_1 \leq t_{m_q}$, $1 \leq q \leq p$,

$$\int_{t_0}^{t_1} x_q(t)^\beta \, \mathrm{d}t = \frac{e^\beta}{\beta^{\beta+1}} \Delta\gamma \left(\beta + 1, \frac{t_1 - t_{m_q}}{\beta \Delta t_{m_q}}, \frac{t_0 - t_{m_q}}{\beta \Delta t_{m_q}} \right) \tag{10.7}$$

with $\Delta t_{m_q} = t_{m_q} - t_{m_{q-1}}$ *and*

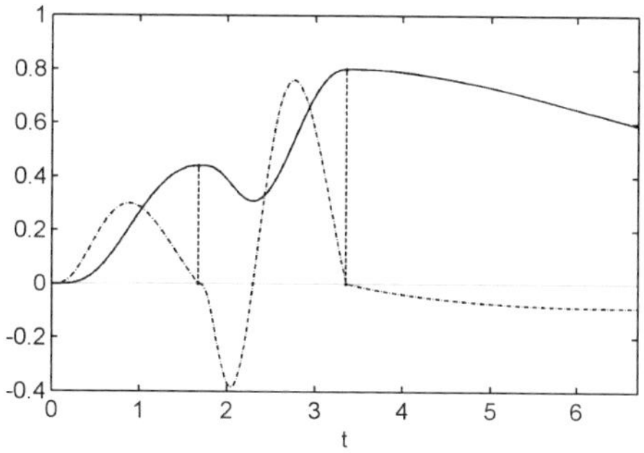

Fig. 10.4 An example of a two-peaked AEF where some of the $\eta_{q,k}$-parameters are negative, so that it has points where the first derivative changes sign between two peaks. The *solid line* is the AEF and the *dashed lines* is the derivative of the AEF

$$\Delta\gamma(\beta, t_0, t_1) = \gamma(\beta + 1, \beta t_1) - \gamma(\beta + 1, \beta t_0),$$

where

$$\gamma(\beta, t) = \int_0^t \tau^{\beta-1} e^{-\tau}\, \mathrm{d}\tau$$

is the lower incomplete Gamma function [1].

If $t_0 = t_{m_{q-1}}$ *and* $t_1 = t_{m_q}$ *then*

$$\int_{t_{m_{q-1}}}^{t_{m_q}} x_q(t)^\beta\, \mathrm{d}t = \frac{e^\beta}{\beta^{\beta+1}} \gamma(\beta + 1, \beta). \tag{10.8}$$

Proof

$$\begin{aligned}\int_{t_0}^{t_1} x_q(t)^\beta\, \mathrm{d}t &= \int_{t_0}^{t_1} \left(\frac{t - t_{m_q}}{\Delta t_{m_q}} \exp\left(1 - \frac{t - t_{m_q}}{\Delta t_{m_q}}\right)\right)^\beta \mathrm{d}t \\ &= \frac{e^\beta}{\beta^{\beta+1}} \int_{t_0}^{t_1} \left(\beta \frac{t - t_{m_q}}{\Delta t_{m_q}}\right)^\beta \exp\left(1 - \beta \frac{t - t_{m_q}}{\Delta t_{m_q}}\right) \mathrm{d}t.\end{aligned}$$

Changing variables according to $\tau = \frac{t - t_{m_q}}{\Delta t_{m_q}}$ gives

$$\begin{aligned}\int_{t_0}^{t_1} x_q(t)^\beta\, \mathrm{d}t &= \frac{e^\beta}{\beta^{\beta+1}} \int_{\tau_0}^{\tau_1} \tau^\beta e^{-\tau}\, \mathrm{d}t \\ &= \frac{e^\beta}{\beta^{\beta+1}} \left(\gamma(\beta + 1, \tau_1) - \gamma(\beta + 1, \tau_0)\right) \\ &= \frac{e^\beta}{\beta^{\beta+1}} \Delta\gamma(\beta + 1, \tau_1, \tau_0) \\ &= \frac{e^\beta}{\beta^{\beta+1}} \Delta\gamma\left(\beta + 1, \beta \frac{t_1 - t_{m_q}}{\Delta t_{m_q}}, \beta \frac{t_0 - t_{m_q}}{\Delta t_{m_q}}\right).\end{aligned}$$

When $t_0 = t_{m_{q-1}}$ and $t_1 = t_{m_q}$ then

$$\int_{t_0}^{t_1} x_q(t)^{\beta} \, \mathrm{d}t = \frac{e^{\beta}}{\beta^{\beta+1}} \Delta\gamma\left(\beta+1, \beta\right)$$

and with $\gamma(\beta+1, 0) = 0$ we get (10.8).

Lemma 10.4 *For any $t_{m_{q-1}} \leq t_0 \leq t_1 \leq t_{m_q}$, $1 \leq q \leq p$,*

$$\int_{t_0}^{t_1} i(t) \, \mathrm{d}t = (t_1 - t_0)\left(\sum_{k=1}^{q-1} I_{m_k}\right) + I_{m_q} \sum_{k=1}^{n_q} \eta_{q,k} \, g_q(t_1, t_0), \tag{10.9}$$

where

$$g_q(t_1, t_0) = \frac{e^{\beta_{q,k}^2}}{\left(\beta_{q,k}^2 + 1\right)^{\beta_{q,k}^2+1}} \Delta\gamma\left(\beta_{q,k}^2 + 2, \frac{t_1 - t_{m_{q-1}}}{\Delta t_{m_q}}, \frac{t_0 - t_{m_{q-1}}}{\Delta t_{m_q}}\right)$$

with $\Delta\gamma(\beta, t_0, t_1)$ defined as in (10.7).

Proof

$$\begin{aligned}
\int_{t_0}^{t_1} i(t) \, \mathrm{d}t &= \int_{t_0}^{t_1} \left(\sum_{k=1}^{q-1} I_{m_k}\right) + I_{m_q} \sum_{k=1}^{n_q} \eta_{q,k} x_q(t)^{\beta_{q,k}^2+1} \, \mathrm{d}t \\
&= (t_1 - t_0)\left(\sum_{k=1}^{q-1} I_{m_k}\right) + I_{m_q} \sum_{k=1}^{n_q} \eta_{q,k} \int_{t_0}^{t_1} x_q(t)^{\beta_{q,k}^2+1} \, \mathrm{d}t \\
&= (t_1 - t_0)\left(\sum_{k=1}^{q-1} I_{m_k}\right) + I_{m_q} \sum_{k=1}^{n_q} \eta_{q,k} \, g_q(t_0, t_1).
\end{aligned}$$

Theorem 10.2 *If $t_{m_{a-1}} \leq t_a \leq t_{m_a}$, $t_{m_{b-1}} \leq t_b \leq t_{m_b}$ and $0 \leq t_a \leq t_b \leq t_{m_p}$ then*

$$\begin{aligned}
\int_{t_a}^{t_b} i(t) \, \mathrm{d}t = {} & (t_{m_a} - t_a)\left(\sum_{k=1}^{a-1} I_{m_k}\right) + I_{m_a} \sum_{k=1}^{n_a} \eta_{a,k} \, g_a(t_a, t_{m_a}) \\
& + \sum_{q=a+1}^{b-1} \left(\Delta t_{m_q}\left(\sum_{k=1}^{q-1} I_{m_k}\right) + I_{m_q} \sum_{k=1}^{n_q} \eta_{q,k} \, \hat{g}\left(\beta_{q,k}^2 + 1\right)\right) \\
& + (t_b - t_{m_b})\left(\sum_{k=1}^{b-1} I_{m_k}\right) + I_{m_b} \sum_{k=1}^{n_b} \eta_{b,k} \, g_b(t_{m_b}, t_b),
\end{aligned} \tag{10.10}$$

where $g_q(t_0, t_1)$ is defined as in Lemma 10.4 and

$$\hat{g}(\beta) = \frac{e^{\beta}}{\beta^{\beta+1}} \gamma\,(\beta + 1, \beta)\,.$$

Proof This theorem follows from integration being linear and Lemma 10.4.

Theorem 10.3 *For $t_{m_p} \leq t_0 < t_1 < \infty$ the integral of the AEF is*

$$\int_{t_0}^{t_1} i(t)\,\mathrm{d}t = \left(\sum_{k=1}^{p} I_{m_k}\right) \sum_{k=1}^{n_{p+1}} \eta_{p+1,k}\, g_{p+1}(t_1, t_0), \tag{10.11}$$

where $g_q(t_0, t_1)$ is defined as in Lemma 10.4.

When $t_0 = t_{m_p}$ and $t_1 \to \infty$ the integral becomes

$$\int_{t_{m_p}}^{\infty} i(t)\,\mathrm{d}t = \left(\sum_{k=1}^{p} I_{m_k}\right) \sum_{k=1}^{n_{p+1}} \eta_{p+1,k}\, \tilde{g}\left(\beta_{p+1,k}^2\right), \tag{10.12}$$

where

$$\tilde{g}(\beta) = \frac{e^{\beta}}{\beta^{\beta+1}} \left(\Gamma(\beta + 1) - \gamma\,(\beta + 1, \beta)\right)$$

with

$$\Gamma(\beta) = \int_0^{\infty} t^{\beta-1} e^{-t}\,\mathrm{d}t$$

is the Gamma function [1].

Proof This theorem follows from integration being linear and Lemma 10.4.

In the next section we will estimate the parameters of the AEF that gives the best fit with respect to some data and for this the partial derivatives with respect to the β_{m_q} parameters will be useful. Since the AEF is a linear function of elementary functions these partial derivatives can easily be found using standard methods.

Theorem 10.4 *The partial derivatives of the p-peak AEF with respect to the β parameters are*

$$\frac{\partial i}{\partial \beta_{q,k}} = \begin{cases} 0, & 0 \leq t \leq t_{m_{q-1}}, \\ 2\, I_{m_q} \eta_{q,k}\, \beta_{q,k}\, h_q(t) x_q(t)^{\beta_{q,k}^2+1}, & t_{m_{q-1}} \leq t \leq t_{m_q},\ 1 \leq q \leq p, \\ 0, & t_{m_q} \leq t, \end{cases} \tag{10.13}$$

$$\frac{\partial i}{\partial \beta_{p+1,k}} = \begin{cases} 0, & 0 \leq t \leq t_{m_p}, \\ 2\, I_{m_{p+1}} \eta_{p+1,k}\, \beta_{p+1,k}\, h_{p+1}(t) x_{p+1}(t)^{\beta_{p+1,k}^2}, & t_{m_p} \leq t, \end{cases} \tag{10.14}$$

where

$$h_q(t) = \begin{cases} \ln\left(\dfrac{t - t_{m_{q-1}}}{\Delta t_{m_q}}\right) - \dfrac{t - t_{m_{q-1}}}{\Delta t_{m_q}} + 1, & 1 \le q \le p, \\ \ln\left(\dfrac{t}{t_{m_q}}\right) - \dfrac{t}{t_{m_q}} + 1, & q = p + 1. \end{cases}$$

Proof Since the $\beta_{q,k}$ parameters are independent, differentiation with respect to $\beta_{q,k}$ will annihilate all terms but one in each linear combination. The expressions (10.13) and (10.14) then follow from the standard rules for differentiation of composite functions and products of functions.

10.3 Least Square Fitting Using MLSM

10.3.1 The Marquardt Least-Squares Method

The Marquardt least-squares method, also known as *the Levenberg-Marquardt algorithm* or *damped least-squares*, is an efficient method for least-squares estimation for functions with non-linear parameters that was developed in the middle of the 20th century (see [9, 14]).

The least-squares estimation problem for functions with non-linear parameters arises when a function of m independent variables and described by k unknown parameters needs to be fitted to a set of n data points such that the sum of squares of residuals is minimized.

The vector containing the independent variables is $\mathbf{x} = (x_1, \ldots, x_n)$, the vector containing the parameters $\boldsymbol{\beta} = (\beta_1, \ldots, \beta_k)$ and the data points

$$(Y_i, X_{1i}, X_{2i}, \ldots, X_{mi}) = (Y_i, \mathbf{X}_i)\,,\; i = 1, 2, \ldots, n.$$

Let the residuals be denoted by $E_i = f(\mathbf{X}_i; \boldsymbol{\beta}) - Y_i$ and the sum of squares of E_i is then written as

$$S = \sum_{i=1}^{n} [f(\mathbf{X}_i; \boldsymbol{\beta}) - Y_i]^2\,,$$

which is the function to be minimized with respect to $\boldsymbol{\beta}$.

The Marquardt least-square method is an iterative method that gives approximate values of $\boldsymbol{\beta}$ by combining the Gauss–Newton method (also known as the inverse Hessian method) and the steepest descent (also known as the gradient) method to minimize S. The method is based around solving the linear equation system

$$\left(\mathbf{A}^{*(r)} + \lambda^{(r)}\mathbf{I}\right)\boldsymbol{\delta}^{*(r)} = \mathbf{g}^{*(r)}, \tag{10.15}$$

where $A^{*(r)}$ is a modified *Hessian matrix* of $\mathbf{E(b)}$ (or $f(\mathbf{X}_i; \mathbf{b})$), $\mathbf{g}^{*(r)}$ is a rescaled version of the gradient of S, r is the number of the current iteration of the method, and λ is a real positive number sometimes referred to as the fudge factor [15]. The Hessian, the gradient and their modifications are defined as follows:

$$\mathbf{A} = \mathbf{J}^\top \mathbf{J},$$

$$\mathbf{J}_{ij} = \frac{\partial f_i}{\partial b_j} = \frac{\partial E_i}{\partial b_j}, \ i = 1, 2, \ldots, m; \ j = 1, 2, \ldots, k,$$

and

$$(\mathbf{A}^*)_{ij} = \frac{a_{ij}}{\sqrt{a_{ii}}\sqrt{a_{jj}}},$$

while

$$\mathbf{g} = \mathbf{J}^\top(\mathbf{Y} - \mathbf{f}_0), \ \mathbf{f}_{0i} = f(\mathbf{X}_i, \mathbf{b}, \mathbf{c}), \ \mathbf{g}_i^* = \frac{\mathbf{g}_i}{a_{ii}}.$$

Solving (10.15) gives a vector which, after some scaling, describes how the parameters $\mathbf{b}$ should be changed in order to get a new approximation of $\boldsymbol{\beta}$,

$$\mathbf{b}^{(r+1)} = \mathbf{b}^{(r)} + \boldsymbol{\delta}^{(r)}, \ \boldsymbol{\delta}^{(r)} = \frac{\boldsymbol{\delta}_i^{*(r)}}{\sqrt{a_{ii}}}. \tag{10.16}$$

It is obvious from (10.15) that $\boldsymbol{\delta}^{(r)}$ depends on the value of the fudge factor λ. Note that if $\lambda = 0$, then (10.15) reduces to the regular Gauss–Newton method [14], and if $\lambda \to \infty$ the method will converge towards the steepest descent method [14]. The reason that the two methods are combined is that the Gauss–Newton method often has faster convergence than the steepest descent method, but is also an unstable method [14]. Therefore, λ must be chosen appropriately in each step. In the Marquardt least-squares method this amounts to increasing λ with a chosen factor v whenever an iteration increases S, and if an iteration reduces S then λ is reduced by a factor v as many times as possible. Below follows a detailed description of the method using the following notation:

$$S^{(r)} = \sum_{i=1}^{n} \left[Y_i - f(\mathbf{X}_i, \mathbf{b}^{(r)}, \mathbf{c})\right]^2, \tag{10.17}$$

$$S\left(\lambda^{(r)}\right) = \sum_{i=1}^{n} \left[Y_i - f(\mathbf{X}_i, \mathbf{b}^{(r)} + \boldsymbol{\delta}^{(r)}, \mathbf{c})\right]^2. \tag{10.18}$$

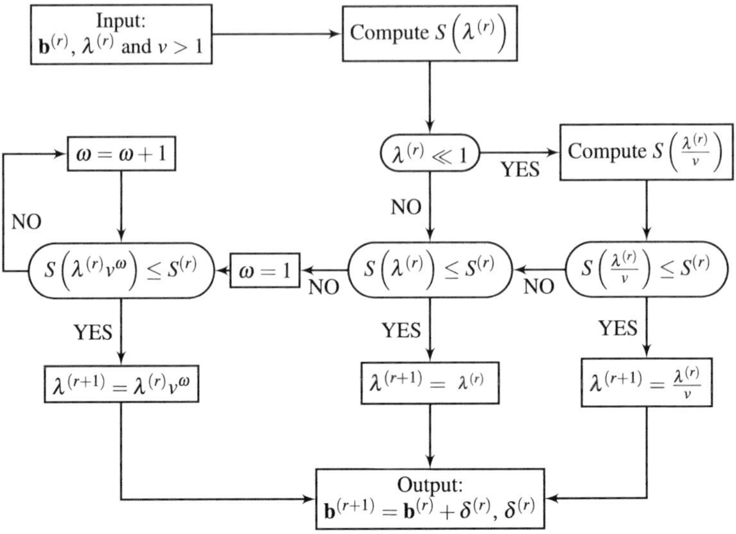

Fig. 10.5 The basic iteration step of the Marquardt least-squares method, definitions of computed quantities are given in (10.16), (10.17) and (10.18)

The iteration step of the Marquardt least-squares method can be described as follows:

- Input: $v > 1$ and $\mathbf{b}^{(r)}$, $\lambda^{(r)}$.
- ◁ Compute $S\left(\lambda^{(r)}\right)$.
- If $\lambda(r) \ll 1$ then compute $S\left(\frac{\lambda^{(r)}}{v}\right)$, else go to ▷.
- If $S\left(\frac{\lambda^{(r)}}{v}\right) \leq S^{(r)}$ let $\lambda^{(r+1)} = \frac{\lambda^{(r)}}{v}$.
- ▷ If $S\left(\lambda^{(r)}\right) \leq S^{(r)}$ let $\lambda^{(r+1)} = \lambda(r)$.
- If $S\left(\lambda^{(r)}\right) > S^{(r)}$ find the smallest integer $\omega > 0$ such that $S\left(\lambda^{(r)}v^{\omega}\right) \leq S^{(r)}$, and then set $\lambda^{(r+1)} = \lambda^{(r)}v^{\omega}$.
- Output: $\mathbf{b}^{(r+1)} = \mathbf{b}^{(r)} + \delta^{(r)}$, $\delta^{(r)}$.

This iteration step is also described in Fig. 10.5. Naturally, some condition for what constitutes an acceptable fit for the function must also be chosen. If this condition is not satisfied the new values for $\mathbf{b}^{(r+1)}$ and $\lambda^{(r+1)}$ will be used as input for the next iteration and if the condition is satisfied the algorithm terminates. The quality of the fitting, in other words the value of S, is determined by the stopping condition and the initial values for $\mathbf{b}^{(0)}$. The initial value of $\lambda^{(0)}$ affects the performance of the algorithm to some extent since after the first iteration $\lambda^{(r)}$ will be self-regulating. Suitable values for $\mathbf{b}^{(0)}$ are challenging to find for many functions f and they are often, together with $\lambda^{(0)}$, found using heuristic methods.

10.3.2 Estimating Parameters for Underdetermined Systems

For the Marquardt least-squares method to work one data point per unknown parameter is needed, $m = k$. It can still be possible to estimate all unknown parameters if there is insufficient data, $m < k$.

Suppose that $k - m = p$ and let $\gamma_j = \beta_{m+j}$, $j = 1, 2, \ldots, p$. If there are at least p known relations between the unknown parameters such that $\gamma_j = \gamma_j(\beta_1, \ldots, \beta_m)$ for $j = 1, 2, \ldots, p$ then the Marquardt least-squares method can be used to give estimates on $\beta_1, \ldots, \beta_m$ and the still unknown parameters can be estimated from these. Denoting the estimated parameters $\mathbf{b} = (b_1, \ldots, b_m)$ and $\mathbf{c} = (c_1, \ldots, c_p)$ the following algorithm can be used:

- Input: $\nu > 1$ and initial values $\mathbf{b}^{(0)}$, $\lambda^{(0)}$.
- $r = 0$
- $\triangleleft$ Find $\mathbf{c}^{(r)}$ using $\mathbf{b}^{(r)}$ together with extra relations.
- Find $\mathbf{b}^{(r+1)}$ and $\boldsymbol{\delta}^{(r)}$ using MLSM.
- Check chosen termination condition for MLSM, if it is not satisfied go to $\triangleleft$.
- Output: $\mathbf{b}$, $\mathbf{c}$.

The algorithm is illustrated in Fig. 10.6.

In order to fit the AEF it is sufficient that $k_q \geq n_q$. Suppose we have some estimate of the β-parameters which is collected in the vector $\mathbf{b}$. It is then fairly simple to calculate an estimate for the η-parameters, see Sect. 10.3.4, which we collect in $\mathbf{h}$. We can then define a residual vector by $(\mathbf{E})_k = i(t_{q,k}; \mathbf{b}, \mathbf{h}) - i_{q,k}$ where $i(t; \mathbf{b}, \mathbf{h})$ is the AEF with the estimated parameters.

The $\mathbf{J}$ matrix can in this case be described as

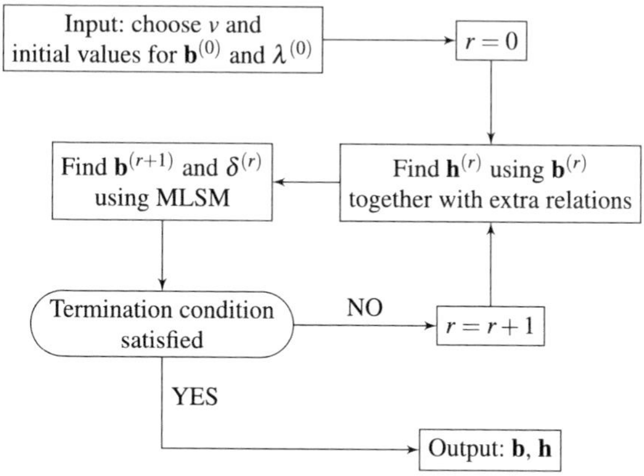

Fig. 10.6 Schematic description of the parameter estimation algorithm

$$\mathbf{J} = \begin{bmatrix} \left.\frac{\partial i}{\partial \beta_{q,1}}\right|_{t=t_{q,1}} & \left.\frac{\partial i}{\partial \beta_{q,2}}\right|_{t=t_{q,1}} & \cdots & \left.\frac{\partial i}{\partial \beta_{q,n_q}}\right|_{t=t_{q,1}} \\ \left.\frac{\partial i}{\partial \beta_{q,1}}\right|_{t=t_{q,2}} & \left.\frac{\partial i}{\partial \beta_{q,2}}\right|_{t=t_{q,2}} & \cdots & \left.\frac{\partial i}{\partial \beta_{q,n_q}}\right|_{t=t_{q,2}} \\ \vdots & \vdots & \ddots & \vdots \\ \left.\frac{\partial i}{\partial \beta_{q,1}}\right|_{t=t_{q,k_q}} & \left.\frac{\partial i}{\partial \beta_{q,2}}\right|_{t=t_{q,k_q}} & \cdots & \left.\frac{\partial i}{\partial \beta_{q,n_q}}\right|_{t=t_{q,k_q}} \end{bmatrix}, \tag{10.19}$$

where the partial derivatives are given by (10.13) and (10.14).

10.3.3 Fitting with Data Points as Well as Charge Transfer and Specific Energy Conditions

By considering the charge transfer at the striking point, Q_0, and the specific energy, W_0, two further conditions need to be considered:

$$Q_0 = \int_0^\infty i(t)\,\mathrm{d}t, \tag{10.20}$$

$$W_0 = \int_0^\infty i(t)^2\,\mathrm{d}t. \tag{10.21}$$

First we will define

$$Q(\mathbf{b},\mathbf{h}) = \int_0^\infty i(t;\mathbf{b},\mathbf{h})\,\mathrm{d}t,$$

$$W(\mathbf{b},\mathbf{h}) = \int_0^\infty i(t;\mathbf{b},\mathbf{h})^2\,\mathrm{d}t.$$

These two quantities can be calculated as follows.

Theorem 10.5

$$Q(\mathbf{b},\mathbf{h}) = \sum_{q=1}^{p}\left(\Delta t_{m_q}\left(\sum_{k=1}^{q-1} I_{m_k}\right) + I_{m_q}\sum_{k=1}^{n_q}\eta_{q,k}\,\hat{g}(\beta_{q,k}^2+1)\right) + \left(\sum_{k=1}^{p} I_{m_k}\right)\sum_{k=1}^{n_{p+1}}\eta_{p+1,k}\,\tilde{g}(\beta_{p+1,k}^2), \tag{10.22}$$

$$
\begin{aligned}
W(\mathbf{b},\mathbf{h}) = \sum_{q=1}^{p}\Bigg(&\left(\sum_{k=1}^{q-1} I_{m_k}\right)^2 + \left(\sum_{k=1}^{q-1} I_{m_k}\right) I_{m_q} \sum_{k=1}^{n_q} \eta_{q,k}\,\hat{g}(\beta_{q,k}^2+1) \\
&+ I_{m_q}^2 \sum_{k=1}^{n_q} \eta_{q,k}^2\,\hat{g}\left(2\,\beta_{q,k}^2+2\right) \\
&+ 2\,I_{m_q}^2 \sum_{r=1}^{n_q-1}\sum_{s=r+1}^{n_q} \eta_{q,r}\,\eta_{q,s}\,\hat{g}\left(\beta_{q,r}^2+\beta_{q,s}^2+2\right)\Bigg) \\
&+ \left(\sum_{k=1}^{p} I_{m_k}\right)^2 \left(\sum_{k=1}^{n_p} \eta_{p,k}^2\,\tilde{g}\left(2\,\beta_{p,k}^2\right) \right.\\
&\left. + 2\sum_{r=1}^{n_{p+1}-1}\sum_{s=r+1}^{n_{p+1}} \eta_{p+1,r}\,\eta_{p+1,s}\,\tilde{g}\left(\beta_{p+1,r}^2+\beta_{p+1,s}^2\right)\right), \quad (10.23)
\end{aligned}
$$

where $\hat{g}(\beta)$ and $\tilde{g}(\beta)$ are defined in Theorems 10.2 and 10.3.

Proof Formula (10.22) is found by combining (10.10) and (10.12). Formula (10.23) is found by noting that

$$
\left(\sum_{k=1}^{n} a_k\right)^2 = \sum_{k=1}^{n} a_k^2 + \sum_{r=1}^{n-1}\sum_{s=r+1}^{n} a_r\,a_s,
$$

and then reasoning analogously to the proofs for (10.10) and (10.12).

We can calculate the charge transfer and specific energy given by the AEF with formula (10.22) and (10.23), respectively, and get two additional residuals $E_{Q_0} = Q(\mathbf{b},\mathbf{h}) - Q_0$ and $E_{W_0} = W(\mathbf{b},\mathbf{h}) - W_0$. Since these are global conditions this means that the parameters $\boldsymbol{\eta}$ and $\boldsymbol{\beta}$ no longer can be fitted separately in each interval. This means that we need to consider all data points simultaneously.

The resulting **J**-matrix is

$$
\mathbf{J} = \begin{bmatrix}
\mathbf{J_1} & & \cdots & & 0 & & \\
\vdots & & \ddots & & \vdots & & \\
0 & & \cdots & & \mathbf{J_{p+1}} & & \\
\frac{\partial E_{Q_0}}{\partial \beta_{1,1}} & \cdots & \frac{\partial E_{Q_0}}{\partial \beta_{1,n_1}} & \cdots & \frac{\partial E_{Q_0}}{\partial \beta_{p+1,1}} & \cdots & \frac{\partial E_{Q_0}}{\partial \beta_{p+1,n_{p+1}}} \\
\frac{\partial E_{W_0}}{\partial \beta_{1,1}} & \cdots & \frac{\partial E_{W_0}}{\partial \beta_{1,n_1}} & \cdots & \frac{\partial E_{W_0}}{\partial \beta_{p+1,1}} & \cdots & \frac{\partial E_{W_0}}{\partial \beta_{p+1,n_{p+1}}}
\end{bmatrix}, \quad (10.24)
$$

where

$$\mathbf{J}_q = \begin{bmatrix} \left.\frac{\partial i}{\partial \beta_{q,1}}\right|_{t=t_{q,1}} & \left.\frac{\partial i}{\partial \beta_{q,2}}\right|_{t=t_{q,1}} & \cdots & \left.\frac{\partial i}{\partial \beta_{q,n_q}}\right|_{t=t_{q,1}} \\ \left.\frac{\partial i}{\partial \beta_{q,1}}\right|_{t=t_{q,2}} & \left.\frac{\partial i}{\partial \beta_{q,2}}\right|_{t=t_{q,2}} & \cdots & \left.\frac{\partial i}{\partial \beta_{q,n_q}}\right|_{t=t_{q,2}} \\ \vdots & \vdots & \ddots & \vdots \\ \left.\frac{\partial i}{\partial \beta_{q,1}}\right|_{t=t_{q,k_q}} & \left.\frac{\partial i}{\partial \beta_{q,2}}\right|_{t=t_{q,k_q}} & \cdots & \left.\frac{\partial i}{\partial \beta_{q,n_q}}\right|_{t=t_{q,k_q}} \end{bmatrix}$$

and the partial derivatives in the last two rows are given by

$$\frac{\partial Q}{\partial \beta_{q,s}} = \begin{cases} 2\, I_{m_q} \eta_{q,s}\, \beta_{q,s} \left.\frac{\mathrm{d}\hat{g}}{\mathrm{d}\beta}\right|_{\beta=\beta_{q,s}^2+1}, & 1 \leq q \leq p, \\ 2\, I_{m_p} \eta_{p+1,s}\, \beta_{p+1,s} \left.\frac{\mathrm{d}\tilde{g}}{\mathrm{d}\beta}\right|_{\beta=\beta_{p+1,s}^2}, & q = p+1. \end{cases}$$

For $1 \leq q \leq p$

$$\frac{\partial W}{\partial \beta_{q,s}} = 2\left(\sum_{k=1}^{q-1} I_{m_k}\right) I_{m_q} \eta_{q,s}\, \beta_{q,s} \left.\frac{\mathrm{d}\hat{g}}{\mathrm{d}\beta}\right|_{\beta=\beta_{q,s}^2+1}$$
$$+4\, I_{m_q}^2 \eta_{q,s} \beta_{q,s} \left(\eta_{q,s} \left.\frac{\mathrm{d}\hat{g}}{\mathrm{d}\beta}\right|_{\beta=2\beta_{q,s}^2+2} + \sum_{\substack{k=1\\k\neq s}}^{n_q} \eta_{q,k} \left.\frac{\mathrm{d}\hat{g}}{\mathrm{d}\beta}\right|_{\beta=\beta_{q,s}^2+\beta_{q,k}^2+2} \right)$$

and

$$\frac{\partial W}{\partial \beta_{p+1,s}} = 4\left(\sum_{k=1}^{p} I_{m_k}\right) \eta_{p+1,s}\beta_{p+1,s}$$
$$\left(\eta_{p+1,s} \left.\frac{\mathrm{d}\tilde{g}}{\mathrm{d}\beta}\right|_{\beta=2\beta_{p+1,s}^2} + \sum_{\substack{k=1\\k\neq s}}^{n_q} \eta_{p+1,k} \left.\frac{\mathrm{d}\tilde{g}}{\mathrm{d}\beta}\right|_{\beta=\beta_{p+1,s}^2+\beta_{p+1,k}^2} \right).$$

The derivatives of $\hat{g}(\beta)$ and $\tilde{g}(\beta)$ are

$$\frac{\mathrm{d}\hat{g}}{\mathrm{d}\beta} = \frac{e^{\beta}}{\beta^{\beta+1}} \left(\Gamma(\beta+1)\big(\Psi(\beta+1) + \ln(\beta)\big) - G(\beta) - \frac{\gamma(\beta+1,\beta)}{\beta} \right) + 1, \tag{10.25}$$

$$\frac{\mathrm{d}\tilde{g}}{\mathrm{d}\beta} = \frac{e^{\beta}}{\beta^{\beta+1}} \left(G(\beta) - \frac{\Gamma(\beta+1) - \gamma(\beta+1,\beta)}{b} \right) - 1, \tag{10.26}$$

where $\Gamma(\beta)$ is the Gamma function, $\Psi(\beta)$ is the digamma function, see [1], and $G(\beta)$ is a special case of the Meijer G-function and can be defined as

$$G(\beta) = G_{2,3}^{3,0}\left(\beta \left| \begin{matrix} 1, 1 \\ 0, 0, \beta + 1 \end{matrix} \right.\right)$$

using the notation from [16]. When evaluating this function it might be more practical to rewrite G using other special functions

$$G(\beta) = G_{2,3}^{3,0}\left(\beta \left| \begin{matrix} 1, 1 \\ 0, 0, \beta + 1 \end{matrix} \right.\right) = \frac{\beta^{\beta+1}}{(\beta+1)^2}\, {}_2F_2(\beta+1, \beta+1;\ \beta+2, \beta+2;\ -\beta) + \left(\ln(\beta) - \Psi(\beta) - \frac{1}{b}\right)\frac{\pi\ \csc(\pi\beta)}{\Gamma(-\beta)},$$

where

$${}_2F_2(\beta+1, \beta+1;\ \beta+2, \beta+2;\ -\beta) = \sum_{k=0}^{\infty}(-1)^k\beta^k\frac{(\beta+1)^2}{(\beta+k+1)^2} = \frac{\beta^2+2\beta+1}{\beta}\left(\frac{1}{\beta^2} - \sum_{k=0}^{\infty}\frac{(-b)^k}{(b+k)^2}\right)$$

is a special case of the hypergeometric function. These partial derivatives were found using software for symbolic computation [13].

Note that all η-parameters must be recalculated for each step, how this is done is detailed in the Sect. 10.3.4.

10.3.4 Calculating the η-Parameters from the β-Parameters

Suppose that we have $n_q - 1$ points $(t_{q,k}, i_{q,k})$ such that

$$t_{m_{q-1}} < t_{q,1} < t_{q,2} < \ldots < t_{q,n_q-1} < t_{m_q}.$$

For an AEF that interpolates these points it must be true that

$$\sum_{k=1}^{q-1} I_{m_k} + I_{m_q}\sum_{s=1}^{n_q}\eta_{q,s}x_q(t_{q,k})^{\beta_{q,s}} = i_{q,k},\ k = 1, 2, \ldots, n_q - 1. \tag{10.27}$$

Since $\eta_{q,1} + \eta_{q,2} + \ldots + \eta_{q,n_q} = 1$ equation (10.27) can be rewritten as

$$I_{m_q} \sum_{s=1}^{n_q-1} \eta_{q,s} \left(x_q(t_{q,k})^{\beta_{q,s}} - x_q(t_{q,k})^{\beta_{q,n_q}} \right) = i_{q,k} - x_q(t_{q,k})^{\beta_{q,n_q}} - \sum_{s=1}^{q-1} I_{m_s} \quad (10.28)$$

for $k = 1, 2, \ldots, n_q - 1$. This can easily be written as a matrix equation

$$I_{m_q} \tilde{X}_q \tilde{\boldsymbol{\eta}}_q = \tilde{\mathbf{i}}_q, \quad (10.29)$$

where

$$\tilde{\boldsymbol{\eta}}_q = \left[\eta_{q,1} \; \eta_{q,2} \; \ldots \; \eta_{q,n_q-1} \right]^\top,$$

$$\left(\tilde{\mathbf{i}}_q \right)_k = i_{q,k} - x_q(t_{q,k})^{\beta_{q,n_q}} - \sum_{s=1}^{q-1} I_{m_s},$$

$$\left(\tilde{X}_q \right)_{k,s} = \tilde{x}_q(k,s) = x_q(t_{q,k})^{\beta_{q,s}} - x_q(t_{q,k})^{\beta_{q,n_q}},$$

with $x_q(t)$ given by (10.4).

When all $\beta_{q,k}, k = 1, 2, \ldots, n_q$ are known then $\eta_{q,k}, k = 1, 2, \ldots, n_q - 1$ can be found by solving (10.29) and $\eta_{q,n_q} = 1 - \sum_{k=1}^{n_q-1} \eta_{q,k}$.

If we have $k_q > n_q - 1$ data points than the parameters can be estimated with the least-squares solution to (10.29), more specifically the solution to

$$I_{m_q}^2 \tilde{X}_q^\top \tilde{X}_q \tilde{\boldsymbol{\eta}}_q = \tilde{X}_q^\top \tilde{\mathbf{i}}_q.$$

If we wish to guarantee monotonicity in an interval by forcing $\eta_{q,k} > 0$, $k \in \{1, 2, \ldots, n_q\}$ (see Lemma 10.2) this becomes a so-called nonnegative least squares problem that can also be solved effectively with well known algorithms, e.g. [8].

10.3.5 Explicit Formulas for a Single-Peak AEF

Consider the case where $p = 1, n_1 = n_2 = 2$ and $\tau = \frac{t}{t_{m_1}}$. Then the explicit formula for the AEF is

$$\frac{i(\tau)}{I_{m_1}} = \begin{cases} \eta_{1,1}\, \tau^{\beta_{1,1}^2+1} e^{(\beta_{1,1}^2+1)(1-\tau)} + \eta_{1,2}\, \tau^{\beta_{1,2}^2+1} e^{(\beta_{1,2}^2+1)(1-\tau)}, & 0 \le \tau \le 1, \\ \eta_{2,1}\, \tau^{\beta_{2,1}^2}\; e^{\beta_{2,1}^2(1-\tau)} + \eta_{2,2}\, \tau^{\beta_{2,2}^2}\; e^{\beta_{2,2}^2(1-\tau)}, & 1 \le \tau. \end{cases} \quad (10.30)$$

Assume that four data points, $(i_k, \tau_k), k = 1, 2, 3, 4$, as well as the charge transfer and specific energy Q_0, W_0 are known.

If we want to fit the AEF to this data using MLSM, then (10.24) gives

$$
\mathbf{J} = \begin{bmatrix}
f_1(\tau_1) & f_2(\tau_1) & 0 & 0 \\
f_1(\tau_2) & f_2(\tau_2) & 0 & 0 \\
0 & 0 & g_1(\tau_3) & g_2(\tau_3) \\
0 & 0 & g_1(\tau_4) & g_2(\tau_4) \\
\frac{\partial}{\partial \beta_{1,1}} Q(\boldsymbol{\beta}, \boldsymbol{\eta}) & \frac{\partial}{\partial \beta_{1,2}} Q(\boldsymbol{\beta}, \boldsymbol{\eta}) & \frac{\partial}{\partial \beta_{2,1}} Q(\boldsymbol{\beta}, \boldsymbol{\eta}) & \frac{\partial}{\partial \beta_{2,2}} Q(\boldsymbol{\beta}, \boldsymbol{\eta}) \\
\frac{\partial}{\partial \beta_{1,1}} W(\boldsymbol{\beta}, \boldsymbol{\eta}) & \frac{\partial}{\partial \beta_{1,2}} W(\boldsymbol{\beta}, \boldsymbol{\eta}) & \frac{\partial}{\partial \beta_{2,1}} W(\boldsymbol{\beta}, \boldsymbol{\eta}) & \frac{\partial}{\partial \beta_{2,2}} W(\boldsymbol{\beta}, \boldsymbol{\eta})
\end{bmatrix},
$$

$$
\begin{aligned}
f_k(\tau) &= 2\,\eta_{1,k}\,\beta_{1,k}\tau^{\beta_{1,k}^2+1} e^{(\beta_{1,k}^2+1)(1-\tau)}\big(\ln(\tau)+1-\tau\big),\\
\eta_{1,1} &= \frac{i_1}{I_{m_1}} - \tau_1^{\beta_{1,2}^2} e^{(\beta_{1,2}^2+1)(1-\tau_1)},\ \eta_{1,2} = 1-\eta_{1,1},\\
g_k(\tau) &= 2\,\eta_{2,k}\,\beta_{2,k}\tau^{\beta_{2,k}^2} e^{\beta_{2,k}^2(1-\tau)}\big(\ln(\tau)+1-\tau\big),\\
\eta_{2,1} &= \frac{i_3}{I_{m_1}} - \tau_3^{\beta_{2,2}^2} e^{\beta_{1,2}^2(1-\tau_3)},\ \eta_{2,2} = 1-\eta_{2,1},\\
\boldsymbol{\beta} &= \left[(\beta_{1,1}^2+1)\ (\beta_{1,2}^2+1)\ \beta_{2,1}^2\ \beta_{2,2}^2\right],\\
\boldsymbol{\eta} &= \left[\eta_{1,1}\ \eta_{1,2}\ \eta_{2,1}\ \eta_{2,2}\right],\\
\frac{Q(\boldsymbol{\beta}, \boldsymbol{\eta})}{I_{m_1}} &= \sum_{s=1}^{2} \eta_{1,s} \frac{e^{\beta_{1,s}^2}}{(\beta_{1,s}^2+1)^{\beta_{1,s}^2+1}} \gamma\left(\beta_{1,s}^2+2, \beta_{2,s}^2+1\right)\\
&\quad + \sum_{s=1}^{2} \eta_{2,s} \frac{e^{\beta_{2,s}^2-1}}{\beta_{2,s}^{2\beta_{2,s}^2}} \left(\Gamma\left(\beta_{2,s}^2+1\right) - \gamma\left(\beta_{2,s}^2+1, \beta_{2,s}^2\right)\right),\\
\frac{\partial Q}{\partial \beta_{q,s}} &= \begin{cases} 2\, I_{m_1} \eta_{1,s}\, \beta_{1,s} \left.\dfrac{\mathrm{d}\hat{g}}{\mathrm{d}\beta}\right|_{\beta=\beta_{1,s}^2+1}, & q=1,\\[2ex] 2\, I_{m_q} \eta_{p,s}\, \beta_{2,s} \left.\dfrac{\mathrm{d}\tilde{g}}{\mathrm{d}\beta}\right|_{\beta=\beta_{2,s}^2}, & q=2, \end{cases}
\end{aligned}
$$

with derivatives of $\hat{g}(\beta)$ and $\tilde{g}(\beta)$ given by (10.25) and (10.26),

$$
\begin{aligned}
\widetilde{\boldsymbol{\beta}} &= \left[(\beta_{1,1}^2+\beta_{1,2}^2+2)\ (\beta_{1,1}^2+\beta_{1,2}^2+2)\ (\beta_{2,1}^2+\beta_{2,2}^2)\ (\beta_{2,1}^2+\beta_{2,2}^2)\right],\\
\widehat{\boldsymbol{\eta}} &= \left[\eta_{1,1}^2\ \eta_{1,2}^2\ \eta_{2,1}^2\ \eta_{2,2}^2\right],\ \widetilde{\boldsymbol{\eta}} = \left[(\eta_{1,1}\eta_{1,2})\ (\eta_{1,1}\eta_{1,2})\ (\eta_{2,1}\eta_{2,2})\ (\eta_{2,1}\eta_{2,2})\right],\\
\frac{\partial}{\partial \beta_{q,s}} W(\boldsymbol{\beta}, \boldsymbol{\eta}) &= 2\,\beta_{q,s} \frac{\partial}{\partial \beta_{q,s}} Q(2\boldsymbol{\beta}, \widehat{\boldsymbol{\eta}}) + \beta_{q,\left((s-1 \bmod 2)+1\right)} \frac{\partial}{\partial \beta_{q,s}} Q(\widetilde{\boldsymbol{\beta}}, \widetilde{\boldsymbol{\eta}}).
\end{aligned}
$$

Remark 10.2 If we only have one datapoint such that (cI_{m_1}, τ_3), $0 < c < 1$, and one term in the decaying part, we can actually interpolate that point using the formula

$$
\beta_2 = \sqrt{\frac{1-\tau_3+\ln(\tau_3)}{\ln(c)}}.
$$

10.3.6 Examples of Fitting a Single-Peak AEF

Here we apply the procedure described in Sect. 10.3.5 to estimate parameters of the single-peaked AEF to fit two different one-peaked waveforms, a fast-decaying waveform [10] and the so called first-positive stroke 10/350 μs from [4]. Each waveform is defined by a Heidler function and all parameters (rise/decay time ratio, T_1/T_2, peak current value, I_{m1}, time to peak current, t_{m1}, charge transfer at the striking point, Q_0, specific energy, W_0, and time to $0.1 I_{m1}$, t_1) are given in Table 10.1. Data points were chosen as follows:

$$(i_1, \tau_1) = (0.1\, I_{m_1}, t_1), \qquad (i_3, \tau_3) = (0.5\, I_{m_1}, t_h = t_1 - 0.1\, T_1 + T_2),$$
$$(i_2, \tau_2) = (0.9\, I_{m_1}, t_2 = t_1 + 0.8\, T_1), \quad (i_4, \tau_4) = (i(1.5\, t_h), 1.5\, t_h).$$

The AEF representation of the fast-decaying waveshape is shown in Fig. 10.7. Rising and decaying parts of the first-positive stroke current in IEC 62305 [4], are shown in Fig. 10.8. Apart from the AEF (solid line), the Heidler function representation of the same waveforms (dashed line), and used data points (red solid circles) are also shown in the figures.

In Fig. 10.7 it can be noticed that the fit in the rising part is very good and the fit in the decaying part is acceptable for many purposes.

From Fig. 10.8 it is clear that the fitting of the AEF can be difficult. In the rising part the fit is poor and this is due to the Heidler function rising steeply in the middle of the interval and when the steepness of the power exponential function is increased it will also move the steepest part of the slope to the right. The charge transfer Q has a low relative accuracy compared to the specific energy W but similar absolute accuracy. This is an example that in some cases a weighted least-square sum is preferable.

For both waveforms the best fit using two terms in each interval for the AEF is not better than the fit that is achieved using only a single term in each interval which can be seen in Table 10.1 since all the η-parameters are either 0 or 1.

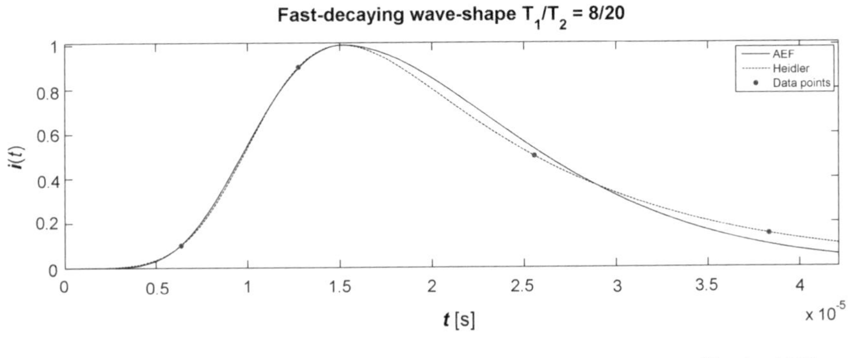

Fig. 10.7 The normalized fast-decaying current waveshape 8/20 μs, represented by the AEF

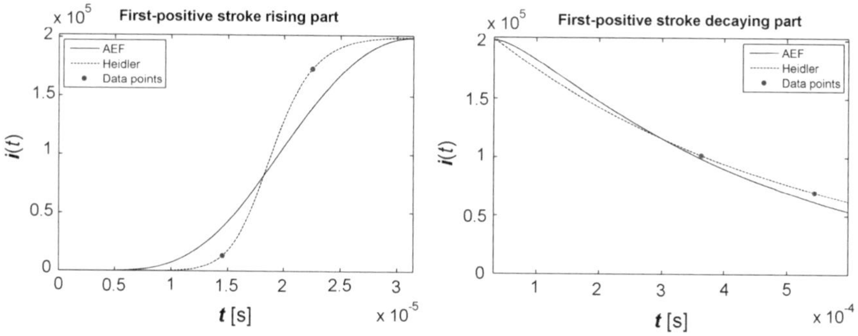

Fig. 10.8 First positive stroke 10/350 μs, for $I_m = 200$ kA, represented by the AEF

Table 10.1 The AEF parameters for the example waveshapes

	First-positive stroke	Fast-decaying
T_1/T_2	10/350	8/20
t_{m1} [μs]	31.428	15.141
I_{m1} [kA]	200	0.001
t_1 [μs]	14.528	6.343
Q_0 [C]	100	/
W_0 [MJ/Ω]	10	/
Q [C]	89.7	/
W [MJ/Ω]	10.000095	/
$\beta_{1,1}$	2.600	2.626
$\beta_{1,2}$	2.477	2.700
$\beta_{2,1}$	0.295	2.500
$\beta_{2,2}$	0.567	1.958
$\eta_{1,1}$	0	1
$\eta_{1,2}$	1	0
$\eta_{2,1}$	1	0
$\eta_{2,2}$	0	1

10.4 Conclusions

We have presented and examined some basic properties of a generalized version of the AEF function intended to be used for approximation of multi-peaked lightning discharge currents. Existence as well as explicit formulas of the analytical solution for the first derivative and the integral of the AEF function has been shown, which is needed in order to perform lightning electromagnetic field (LEMF) calculations based on it.

A method for finding a least square approximation using the Marquardt least square method (MLSM) that works for any number of peaks has been presented.

Two examples of parameter estimation for single-peaked waveforms, the Standard IEC 62305 first-positive stroke 10/350 μs function and a fast-decaying waveform 8/20 μs, have been shown. An estimation of their parameters using MLSM was performed using two pairs of data points for each waveform (one pair for the rising part and one pair for the decaying part). As it can be observed from the results a good approximation is achievable but not under all circumstances.

References

1. Abramowitz, M., Stegun, I.: Handbook of Mathematical Functions with Formulas, Graphs, and Mathematical Tables. Dover, New York (1964)
2. Heidler, F.: Travelling current source model for lemp calculation. In: Proceedings of papers, 6th International Zurich Symposium EMC, Zurich, pp. 157–162 (1985)
3. Heidler, F., Cvetić, J.: A class of analytical functions to study the lightning effects associated with the current front. Trans. Electr. Power **12**(2), 141–150 (2002)
4. IEC 62305-1 Ed.2: Protection Against Lightning - Part I: General Principles (2010)
5. Javor, V.: Multi-peaked functions for representation of lightning channel-base currents. In: Proceedings of Papers, 2012 International Conference on Lightning Protection - ICLP, Vienna, Austria, pp. 1–4 (2012)
6. Javor, V.: New functions for representing IEC 62305 standard and other typical lightning stroke currents. J. Light. Res. **4**(Suppl 2: M2), 50–59 (2012)
7. Javor, V., Rančić, P.D.: A channel-base current function for lightning return-stroke modeling. IEEE Trans. EMC **53**(1), 245–249 (2011)
8. Lawson, C.L., Hanson, R.J.: Solving Least Squares Problems. Prentice Hall Inc., Englewood Cliffs (1974)
9. Levenberg, K.: A method for the solution of certain non-linear problems in least squares. Q. J. Appl. Math. I **I**(2), 164–168 (1944)
10. Lovrić, D., Vujević, S., Modrić, T.: On the estimation of Heidler function parameters for reproduction of various standardized and recorded lightning current waveshapes. Int. Trans. Electr. Energy Syst. **23**, 290–300 (2013)
11. Lundengård, K., Rančić, M., Javor, V., Silvestrov, S.: Application of the Marquardt least-squares method to the estimation of Pulse function parameters. In: AIP Conference Proceedings 1637, ICNPAA, Narvik, Norway, pp. 637–646 (2014)
12. Lundengård, K., Rančić, M., Javor, V., Silvestrov, S.: Estimation of Pulse function parameters for approximating measured lightning currents using the Marquardt least-squares method. In: Conference Proceedings, EMC Europe, Gothenburg, Sweden, pp. 571–576 (2014)
13. Maple 18.02: Maplesoft, a division of Waterloo Maple Inc., Waterloo, Ontario
14. Marquardt, D.W.: An algorithm for least-squares estimation of nonlinear parameters. Soc. Ind. Appl. Math. J. Soc. Ind. Appl. Math. **11**(2), 431–11 (1963)
15. Press, W.H., Teukolsky, S.A., Vetterling, W.T., Flannery, B.P.: Numerical Recipes: The Art of Scientific Computing, 3rd edn. Cambridge University Press, Cambridge (2007)
16. Prudnikov, A.P., Brychkov, Y.A., Marichev, O.I.: Integrals and Series, Volume 3: More Special Functions. Gordon and Breach Science Publishers, New York (1990)

Chapter 11
Mathematical Modelling of Cutting Process System

Jüri Olt, Olga Liivapuu, Viacheslav Maksarov, Alexander Liyvapuu and Tanel Tärgla

Abstract The mathematical modelling of the process system allows carrying out research into the selection and optimisation of machining conditions. The conceptualization of the operator that represents the dynamic characteristics of the cutting and friction process is an important issue in the development of the mathematical formulation of the interaction between subsystems in the cutting process. Currently, different approaches exist to the description of cutting and friction processes with the use of dynamic and quasi-static concepts, which results in the different studies using the machining process system models that are essentially distinct from each other. The subject of this paper is the method of dynamic process approximation, which allows analysing the behaviour of the machining process system in the process of chip formation at a sufficient level of accuracy.

Keywords Cutting · Elastic and plastic deformation · Chip formation · Oscillation · Dynamic process

---

J. Olt (✉) · O. Liivapuu · A. Liyvapuu · T. Tärgla
Institute of Technology, Estonian University of Life Sciences,
Fr. R. Kreutzwaldi 56, 51014 Tartu, Estonia
e-mail: jyri.olt@emu.ee

O. Liivapuu
e-mail: olga.liivapuu@emu.ee

A. Liyvapuu
e-mail: alexander.liyvapuu@emu.ee

T. Tärgla
e-mail: tanel.targla@emu.ee

V. Maksarov
Saint-Petersburg Mining University,
21 Line, 2, Vasilevsky Island, St. Petersburg, Russia
e-mail: maks78.54@mail.ru

S. Silvestrov and M. Rančić (eds.), *Engineering Mathematics I*,
Springer Proceedings in Mathematics & Statistics 178,
DOI 10.1007/978-3-319-42082-0_11

11.1 Mathematical Model of Technological System

One of the ways of analysing a process system is its mathematical modelling, which allows to perform studies on the selection of the optimum machining conditions. The mathematical formulation of the behaviour of a process system at the phenomenological level shall be regarded as the approximation of dynamic processes based on the results of experimental studies and allowing to identify, within the framework of the assumed concept, the time values of the cutting force lagging behind the movement of the tool and the friction force lagging behind the cutting force, the time constants and the transfer factors of the respective approximation elements.

In the light of today's views, the process of chip formation in cutting has discrete nature [2], defined by the presence of several stages in the chip formation. Let us consider the dynamical model of four-loop manufacturing process system, which is divided to workpiece sub-system with generalized coordinates X and Y and tool sub-system with generalized coordinates x and y (Fig. 11.1). The analysis that has been carried out in [1] allows to approximate the discrete cutting process by a continuous one with a sufficient accuracy. In accordance with this supposition it is assumed that the displacements of the cutting tool along the lines of cutting and friction forces (respectively, along the tangent direction y and the normal direction x) for the lag times τ_P and τ_Q of the cutting force P behind the displacement x and the friction force Q behind the cutting force P, respectively, are equal to l_P and l_Q. The values l_P and l_Q are supposed to be constant, they depend on the properties of the machined material (cutting coefficient k, chip contraction coefficient ξ) as well as the coefficient f of the friction of the chip on the cutting tool's rake face.

Denoting the nominal cutting speed by v_S and taking into account the adopted supposition about l_P and l_Q, we obtain

$$l_P = \int_0^{\tau_P} (v_S + \dot{y})dt = v_S\tau_P + \int_0^{\tau_P} \dot{y}dt = \text{const}, \tag{11.1}$$

$$l_Q = \int_0^{\tau_Q} (v_S + \dot{y} + \xi\dot{x})dt = v_S\tau_Q + \int_0^{\tau_Q} (\dot{y} + \xi\dot{x})dt = \text{const}, \tag{11.2}$$

where $\dot{x}$ and $\dot{y}$ are time derivatives of the coordinates; ξ is the chip contraction coefficient defined by the combined action of the forces P and Q. The constant components of time lag parameters τ_P and τ_Q are found using the formulae $T_P = \frac{l_P}{v_S}$ and $T_Q = \frac{l_Q}{v_S}$ and referred as cutting process lag constants. The static cutting force P_S and friction force Q_S can be represented in the following form [1, 5]

$$P_S = kb_c\delta^{\varepsilon}, \quad Q_S = fP, \tag{11.3}$$

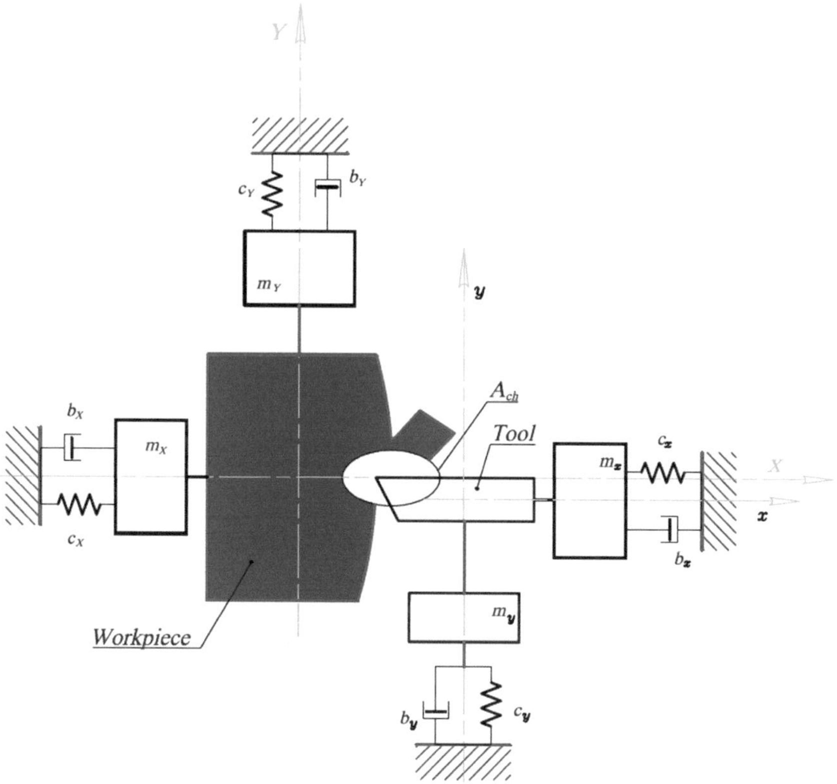

Fig. 11.1 Dynamic model of four-loop process system; WP - work piece sub-system; T - tool; A_{ch} - chip formation operator

where f is the coefficient of the friction of the chip on the cutting tool's rake face; k is the cutting factor; b_c is the width of the sheared-off layer; δ is the thickness of the sheared-off layer; ε is an empirical index (it is provisionally assumed to be equal to 0.75 for steel).

In case of small displacements of the cutter the cutting force $P_x(t)$ can be sufficiently closely approximated using the expression [7]

$$P_x(t) = -kb_c\delta^{\varepsilon-1}x(t), \tag{11.4}$$

subsequently

$$Q_P(t) = f P_x(t). \tag{11.5}$$

Without passing on to the time lagging mechanism, we assume in accordance with [7], that at any instant of time the following relations hold

$$P(t + \tau_P) = -Bx(t) \equiv P_x(t), \tag{11.6}$$

$$Q(t + \tau_Q) = fP(t) \equiv Q_P(t), \tag{11.7}$$

where B is is the disturbance factor in the loop x.

Expanding each of the previous functions in the Taylor series about the operating point and keeping the linear part of the series, we obtain the formulae for the determination of τ_P and τ_Q

$$\tau_P = \frac{P_x(t) - P(t)}{\dot{P}(t)}, \quad \tau_Q = \frac{Q_P(t) - Q(t)}{\dot{Q}(t)}. \tag{11.8}$$

Taking into consideration (11.1) (11.2) and (11.8), after a number of transformations we obtain the lag equations as follows

$$\begin{cases} T_P\dot{P} + (1 + \frac{\dot{y}}{v_S})(P + Bx) = 0, \\ \\ T_Q\dot{Q} + (1 + \frac{\dot{y}+\xi\dot{x}}{v_S})(Q - fP) = 0. \end{cases} \tag{11.9}$$

The assumptions of the invariability of the values l_P, l_Q and the satisfaction of the relations (11.6) and (11.7) are essential for the system of Eq. (11.9).

The two-loop dynamic model of the machining process system for the loops x and y (Fig. 11.1) in the case of small perturbations can be represented by the following differential equation system

$$\begin{cases} m_x\ddot{x} + b_x\dot{x} + c_x x = Q, \\ m_y\ddot{y} + b_y\dot{y} + c_y y = P, \end{cases} \tag{11.10}$$

where m_x, m_y are reduced inertial parameters of the loops x and y of the system; b_x, b_y are factors, which take account of the energy dissipation in the loops x and y of the system; c_x, c_y are stiffness factors of the loops x and y.

Basing on simultaneous analysis of the Eqs. (11.9) and (11.10), the behaviour of the two-loop model of the machining process system can be represented by the following system of differential equations

$$\begin{cases} m_x\ddot{x} + b_x\dot{x} + c_x x = Q, \\ m_y\ddot{y} + b_y\dot{y} + c_y y = P, \\ T_P\dot{P} + (1 + \frac{\dot{y}}{v_S})(P + Bx) = 0, \\ T_Q\dot{Q} + (1 + \frac{\dot{y}+\xi\dot{x}}{v_S})(Q - fP) = 0. \end{cases} \tag{11.11}$$

This system of equations is nonlinear. That is a crucial statement for a self-oscillating cutting machining process system. For the further research into it we are going to use the approximation of the nonlinear differential equations by a system of quasi-linear differential equations with piecewise constant coefficients [6]. For that purpose, it would be more convenient to reduce the system of differential equation (11.11) to its canonical representation. We introduce the variables of the canonical system of equations with the following formulae

$$q_1 = x, \; q_2 = \dot{x}, \; q_3 = y, \; q_4 = \dot{y}, \; q_5 = P, \; q_6 = Q. \tag{11.12}$$

Then the system (11.11) takes the following form

$$\begin{cases} \dot{q}_1 - q_2 = 0, \\ m_x \dot{q}_2 + b_x q_2 + c_x q_1 - q_6 = 0, \\ \dot{q}_3 - q_4 = 0, \\ m_y \dot{q}_4 + b_y q_4 + c_y q_3 - q_5 = 0, \\ T_P \dot{q}_5 + (1 + \frac{q_4}{v_S})(q_5 + B q_1) = 0, \\ T_Q \dot{q}_6 + (1 + \frac{q_4 + \xi q_2}{v_S})(q_6 - f q_5) = 0. \end{cases} \tag{11.13}$$

Thus, the differential equation system (11.13) can be represented in terms of state variables in the following vector-matrix form

$$T\dot{q} + Sq = 0, \tag{11.14}$$

where

- $q = (q_1, \ldots, q_6)^T$ is a six-component vector of the system's state variables;
- $T = \text{diag}(t_1, \ldots, t_6)$ is a diagonal matrix of the following constants:

$$T = \begin{pmatrix} 1 & 0 & 0 & 0 & 0 & 0 \\ 0 & m_x & 0 & 0 & 0 & 0 \\ 0 & 0 & 1 & 0 & 0 & 0 \\ 0 & 0 & 0 & m_y & 0 & 0 \\ 0 & 0 & 0 & 0 & T_P & 0 \\ 0 & 0 & 0 & 0 & 0 & T_Q \end{pmatrix}. \tag{11.15}$$

- $S = (s_{ij}), \; (i, j = 1, \ldots, 6)$, is a square matrix of order 6, where the elements $s_{51} = s_{55}, \; s_{56} = -f s_{66}$ are coordinate functions:

$$S = \begin{pmatrix} 0 & -1 & 0 & 0 & 0 & 0 \\ c_x & b_x & 0 & 0 & 0 & -1 \\ 0 & 0 & 0 & -1 & 0 & 0 \\ 0 & 0 & c_y & b_y & -1 & 0 \\ 1 + \frac{q_4}{v_S} & 0 & 0 & 0 & 1 + \frac{q_4}{v_S} & 0 \\ 0 & 0 & 0 & 0 & -f\left(1 + \frac{q_4 + \xi q_2}{v_S}\right) & 1 + \frac{q_4 + \xi q_2}{v_S} \end{pmatrix}. \tag{11.16}$$

Since the matrix S contains the elements s_{51}, s_{55}, s_{56} and s_{66}, which are coordinate functions, i.e. it includes nonlinear elements in the form of the products of state variables, the system of Eq. (11.14) is nonlinear.

The nonlinear parts of the system represented by the Eq. (11.14) can be transformed by the approximate substitution of the products of variables by the corresponding linear approximants

$$\left(1+\frac{q_4}{v_S}\right)(q_5+Bq_1)=q_5+Bq_1+\frac{q_4}{v_S}(a_P+Ba_x)+\frac{a_y}{v_S}(q_5+Bq_1)=$$
$$=(q_5+Bq_1)\left(1+\frac{a_y}{v_S}\right)+\frac{q_4}{v_S}(a_P+Ba_x)=$$
$$=q_5+Bq_1+\frac{q_4}{v_S}(a_P+Ba_x), \qquad (11.17)$$

where a_x, a_y and a_P are some average small perturbations of parameters in the neighbourhood of the stability region boundary, which is brought into coincidence with the supposed stable limit cycle. The product of the variables containing the value a_y will be dropped out because of the smallness of the latter compared to unity.

The product of other parameters is linearized similarly:

$$\left(1+\frac{q_4+\xi q_2}{v_S}\right)(q_6-fq_5)=\left(1+\frac{a_{yx}}{v_S}\right)(q_6-fq_5)+\left(\frac{q_4+\xi q_2}{v_S}\right)a_{PQ}=$$
$$=(q_6-fq_5)+\left(\frac{q_4+\xi q_2}{v_S}\right)a_{PQ}. \qquad (11.18)$$

To determine the values a_{yx} and a_{PQ} the equation of the energy balance for a period of oscillation is set up following the method described in the work [1].

Now we present the initial system of differential equation (11.14) in the linearized form in the neighbourhood of the limit cycle. This will be the system of linear homogeneous differential equations. We present it in the matrix form, where all matrices are with constant elements

$$\tilde{T}\dot{q}+\tilde{S}q=0. \qquad (11.19)$$

Let us introduce the relation formulae between coefficients of the system of Eq. (11.19), which facilitate reducing the components of the system of differential equation (11.14) to the dimensionless form

- $\omega_x=\sqrt{\frac{c_x}{m_x}}$ and $\omega_y=\sqrt{\frac{c_y}{m_y}}$ are angular frequencies of the loops x and y;
- $T_{x2}=\frac{1}{\omega_x}$ and $T_{y2}=\frac{1}{\omega_y}$ are time constants of the loops x and y;
- $d_x=\frac{\psi_x}{2\pi}$ and $d_y=\frac{\psi_y}{2\pi}$, where ψ_x and ψ_y are energy dissipation coefficients (absorption coefficients) in the loops x and y;
- $T_{x1}=\frac{b_x}{c_x}\approx\frac{d_x}{\omega_x}$ and $T_{y1}=\frac{b_y}{c_y}\approx\frac{d_y}{\omega_y}$ are damping time constants of the loops x and y;

- $k_x = \frac{fB}{c_x}$ is the system transfer factor in the loop x, B is the disturbance factor in the loop x;
- $\gamma = \sqrt{\frac{1+k_x}{T_{x2}^2+T_{x1}(T_P+T_Q)}}$ is the angular frequency of the system.

After the transformations the diagonal matrix $\tilde{T}$ of time constants will take the following form

$$\tilde{T} = \begin{pmatrix} 1 & 0 & 0 & 0 & 0 & 0 \\ 0 & T_{x2}^2 & 0 & 0 & 0 & 0 \\ 0 & 0 & 1 & 0 & 0 & 0 \\ 0 & 0 & 0 & T_{y2}^2 & 0 & 0 \\ 0 & 0 & 0 & 0 & T_P & 0 \\ 0 & 0 & 0 & 0 & 0 & T_Q \end{pmatrix}. \tag{11.20}$$

In the matrix $\tilde{S}$ the elements $\tilde{s}_{51}$, $\tilde{s}_{55}$, $\tilde{s}_{56}$ and $\tilde{s}_{66}$ are no more coordinate functions, while the elements $\tilde{s}_{51} = k_x$ and $\tilde{s}_{55} = \tilde{s}_{56} = -\tilde{s}_{66} = 1$. The elements $\tilde{s}_{54}$, $\tilde{s}_{62}$ and $\tilde{s}_{64}$ are equal to the cutting speed oscillation time constants $\tilde{s}_{54} = T_{ky1}$, $\tilde{s}_{62} = T_{kx}$, $\tilde{s}_{64} = T_{kx2}$. The matrix $\tilde{S}$ will look as follows

$$\tilde{S} = \begin{pmatrix} 0 & -1 & 0 & 0 & 0 & 0 \\ 1 & T_{x1} & 0 & 0 & 0 & -1 \\ 0 & 0 & 0 & -1 & 0 & 0 \\ 0 & 0 & 1 & T_{y1} & -1 & 0 \\ k_x & 0 & 0 & T_{ky1} & 1 & 0 \\ 0 & T_{kx} & 0 & T_{ky2} & -1 & 1 \end{pmatrix}, \tag{11.21}$$

where

$$\begin{aligned} \tilde{s}_{54} &= T_{ky1} = AT_P\psi_y L_P, \\ \tilde{s}_{64} &= T_{ky2} = \frac{AT_P\psi_{xy}L_P}{N_Q}, \\ \tilde{s}_{62} &= T_{kx} = \frac{Af\xi T_Q L_P \psi_{xy} k_y}{N_Q}, \\ L_P &= \frac{B\gamma}{v_S c_y\sqrt{T_P^2\gamma^2+1}}, \\ N_Q &= \sqrt{T_Q^2\gamma^2+1}, \\ k_y &= \frac{c_x}{c_y}. \end{aligned} \tag{11.22}$$

The amplitude A is entered into this expressions as a parameter. Its magnitude is defined by the requirements of the soft limitation of the contact between the clearance face of the cutting tool and the machined surface [7]

$$A = \frac{v_s \tan(\alpha_P)}{\gamma\sqrt{1 + a_y^2 \tan(\alpha_P) + 2a_y \tan(\alpha_P)\cos(\varphi_{xy})}}, \quad (11.23)$$

where

- α_P is the cutting tool clearance angle;
- $a_y = \frac{A_y}{A_x}$, where A_x and A_y are amplitudes of the oscillation of x and y with relation to the oscillation of the force P;
- φ_{xy} is the phase shift of the oscillation of x and y with relation to the oscillation of the force P;
- γ is the angular frequency of the closed-loop system.

The functions ψ_y and ψ_{xy} are averaged over a period of oscillation and represent the effect of the phase shift between variables

$$\psi_y = \frac{1}{2}\left[\left(\frac{\pi}{2} - \varphi_{P_y}\right)\cos(\varphi_{P_y}) + \sin(\varphi_{P_y})\right], \quad (11.24)$$

$$\varphi_{P_y} = \begin{cases} \pi + \arctan\left(\frac{d_y}{1-v_y^2}\right), & \text{when } v_y \geq 1, \\ \arctan\left(\frac{d_y}{1-v_y^2}\right), & \text{when } v_y < 1, \end{cases} \quad (11.25)$$

$$\psi_{xy} = \frac{1}{2}\left[\left(\frac{\pi}{2} - |\theta_{xy} - \theta_{PQ}|\right)\cos(|\theta_{xy} - \theta_{PQ}|) + \sin(|\theta_{xy} - \theta_{PQ}|)\right], \quad (11.26)$$

where

$$\theta_{xy} = \arctan\left[\frac{(B\cos(\varphi_{P_y}) - \xi c_y\sqrt{N^2 + d_y^2}\sqrt{T_P^2\gamma^2 + 1})}{(B\sin(\varphi_{P_y}) + \xi c_y\sqrt{N^2 + d_y^2}\sqrt{T_P^2\gamma^2 + 1})} \cdot \frac{\cos(\varphi_{xP})}{\sin(\varphi_{xP})}\right],$$

$$\theta_{PQ} = \arctan\left(\frac{1}{T_Q\gamma}\right),$$

$$N = 1 - v_y^2,$$

$$v_y = \frac{\gamma}{\omega_y},$$

$$\varphi_{xP} = \arctan(T_P\gamma),$$

$$\varphi_{xy} = \varphi_{xP} + \varphi_{P_y} = \arctan(T_P\gamma) + \arctan\left(\frac{d_y}{N}\right). \quad (11.27)$$

Thus, the initial nonlinear system of differential equation (11.11) can be represented in the form of a linearized two-loop model as follows

$$T_{x2}^2\ddot{x} + T_{x1}\dot{x} + x = Q, \tag{11.28}$$

$$T_{y2}^2\ddot{y} + T_{y1}\dot{y} + y = P, \tag{11.29}$$

$$T_P\dot{P} + P = -k_x x - T_{ky1}\dot{y}, \tag{11.30}$$

$$T_Q\dot{Q} + Q = P - T_{kx}x - T_{ky2}\dot{y}. \tag{11.31}$$

The system of equations in the form of (11.28)–(11.31) defines the behaviour of the process system when describing the friction between the cutting face of the cutting tool and the chip in the process of cutting in the form of quasi-static performance like in the Amontons–Coulomb model. In the studies [3, 7] these equations were modified taking into account the molecular and mechanical conception of the contact interaction between the cutting tool and the chip on the basis of the binomial friction law that includes the stages of slipping and adhesion. Their influence on the behaviour of the process system was taken into consideration. The dynamic equations for the four-loop model of the machining system were generated with due account for the mentioned stages. In the state of slipping the behaviour of the process system was defined by the system of equations (11.28)–(11.31).

11.2 Mathematical Simulation of Plastic Deformation and Destruction in the Process of Chip Formation

The results of the analysis of mathematical simulation of the process of chip formation suggest that the adequate representation of this process is a rheological model constructed of the elastic-ductile-plastic relaxing medium of Ishlinskiy and the medium of Voigt with the delay combined in serial [3, 4, 7].

In the state of adhesion the contact interaction during machining was represented by the Voigt model [7], therefore, the equations of motion took the following form

$$m_x\ddot{x} + b_x\dot{x} + c_x x + \beta_\tau\dot{x} + c_\tau x = Q, \tag{11.32}$$

$$m_y\ddot{y} + b_y\dot{y} + c_y y + \beta_n\dot{y} + c_n y = P, \tag{11.33}$$

where c_τ, c_n, β_τ, β_n were quasi-elastic and dissipative factors for the tangent and normal directions of the chip formation area.

The dynamic model presented in the paper [4] incorporates the relation between the work piece and cutting tool sub-systems through the cutting process, but the elastic and dissipative characteristics c_τ, c_n, β_τ, β_n are given consideration there only in the adhesion zone, i.e. during the adhesive interaction between the already departing chip and the tool. Thence, effectively only the secondary deformation process and its influence on the dynamic behaviour of the machining process system are taken into account, while the process of the active primary deformation of metal in the chip formation area, which continues throughout the whole cutting process irrespective of the current stage of the chip progression over the tool's rake face, is disregarded.

The transition from one stage to the other affects the primary plastic deformation only on account of the effective force vector change. Thus, for the formulation of a rheological model of the machining process system we have to take into account as the primary plastic deformation process in the area of the sheared-off layer, so the secondary deformation and friction processes during the advancement of the chip over the cutting tool's rake face.

Figure 11.2 shows the rheological model of chip formation that represents the relations between the subsystems in cutting. The rheological model depicts the chip formation process as a series combination of the elastoviscoplastic relaxing Ishlinsky medium and the Voigt medium with two elastic and dissipative elements in the tangent and normal directions [7]. The properties of the elastic elements of the mechanical system are determined by stiffness coefficients c_i, $(i = 1, 2, 3)$, (or pliability $e_i = 1/c_i$), they characterize the ability of the formed links to accumulate strain energy. The elements of damping in the mechanical system are interpreted as the ideal linear elements creating resistant forces proportional to the relative deformation rate. The

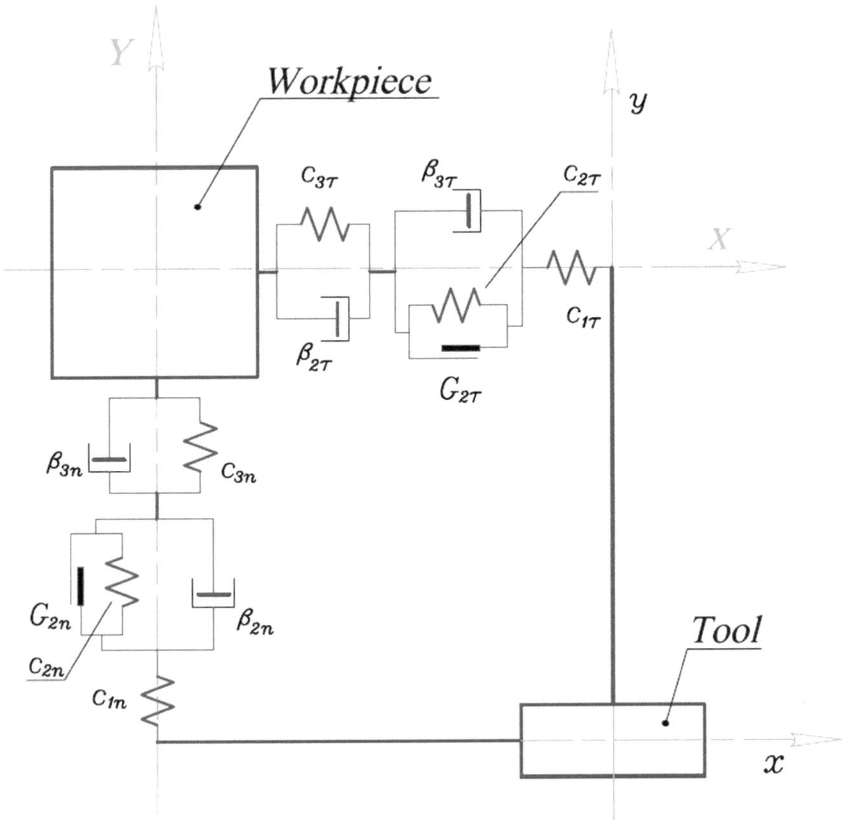

Fig. 11.2 Rheological model of chip formation in the cutting process

elements of damping characterize irreversible losses, connected with the dissipation of energy both as a result of internal and external viscous resistances. The element of damping is generally characterized by the coefficient of the linear resistance β_i.

In accordance with the rheological model of chip formation (Fig. 11.2), the Ishlinsky equation can be formulated as follows

$$\beta_2\dot{\sigma} + (c_1 + c_2)\sigma = c_1\beta_2\dot{\varepsilon} + c_1c_2\varepsilon \pm c_1\sigma_y, \tag{11.34}$$

where σ_y is a maximum load of a plastic element.

After reducing to unity the coefficients of the output variable, the Eq. (11.34) assumes the following form

$$\frac{\beta_2}{(c_1 + c_2)} + \sigma = \frac{c_1c_2}{(c_1 + c_2)}\left(\frac{\beta_2}{c_2}\dot{\varepsilon} + \varepsilon\right) \pm \frac{c_1}{(c_1 + c_2)}\sigma_y. \tag{11.35}$$

Assuming $\frac{\beta_2}{(c_1+c_2)} = n$ – relaxation time, $\frac{c_1c_2}{(c_1+c_2)} = H$ – continuous elasticity modulus and $c_1 = E$ – instantaneous elasticity modulus, we obtain

$$n\dot{\sigma} + \sigma = H\left(\frac{En}{H}\dot{\varepsilon} + \varepsilon\right), \tag{11.36}$$

since $c_1 \gg c_2$, then $\frac{c_1}{c_1+c_2} \approx 1$, and σ_y is assumed to have a constant value.

After transforming the rheological equation and its coefficients, which represent the time constants

$$T_{P1} = n = \frac{\beta_2}{(c_1 + c_2)},$$

$$T_{P2} = \frac{En}{H} = \frac{\beta_2}{c_2},$$

we come to

$$T_{P1}\dot{\sigma} + \sigma = k_P(T_{P2}\dot{\varepsilon} + \varepsilon), \tag{11.37}$$

where k_P is a factor representing the rheological features of the chip formation process. Employing the relations that enable the transition from stresses and strains to forces and displacements, we obtain

$$\sigma = \frac{P_S}{\delta b_c},$$

$$\varepsilon = \frac{\Delta y\xi + \Delta x}{l_0},$$

where P_S is the static cutting force, δ and b_c are the thickness and width of the sheared-off layer, $l_0 = v_S\tau$, Δx and Δy are the displacement increments. Thereupon, we come to the following expression for k_P

$$k_P = \frac{\delta b_c H}{l_0 c_x}. \tag{11.38}$$

Thence, the final appearance of the differential equation in the dimensionless form that represents the rheological processes in chip formation will be

$$T_{P1}\dot{P} + P = k_{Px}(T_{P2}\dot{x} + x) + k_{Py}(T_{P2}\dot{y} + y), \tag{11.39}$$

where $k_{Px} = \frac{1}{\xi}k_P$ and $k_{Py} = k_P$. Now our aim is to present the Eq. (11.39) that represents the rheological processes in chip formation in the operator form

$$(T_{P1}p + 1)P(p) = k_{Px}\frac{1}{\xi}(T_{P2}p + 1)x(p) + k_{Py}(T_{P2}p + 1)y(p), \tag{11.40}$$

where p is differential operator.

The lagging Eq. (11.30) in the operator form appears as follows

$$(T_P p + 1)P(p) = -k_x x(p) - T_{ky1} p y(p). \tag{11.41}$$

Analysing simultaneously the rheological Eq. (11.40) and the lagging Eq. (11.41), we obtain, for the x-direction

$$\frac{P(p)}{x(p)} = \frac{[k_{Px}(T_P + T_{P2}) - k_x T_{P1}]p + k_{Px} - k_x}{(T_P + T_{P1})p + 1}, \tag{11.42}$$

and for the y-direction

$$\frac{P(p)}{y(p)} = \frac{[k_{Py}(T_P + T_{P2}) - T_{ky1}]p + k_{Py}}{(T_P + T_{P1})p + 1}. \tag{11.43}$$

After the transformation of the previous equations, we obtain the general equation in the operator form

$$\begin{aligned}[(T_P + T_{P1})p + 1]P(p) = (k_{Px}T_P p + k_{Px}T_{P2}p + k_{Px} - k_x T_{P1} p - k_x)x(p) + \\ +(k_{Py}T_P p + k_{Py}T_{P2}p - T_{ky1} p + k_{Py})y(p). \end{aligned} \tag{11.44}$$

Finally, the differential equation representing the lagging process and rheological features of chip formation looks as follows

$$\begin{aligned}(T_P + T_{P1})\dot{P} + P = -(k_x - k_{Px})x - (T_{ky1} - k_{Py}(T_P + T_{P2}))\dot{y} + \\ +k_{Py}y - (k_x T_{P1} - k_{Px}(T_P + T_{P2}))\dot{x}. \end{aligned} \tag{11.45}$$

The modelling on the basis of the piecewise linear approximation of the chip formation process with due consideration of the slipping and adhesion (slip and stick) stages facilitates establishing the foundation for the generation of the differential equations of the machining process system. Accordingly, the behaviour of the adopted dynamic four-loop model in line with the assumed rheological chip formation model is represented in general by the following differential equation system

$$
\begin{aligned}
T_{x2}^2\ddot{x} + (T_{x1} + T_{x3})\dot{x} - T_{x3}\dot{X} + 2x - X &= Q,\\
T_{y2}^2\ddot{y} + (T_{y1} + T_{y3})\dot{y} - T_{y3}\dot{Y} + 2y - Y &= P,\\
T_{X2}^2\ddot{X} + (T_{X1} + T_{X3})\dot{X} - T_{X3}\dot{x} + 2X - x &= -Q,\\
T_{Y2}^2\ddot{Y} + (T_{Y1} + T_{Y3})\dot{Y} - T_{Y3}\dot{y} + 2Y - y &= -P,\\
(T_P + T_{P1})\dot{P} + P = -(k_x - k_{Px})(x - X) - (T_{ky1} - k_{Py}(T_P + T_{P2}))(\dot{y} - \dot{Y}) &+\\
+k_{Py}(y - Y) - (k_x T_{P1} - k_{Px}(T_P + T_{P2}))(\dot{x} - \dot{X})&,\\
T_Q\dot{Q} + Q = P - T_{kx}(\dot{x} - \dot{X}) - T_{ky2}(\dot{y} - \dot{Y})&,
\end{aligned}
$$

where

- T_{x2}, T_{y2}, T_{X2}, T_{Y2} are time constants of the loops x, y, X and Y;
- T_{x1}, T_{y1}, T_{X1}, T_{Y1}, T_{x3}, T_{y3}, T_{X3}, T_{Y3} are damping time constants of the loops x, y, X and Y;
- k_x is the transfer factor of the system's loop x;
- T_P and T_Q are time constants;
- k_{Px}, k_{Py}, T_{P1}, T_{P2} are factors and time constants taking into account the rheological features of the chip formation process.

The last system of differential equations represents the dynamic processes in the machining process system taking into account the elastoplastic properties involved in the dynamics of the contact interaction between the cutting tool and the work piece and the rheological features of the chip formation process in the area of active plastic deformation. The obtained system of equations provides the basis for solving the problem of analysing the behaviour of the machining process system during the chip segmentation.

Acknowledgements The authors are grateful for partial support from Linda Peetre's Foundation for cooperation between Sweden and Estonia provided by Swedish Mathematical Society. The authors are grateful to the international research project of the Institute of Technology at Estonian University of Life Sciences: 8-2/TI13002TEDT. The research was partially financially supported by institutional research funding IUT20-57 of the Estonian Ministry of Education and Research..

References

1. Elyasberg, M.E.: Oscillations of Machine Tools. OKBS, St. Petersburg (1993). (in Russian)
2. Gregg, B.C., Suh, C.S., Luo, A.C.J.: Machine Tool Vibrations and Cutting Dynamics. Springer, New York (2011)
3. Maksarov, V., Olt, J.: Analysis of the reological model of the process of chip formation with metal machining. Agron. Res. **6**, 249–263 (2008)

4. Olt, J., Liyvapuu, A., Madissoo, M., Maksarov, V.: Dynamic simulation of chip formation in the process of cutting. Int. J. Mater. Prod. Technol. **53**(1) (2016)
5. Schmitz, T.L., Smith, K.S.: Machining Dynamics: Frequency Response to Improved Productivity. Springer, New York (2009)
6. Schmitz, T.L., Smith, K.S.: Mechanical Vibrations: Modeling and Measurement. Springer, New York (2012)
7. Veitz, V.L., Maksarov, V.V.: Dynamics of Technological Systems Machining. SZTU, St. Petersburg (2001). (in Russian)

Chapter 12
Mixed Convection Heat Transfer in MHD Non-Darcian Flow Due to an Exponential Stretching Sheet Embedded in a Porous Medium in Presence of Non-uniform Heat Source/Sink

Prashant G. Metri, Veena M. Bablad, Pushpanjali G. Metri, M. Subhas Abel and Sergei Silvestrov

Abstract A mathematical analysis has been carried out to describe mixed convection heat transfer in MHD non-Darcian flow due to an exponential stretching sheet embedded in a porous medium in presence of non-uniform heat source/sink. Approximate analytical similarity solutions of the highly non-linear momentum and energy equations are obtained. The governing system of partial differential equations is first transformed into a system of non-linear ordinary differential equations using similarity transformation. The transformed equations are non-linear coupled differential equations and are solved very efficiently by using fifth order Runge–Kutta–Fehlberg method with shooting technique for various values of the governing parameters. The numerical solutions are obtained by considering an exponential dependent stretching velocity and prescribed boundary temperature on the flow directional coordinate. The computed results are compared with the previously published work on various special cases of the problem and are in good agreement with the earlier studies. The effect of various physical parameters, such as the Prandtl number,

P.G. Metri (✉) · S. Silvestrov
Division of Applied Mathematics, The School of Education, Culture and Communication, Mälardalen University, Box 883, 721 23 Västerås, Sweden
e-mail: prashant.g.metri@mdh.se

S. Silvestrov
e-mail: sergei.silvestrov@mdh.se

M.S. Abel
Department of Mathematics, Gulbarga University, Gulbarga, Karnataka, India
e-mail: msabel2001@yahoo.co.uk

V.M. Bablad
Department of Mathematics, PDA College of Engineering, Gulbarga, Karnataka, India
e-mail: 6666veena@gmail.com

P.G. Metri
Department of Physics, Sangameshwar College,
Solapur, Maharashtra, India
e-mail: pushpa22metri@gmail.com

S. Silvestrov and M. Rančić (eds.), *Engineering Mathematics I*,
Springer Proceedings in Mathematics & Statistics 178,
DOI 10.1007/978-3-319-42082-0_12

the Grashoff number, the Hartmann number, porous parameter, inertia coefficient and internal heat generation on flow and heat transfer characteristics are presented graphically to show some interesting aspects of the physical parameter.

Keywords Exponential stretching · Magnetohydrodynamics (MHD) · Porous medium · Similarity solutions

12.1 Introduction

Analysis of fluid flow in a boundary layer in a stretching sheet is an important part in the fluid mechanics and heat transfer occurring in a number of engineering processes. Few examples of such technological processes are the extrusion of plastic sheet, hot rolling, wire drawing, glass-fiber and paper production, drawing of plastic films and the cooling of a metallic plate in a cooling bath. A class of flow problems with obvious relevance to polymer extrusion is the flow induced by the stretching motion of a flat elastic sheet. For example, in a metal spinning process, the extradite from the die is generally drawn and simultaneously stretched into a filament or sheet, which is there after solidified through rapid quenching or gradual cooling by direct contact with water or chilled metal rolls. Annealing and thinning of copper wires is another example in which the final product depends on the rate of heat transfer at the stretching continuous surface with power-law and exponential variations of stretching velocity and temperature distribution. By drawing the strips in an electrically conducting fluid subjected to a magnetic field the rate of cooling can be controlled and the final products of desired characteristic might be achieved. Both the kinematics of stretching and the simultaneous heating or cooling during such processes have a decisive influence on the quality of the final products. Considering their importance, those flows have been studied by several research groups [22, 24, 28, 30]. The continuing interest in heat transfer and fluid flow through porous media is mainly due to several engineering and geophysical fields such as cooling of nuclear reactors, enhanced oil recovery, thermal insulation drying of porous solids, solid matrix heat exchanges, geothermal and petroleum resources, ceramic processing, filtration processes, chromatography, etc. Ali [6] investigated the effect of variable viscosity on mixed convection heat transfer along a moving surface. The study of magnetohydrodynamic (MHD) flow of an electrically conducting fluid is of considerable interest in modern metallurgical and metal-working process such as drawing of continuous filaments through quiescent fluids, and annealing and tinning of copper wires, the properties of the end product depend greatly on the rate of cooling involved in these processes. This type of flow has also attracted many investigators due to its application in various engineering problems such as MHD generators, nuclear reactors, geothermal energy extraction.

Numerous attempts have been made to analyse the effect of transverse magnetic field on boundary layer flow characteristics. Abel et al. [4] studied viscoelastic MHD flow and heat transfer over a stretching sheet with viscous with Ohmic dissipations in the presence of electric field. Pal and Chatterjee [16] investigated similar problem by considering micropolar fluid. Sharma and Singh [27] analyzed the effects of variable thermal conductivity, viscous dissipation on steady MHD natural convection flow of low Prandtl number fluid on an inclined porous plate with Ohmic dissipation. Singh and Tewari [29] studied the effect of thermal stratification on Non-Darcian free convection flow by using the Ergun model [10] to include the inertia effect. It is well known that there exists non-Darcian flow phenomena bodies inertia effect and solid-boundary viscous resistance. Seddeek [26] analyzed Non-Darcian effect on forced convection heat transfer over a flat plate in a porous medium with temperature-dependent viscosity. Recently, Pal and Mondal [18] analyzed the effect of variable viscosity on MHD non-Darcy mixed convective heat transfer in porous medium with non-uniform heat source/sink. Abel et al. [3] and Bataller [8] investigated the effects of non-uniform heat source on viscoelastic fluid flow and heat transfer over a stretching sheet.

In certain porous media applications such as those involving heat removal from nuclear fuel debris, underground disposal of radiative waste material, storage of food stuffs, the study of heat transfer is of much importance. Comprehensive reviews of the convection through porous media have been reported by Nield and Bejan [15] and by Ingham and Pop [12]. Ali [7] analyzed the effect of lateral mass flux on the natural convection boundary layer induced by a heated vertical plate embedded in a saturated porous medium with an exponential decaying heat generation. It is worth mentioning that Non-Darcian forced flow boundary layers from a very important group of flows, the solution of which is of great importance in many practical applications such as in biomechanical problems, in filtration transpiration cooling and geothermal. In all the above studies, the thermal-diffusion effects are negligible. However, the thermal-diffusion effects, which is caused by temperature gradient is an interesting macroscopically physical phenomenon in fluid mechanics. Usually, in heat and mass transfer problems the variation of density with temperature give rise to combined buoyancy effect under natural convection and hence the temperature will influence the diffusion of species. Recently, Pal and Chatterjee [17] analyzed the effect of mixed convection magnetohydrodynamic heat and mass transfer past a stretching surface in a micropolar fluid-saturated porous medium under the influence of Ohmic heating, Soret and Dufour effects. Alam et al. [5] have studied the Dufour and Soret effects on steady free convection and mass transfer flow past a semi-infinite vertical porous plate in a porous medium. Pal and Mondal [19] examined the effect of Soret and Dufour on MHD non-Darcy unsteady mixed convection heat and mass transfer over a stretching sheet.

In all above investigations porous medium is excluded. But the study of Non-Newtonian fluid flow through porous medium gained momentum as some particular polymer solutions while injected into oil reservoir attain better volumetric sweep efficiency in oil displacement mechanism, which is very important. Abel and Veena [1] studied the flow and heat transfer characteristics in viscoelastic boundary layer flow in porous medium over a stretching surface, Abel et al. [2] studied the hydromagnetic viscoelastic fluid flow and heat transfer over a non-isothermal stretching sheet embedded in porous media. Pillai et al. [20] investigated the effects of work done by deformation in viscoelastic fluid in porous medium with uniform heat source.

Chung [11] also studied the heat transfer characteristics in porous medium in presence of transverse magnetic field.

Rohni et al. [23] investigated steady laminar two-dimensional flow and heat transfer of an incompressible viscous fluid in presence of buoyance force over an exponentially shrinking vertical sheet with suction. The shrinking velocity and wall temperature are assumed to have exponential functions form. Swati et al. [13] studied boundary layer flow and mass transfer an exponential stretching sheet embedded in porous medium. A first order constructive/destructive chemical reaction is also considered. Chetan et al. [9] investigated viscoelastic flow and heat transfer over an exponential stretching sheet with Navier slip boundary condition, here two types of different heating process are considered, namely prescribed exponential order surface and prescribed exponential order heat flux. Sandeep et al. [25] analyzed the unsteady magneto hydrodynamic radiative flow and heat transfer characteristics of a dusty nano fluid over an exponentially permeable stretching surface in presence volume fraction of nano particles. Raju et al. [21] investigated the flow and heat transfer behavior of Casson fluid past an exponential permeable stretching sheet in presence of thermal radiation, magnetic field, viscous dissipation and chemical reaction. Also, dual solutions are presented by comparing the results of the Casson fluid with the Newtonian fluid.

In view of above discussion authors envisage to investigate the effect of mixed convection heat transfer in the MHD non-Darcian flow due to an exponential stretching sheet embedded in a porous medium in presence of non-uniform heat source/sink. The Darcy-Forchheimer model is used to describe the flow in the porous medium. Highly non-linear momentum and heat transfer equations are solved numerically using the fifth order Runge-Kutta-Fehlberg method with shooting technique, (Na [14]), since the governing equations are solved analytically. The novelty of present investigation is to consider temperature dependent and MHD Non-Darcian saturated porous medium in the presence of effect of non-uniform heat source/sink. The effect of various parameters on the velocity and temperature profiles as well as various physical parameters, such as the Prandtl number, the Grashoff number, the Hartmann number, porous parameter, inertia coefficient and internal heat generation on flow and heat transfer characteristics are presented graphically to show some interesting aspects of the physical parameter. It is hoped that the results obtained from the present investigation will provide useful information for application and also serve as a complement to the previous study.

12.2 Mathematical Formulation

Consider a two-dimensional flow of an electrically conducting and incompressible viscous fluid near an impermeable plane wall stretching with velocity U_w and a given temperature distribution T_w. The x-axis is directed along the continuous stretching surface and points in the direction of motion. The y-axis is perpendicular to x-axis whence the continuous stretching plane surface issues (see Fig. 12.1). A uniform

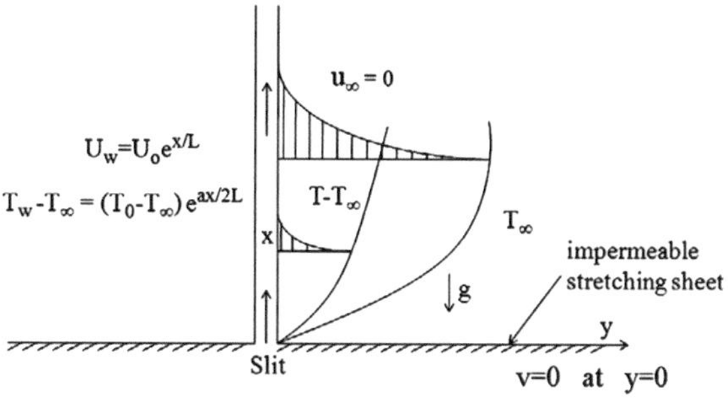

Fig. 12.1 Schematic of the boundary layer induced by stretching sheet

magnetic field B_0 is assumed to be applied in the y-direction. It is assumed that the induced magnetic field of the flow is negligible in comparison with the applied one which corresponds to a very small magnetic Reynolds number. Under boundary layer along with the Boussinesq approximation, the continuity, momentum, and energy equations can be written as,

$$\frac{\partial u}{\partial x} + \frac{\partial u}{\partial y} = 0, \tag{12.1}$$

$$u\frac{\partial u}{\partial x} + v\frac{\partial u}{\partial y} = \nu\frac{\partial^2 u}{\partial y^2} - \frac{\nu}{k}u - \frac{C_b}{\sqrt{k}}u^2 + g\beta(T - T_\infty) - \frac{\sigma B_0^2}{\rho}u, \tag{12.2}$$

$$u\frac{\partial T}{\partial x} + v\frac{\partial T}{\partial y} = \alpha\left(\frac{\partial^2 T}{\partial y^2}\right) + \frac{\sigma B_0^2}{\rho C_p}u^2 + \frac{\mu}{\rho C_p}\left(\frac{\partial T}{\partial y}\right)^2 + \frac{q'''}{\rho C_p}, \tag{12.3}$$

$$q''' = \left(\frac{k u_w(x)}{x\nu}\right)[A^*(T_w - T_\infty)e^{-\alpha\eta} + B^*(T - T_\infty)], \tag{12.4}$$

where ρ is density, T is temperature, ν kinematic viscosity, k is the coefficient of thermal conductivity, C_p is the specific heat at constant pressure, q''' is non-uniform heat generation.

The parameters A^* and B^* are parameters of space temperature-dependent internal heat generation/absorption. It is to be noted that $A^* > 0$ and $B^* > 0$ corresponds to internal heat generation while $A^* < 0$ and $B^* < 0$ corresponds to internal heat absorption.

The associated boundary conditions are:

$$U = U_w(x), \quad v = 0, \quad T = T_w(x) \quad \text{at} \quad y = 0, \tag{12.5}$$

$$u = 0, \quad T \to T_\infty \quad \text{as} \quad y \to \infty, \tag{12.6}$$

where $(u,\ v)$ and $(x,\ y)$ are the components of the velocity field of the steady plane boundary flow, α is Thermal diffusivity of the ambient fluid. σ is the electrical conductivity, and B_0 is the magnetic flux density. The fluid flow is under the effect of the temperature field, where T_∞ is the temperature of the ambient fluid, C_b is the drag coefficient. The stretching velocity U_w and exponential temperature distribution T_w are defined as follows,

$$U_w(x) = U_o e^{\frac{x}{L}}, \tag{12.7}$$

$$T_w(x) = T_\infty + (T_0 - T_\infty) e^{\frac{ax}{2L}}, \tag{12.8}$$

where T_0 and a are parameters of temperature distribution over the stretching surface. We now introduce the following non-dimensional parameters,

$$\eta = \sqrt{\frac{Rey}{2L}} e^{\frac{x}{2L}}, \ \psi(x, \eta) = \sqrt{2Rev} e^{\frac{x}{2L}} f'(\eta), \tag{12.9}$$

$$T(x, y) = T_\infty + (T_0 - T_\infty) e^{\frac{ax}{2L}} \theta(\eta), \tag{12.10}$$

where ψ is the stream function which is defined in the usual form as

$$u = \frac{\partial \psi}{\partial y} \text{ and } v = -\frac{\partial \psi}{\partial x}. \tag{12.11}$$

Thus, substituting (12.9) and (12.10) into (12.11), we obtain u and v as follows

$$u(x, y) = u_0 e^{\frac{x}{L}} f'(\eta) \text{ and } v(x, y) = -\frac{v}{L}\sqrt{\frac{Re}{2}} e^{\frac{x}{2L}} [f(\eta) + \eta f'(\eta)]. \tag{12.12}$$

Equations (12.1)–(12.6) are transformed into ordinary differential equation with the aid of Eqs. (12.9)–(12.11). Thus the governing equations are,

$$f''' + f'' - (2 + N2) f'^2 + 2Gr e^{\frac{aX}{2}} e^{-2X} \theta - 2e^{-X} f' \left(\frac{H^2}{Re} + N1 \right) = 0, \tag{12.13}$$

$$Pr^{-1}\theta'' + f\theta' - af'\theta + e^{\frac{X(2-a)}{2}} Ec \left(2\frac{H^2}{Re} f'^2 + f''^2 e^X \right) + 2e^{-X} [A^* f' e^{-\alpha\eta} + B^* \theta] = 0. \tag{12.14}$$

The boundary condition given in (12.5) and (12.6) reduce to

$$f(0) = 0, \ f'(0) = 1, \ \theta(0) = 1, \text{ at } \eta \to 0, \tag{12.15}$$

$$f'(\infty) = 0, \ \theta(\infty) = 0, \text{ as } \eta \to \infty, \tag{12.16}$$

where $X = \frac{x}{L}$, $H = (\frac{\sigma B_0^2 L^2}{\rho} \nu)^{\frac{1}{2}}$ is the Hartman number, $Ec = \frac{U_0^2}{C_p(T_0 - T_\infty)}$ is the Eckert number, $Gr = \frac{g\beta(T_0 - T_\infty)L^3}{\nu^2}$ is the Grashof number, $Re = \frac{U_0 L}{\nu}$ is the Reynolds number, $\lambda = \frac{Gr}{Re^2}$ is the thermal Buoyancy parameter, $Pr = \frac{\nu}{\alpha}$ is the Prandtl number, $N1 = \frac{L^2}{kRe}$ is porous parameter and $N2 = \frac{2C_b L}{\sqrt{k}}$ is the inertia coefficient. In the above system of local similarity equations, the effect of the magnetic field is included as a ratio of the Hartman number to the Reynolds number.

The physical quantities of interest in the problem are the local skin friction acting on the surface in contact with the ambient fluid of constant density, which is defined as follows

$$\tau_{wx} = \rho\nu\left(\frac{\partial u}{\partial y}\right)_{y=0} = \left(\frac{\rho\nu U_0}{L}\right)\frac{Re^{\frac{1}{2}}}{2}e^{\frac{x}{2}}f''(0), \tag{12.17}$$

and non-dimensional skin friction coefficient, C_f can be written as,

$$C_f = \frac{2\tau_{wx}}{\rho U_w^2} \text{ or } C_f\sqrt{Re_x} = \sqrt{2X}f''(0). \tag{12.18}$$

The local surface heat flux through the wall with k as the thermally conductivity of the fluid is given by,

$$q_w = -k\left(\frac{\partial T}{\partial y}\right)_{y=0} = \frac{k(T_0 - T_\infty)}{L}\left(\frac{Re}{2}\right)^{\frac{1}{2}}e^{\frac{(a+1)}{2}}\theta'(0). \tag{12.19}$$

The local Nusselt number, Nu_x is defined as

$$Nu_x = \frac{xq_w(x)}{k(T_w - T_\infty)}, \tag{12.20}$$

$$\frac{Nu_x}{\sqrt{Re_x}} = -\left(\frac{X}{2}\right)^{\frac{1}{2}}\theta'(0), \tag{12.21}$$

where Re_x is the local Reynolds number based on the surface velocity and is given by,

$$Re_x = \frac{U_{(x)}}{\nu}. \tag{12.22}$$

12.3 Numerical Solution

The system of non-linear differential equations (12.13) and (12.14) together with the boundary condition (12.8)–(12.10) have been solved numerically using the Runge-Kutta-Fehlberg and the Newton-Raphson schemes based on the shooting technique. The most important step in this method is to choose an appropriate finite value of $\eta \to \infty$ for the boundary value problem described by Eqs. (12.13) and (12.14). We start with initial guess values for a particular set of physical parameters to obtain $f''(0), \theta'(0)$. The solution procedure is repeated with another large value of $\eta \to \infty$ until two successive values of $f''(0), \theta'(0)$ differ only by a specified significant digit.

In this method, the third order nonlinear differential equation (12.13) and the second order equation (12.14) have been reduced to 5 ordinary differential equations as follows:

$$\frac{dy_1}{d\eta} = y_2, \tag{12.23}$$

$$\frac{dy_2}{d\eta} = y_3, \tag{12.24}$$

$$\frac{dy_3}{d\eta} = -y_3 - (2 + N2)y_2^2 - 2Gre^{\frac{aX}{2}}e^{-2X}\theta - 2e^{-X}y_2\left(\frac{Ha^2}{Re} + N1\right), \tag{12.25}$$

$$\frac{dy_4}{d\eta} = y_5, \tag{12.26}$$

$$\frac{dy_5}{d\eta} = Pr\left[-y_1y_5 + ay_2y_4 - e^{\frac{X(2-a)}{2}}Ec\left(2\frac{Ha^2}{Re}y_2^2 + y_3^2e^X\right) - 2e^{-X}[A^*y_2e^{-\alpha\eta} + B^*y_4]\right]. \tag{12.27}$$

Boundary conditions are:

$$y_1(0) = 0,\ y_2(0) = 1,\ y_3(0) = S_1,\ y_4(0) = 1,\ y_5(0) = S_2 \text{ as } \eta \to 0, \tag{12.28}$$

$$y_2(\infty) = 0,\ y_4(\infty) = 1 \text{ as } \eta \to \infty, \tag{12.29}$$

where S_1 and S_2 are determined such that $y_2(\infty) = 0$, $y_4(\infty) = 0$. Thus we have to solve this system, we require five initial conditions. However, since we have two initial conditions for f, and one initial condition for θ, the conditions $f''(0)$, $\theta'(0)$ are to be determined by the shooting method using initial guess values S_1 and S_2 until the conditions $y_2(\infty) = 0$, $y_4(\infty) = 0$ are satisfied. In this paper we employed the shooting technique with the Runge-Kutta-Fehlberg and Newton-Raphson schemes to determine unknowns in order to convert the boundary value problem into an initial value problem. Once all initial conditions are determined, the resulting differential equations were integrated using an initial value solver. For this purpose Runge-Kutta-Fehlberg scheme was used.

12.4 Results and Discussion

In this work MHD mixed convective flow and heat transfer characteristics over an exponential stretching sheet embedded in porous medium in presence of viscous dissipation and non-uniform heat source/sink are investigated. Both numerical and analytical solutions are presented. The similarity transformations were used to transform the governing partial differential equations of flow and heat transfer into a system of non-linear ordinary differential equations. The accuracy of the method was established by comparing analytical solution with the numerical solution obtained by the shooting method together with the Runge-Kutta-Fehlberg and the Newton-Raphson schemes. The effects of magnetic field, porous parameter, Grashoff number, Eckert number, and space-temperature dependent heat source/sink parameters on the velocity and temperature profiles are shown in figures (Figs. 12.2, 12.3, 12.4, 12.5, 12.6, 12.7 and 12.8). In Table 12.1, we listed some particular parameter and physical quantities name.

12.5 Conclusion

The flow and heat transfer for MHD non-Darcy boundary layer flow and heat transfer characteristics in an incompressible electrically conducting fluid over an exponential stretching sheet in presence of non-uniform heat source/sink has been analyzed and discussed. The similarity transformations were used to transform the governing partial differential equations of flow and heat transfer into a system of non-linear ordinary differential equations. The accuracy of the method was established by comparing analytical solution with the numerical solution obtained by the shooting method together with the Runge-Kutta-Fehlberg and the Newton-Raphson schemes. The effect of various physical parameters like non uniform heat source/sink parameter,

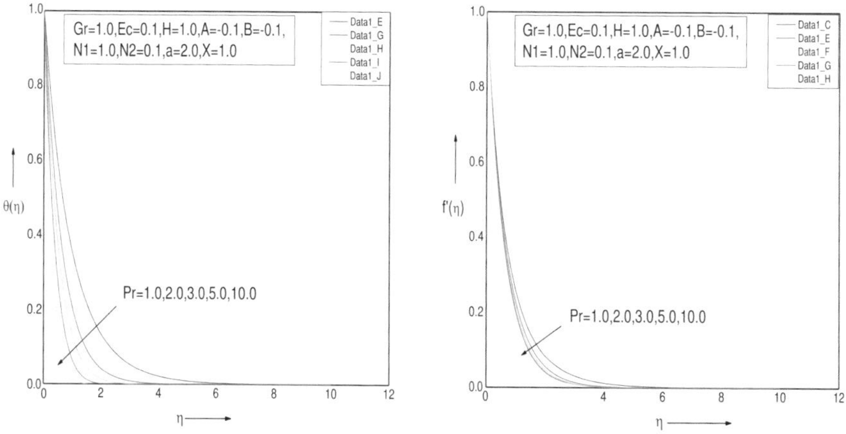

(a) Temperature profile for different values of *Pr*

(b) Velocity profile for different values of *Pr*

Fig. 12.2 **a** and **b** Shows the effect of Prandtl number on the heat transfer is shown in temperature and velocity profiles for different values of Pr. We infer from the figures that temperature decreases with increase in Pr which implies that viscous boundary layer is thicker than the thermal boundary layer. This is in contrast to the effects of other parameters on heat transfer

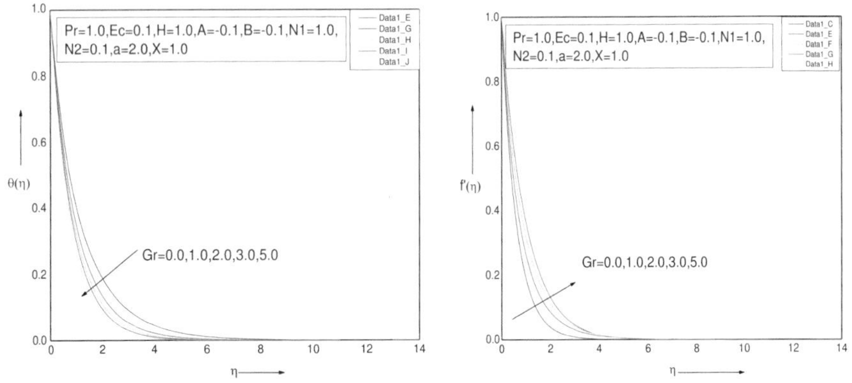

(a) Temperature profile for different values of *Gr*

(b) Velocity profile for different values of *Gr*

Fig. 12.3 **a** Depicts the variation of temperature profile for different values of Grashoff number Gr. From this figure, it is noticed that the temperature decreases with increase in the value of Gr in the boundary layer. Temperature shows increase in the temperature difference between the stretched wall and adjacent fluid which is the reason for enhancing the heat transfer process from surface to the ambient fluid. Increase in the value of the Gr results in the decreasing of the thermal boundary layer thickness. **b** Depicts the velocity response to distinct values of Grashoff number Gr. It is found that the increase in Grashoff number results in rise in the values of velocity due to enhancement in buoyancy force. Here, increase in the values of the Grashoff number correspond to cooling of the surface

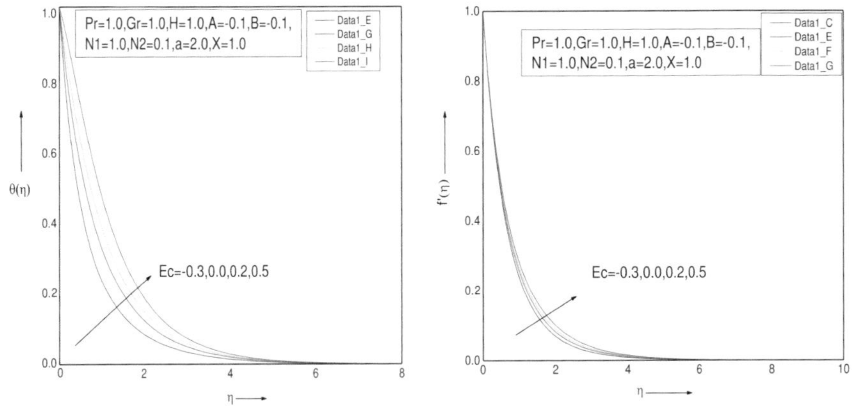

(a) Temperature profile for different values of *Ec*

(b) Velocity profile for different values of *Ec*

Fig. 12.4 **a** and **b** Shows the effect of the Eckert number Ec in both temperature and velocity respectively. It is evident that thermal boundary layer is broadened due to increase in Ec, the energy dissipation exhibits an appreciable increase in the wall temperature in both temperature and velocity. This is quite consistent with the physical situation as the dissipative energy due to elastic deformation work, frictional and ohmic heating are considered, which results in the increase in the temperature and velocity profiles

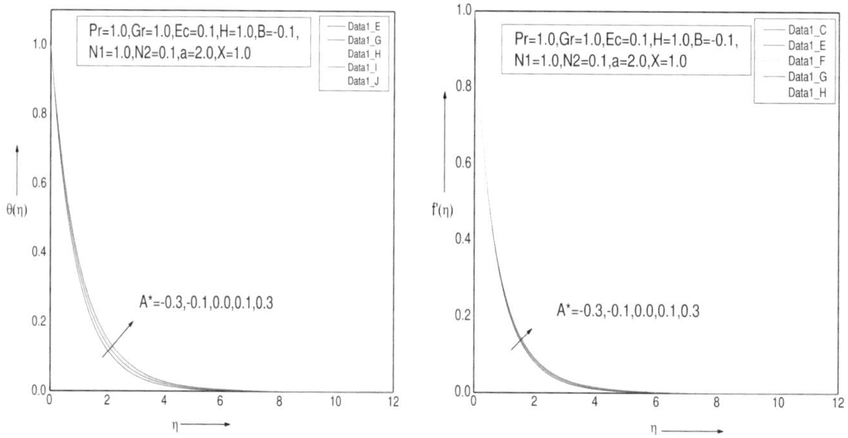

(a) Effect of space dependent heat source/sink parameter A^* on heat transfer in Temperature profile

(b) Effect of space dependent heat source/sink parameter A^* on heat transfer in Temperature profile

Fig. 12.5 **a** and **b** Show temperature and velocity profiles, respectively, for different values of A^*. For $A^* > 0$, it can be seen that the thermal boundary layer generates the energy, and this causes the temperature of the fluid to increase with increase in the value of $A^* > 0$, where as in the case of the $A^* < 0$ the boundary layer absorbs the energy resulting the temperature to fall considerably with decreasing values of A^*

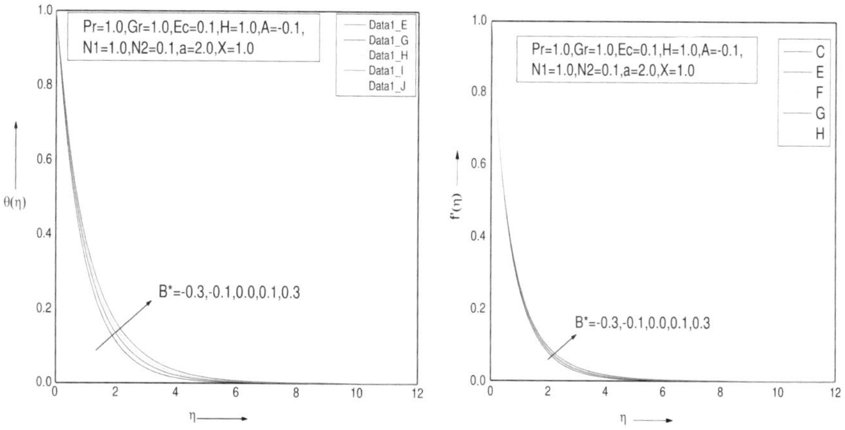

(a) Effect of temperature dependent heat source/sink parameter B^* on heat transfer in temperature profile

(b) Effect of temperature dependent heat source/sink parameter B^* on heat transfer in velocity profile

Fig. 12.6 **a** and **b** Shows the effect of temperature dependent heat source/sink parameter B^* on temperature and velocity profiles, is demonstrated in the figure. The graphs illustrate that energy is released for increasing $B^* > 0$, which causes the temperature to increase both temperature and velocity profiles, where as energy is absorbed for decreasing values of $B^* < 0$, resulting temperature to drop significantly near the boundary layer

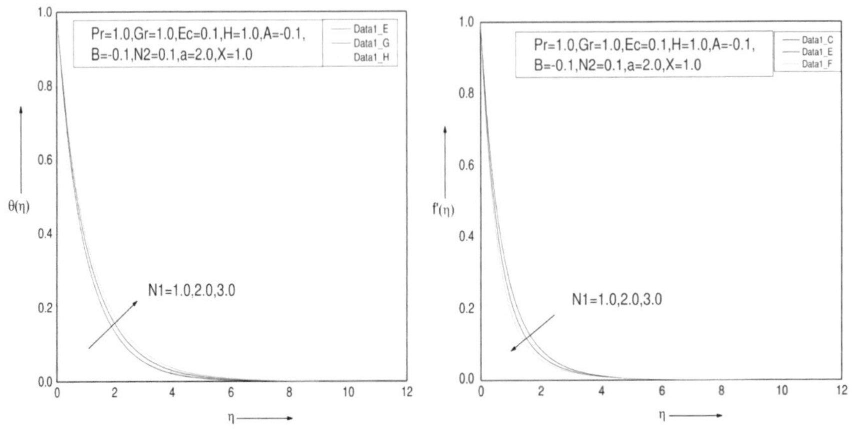

(a) Temperature profile for different values of $N1$

(b) Velocity profile for different values of $N1$

Fig. 12.7 **a** The variation of porous parameter in temperature profile is shown. The increase in the porous parameter decreases permeability which results in obstruction in motion of the fluid due to which there is an increase in the temperature in the thermal boundary layer. **b** The effect of porous parameter on velocity profile is shown. It is observed that the effect of temperature distribution decreases with increase in the porous parameter. This is due to increase in the obstruction of the fluid motion with increase in the porous parameter, thereby increase in the porous parameter indicates decreases in the permeability of the porous medium so the fluid velocity decreases

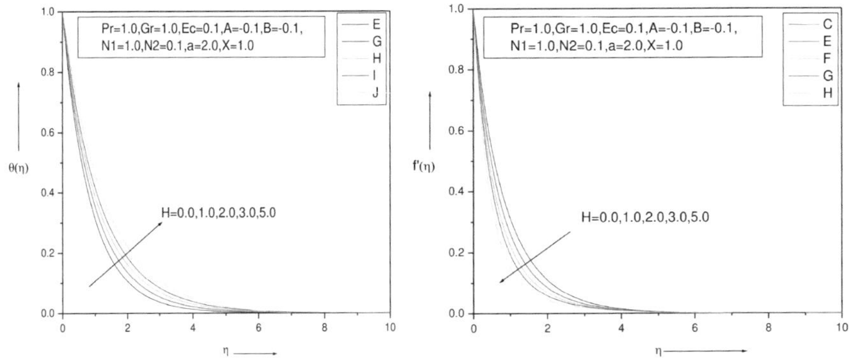

(a) Temperature profile for different values of H

(b) Velocity profile for different values of H

Fig. 12.8 **a** The thermal boundary layer thickness increases with increasing values of the magnetic parameter. The opposing force introduced in the form of the Lorentz drag contributes in increasing the frictional heating between the fluid layers, and hence energy is released in the form of heat. This results in thickening of the thermal boundary layer. **b** The variation of the velocity profile against the magnetic parameter is shown. We notice that the effect of the magnetic parameter is to reduce the velocity of the fluid in the boundary layer region. This is due to an increase in the Lorentz force, similar to Darcy's drag observed in the case of flow through a porous medium. This adverse force is responsible for slowing down the motion of the fluid in the boundary layer region

Table 12.1 Nomenclature

A^*, B^*	Coefficients of space and temperature dependent heat source/sink	a	Temperature distribution on the stretching sheet
C_b	Drag coefficient which is independent of viscosity	k	Permeability of the porous medium
C_p	Specific heat at constant pressure	Ec	Eckert number
Gr	Grashof number	H	Hartmann number
L	Characteristic length	Nu_x	Nusselt number
Pr	Prandtl number	q_w	Local heat flux
q'''	Non-uniform heat source/sink	Re_x	Local Reynolds number
T_w	Stretching sheet temperature	B_o	Uniform transverse magnetic field
T_∞	Temperature for away from the stretching sheet	T	Temperature of the fluid
u	Velocity of the fluid in x direction	v	Velocity of the fluid y directions
x	Flow directional coordinate along the stretching sheet	y	Distance normal to the stretching sheet
C_f	Skin friction coefficient	f	Non-dimensional stream function
Greek symbols			
α	Thermal diffusivity	η	Similarity variable
θ	Dimensionless temperature	τ_{wx}	Local shear stress
μ	Absolute viscosity of the base fluid	υ	Kinematic viscosity of the base fluid
λ	Buoyancy parameter	σ	Fluid electrical conductivity
ρ	Density of the fluid	ψ	Stream function
Subscripts			
w, ∞	Conditions at the wall and at infinity, respectively		

magnetic field parameter, viscous dissipation parameter and porous parameter on velocity and temperature profiles are analyzed. Some of important findings of our analysis obtained by graphical representations are listed below:

1. The effect of temperature dependent heat source/sink parameter, A^* and B^* leads to increase in both temperature and velocity profiles.
2. The effect of convection parameter is to decrease the temperature distribution in the momentum boundary layer.
3. The effect of porous parameter is to increase in the temperature distribution.
4. The effect of Prandtl number is to decrease in both velocity and temperature profiles.

Acknowledgements The first author is grateful to Erasmus Mundus project FUSION (Featured Europe and south-east Asia mobility Network) for support and to the Division of Applied Mathematics, School of Education, Culture and Communication at Mälardalen University for creating excellent research environment during his visit and work on this paper. The author is also thankful to Prof. M. Subhas Abel and Prof. Sergei Silvestrov for making useful suggestions as contributing editors. Further the authors wish to express their thanks to Gulbarga University Gulbarga, Karnataka, India and Mälardalen University, Västerås, Sweden.

References

1. Abel, M.S., Veena, P.: Viscoelastic fluid flow and heat transfer in a porous medium over a stretching sheet. Int. J. Non-linear Mech. **33**(3), 531–540 (1998)
2. Abel, M.S., Joshi, A., Prasad, K.V., Mahaboob, A.: Hydromagnetic viscoelastic fluid flow and heat transfer over a non-isothermal stretching surface embedded in a porous medium. Int. J. Trans. Phenom. **4**, 225–233 (2002)
3. Abel, M.S., Siddheshwar, P.G., Nandeppanavar, M.M.: Heat transfer in a viscoelastic boundary layer flow over a stretching sheet with viscous dissipation and non-uniform heat source. Int. J. Heat. Mass. Transf. **50**, 960–966 (2007)
4. Abel, M.S., Sanjayanand, E., Nandeppanavar, M.M.: Viscoelastic MHD flow and heat transfer over a stretching sheet with viscous and Ohmic dissipation. Commun. Nonlinear. Sci. Numer. Simulat. **13**, 1808–1821 (2008)
5. Alam, M.S., Rahman, M.M., Maleque, A.M., Ferdows, M.: Dufour and Soret effects on steady MHD combined free-forced convective and mass transfer flow past a semi-infinite vertical plate. Int. J. Sci. Tech. **11**, 1–12 (2006)
6. Ali, M.E.: The effect of variable viscosity on mixed convection heat transfer along a vertical moving surface. Int. J. the. Sci. **45**, 60–69 (2006)
7. Ali, M.E.: The effect of lateral mass flux on the natural convection boundary layers induced by a heated vertical plate embedded in a saturated porous medium with internal heat generation. Int. J. Therm. Sci. **46**, 157–163 (2007)
8. Bataller, R.C.: Viscoelastic fluid flow and heat transfer over a stretching sheet under the effects of a non-uniform heat source, viscous dissipation and thermal radiation. Int. J. Heat. Mass. Transf. **50**, 3152–3162 (2007)
9. Chetan, A.S., Shekhar, G.N., Siddeshwar, P.G.: Flow and heat transfer over an exponentially stretching sheet in a viscoelastic liquid with Navier slip boundary conditions. JAFM. **8**, 223–229 (2015)
10. Ergun, S.: Fluid flow through packed columns. Chem. Eng. Prog. **48**, 89–94 (1952)

11. I-Chung, L.: Flow and heat transfer of an electrically conducting fluid of a second grade in a porous medium over a stretching sheet subject to a transverse magnetic field. Int. J. Non. Linear. Mech. **40**, 465–474 (2005)
12. Ingham, D.B., Pop, I.: Transfer Phenomena in Porous Media. Perg, Oxford (1998)
13. Mukhopadhyay, S., Bhattacharyya, K., Layek, G.C.: Mass transfer over an exponentially stretching porous sheet embedded in a stratified medium. Chemical Eng. Communications. **201**, 272–286 (2014)
14. Na, T.Y.: Computational Method in Engineering Boundary Value Problems. Academic Press, New York (1979)
15. Nield, D.A., Bejan, A.: Convection in Porous Media. Springer, New York (1992)
16. Pal, D., Chatterjee, S.: Heat and mass transfer in MHD Non-Darcian flow of a micropolar fluid over a stretching sheet embedded in a porous media with non-uniform heat source and thermal radiation. Commun. Nonlinear. Sci. Numer. Simulat. **15**, 1843–1857 (2010)
17. Pal, D., Chatterjee, S.: Mixed convection magnetohydrodynamic heat and mass transfer past a stretching surface in a micropolar fluid-saturated porous medium under the influence of Ohmic heating, Soret and Dufour effects. Commun. Nonlinear. Sci. Numer Simulat. **16**, 1329–1346 (2011)
18. Pal, D., Mondal, H.: Effect of variable viscosity on MHD Non-Darcy mixed convective heat transfer over a stretching sheet embedded in a porous medium with non-uniform heat source/sink. Commun. Nonlinear. Sci. Numer. Simulat. **15**, 1553–1564 (2010)
19. Pal, D., Mondal, H.: Effects of Soret Dufour, chemical reaction and thermal radiation on MHD Non-Darcy unsteady mixed convective heat and mass transfer over a stretching sheet. Commun. Nonlinear. Sci. Numer. Simulat. **16**, 1942–1958 (2011)
20. Pillai, K.M.C., Sai, K.S., Swamy, N.S., Nataraja, H.R., Tiwari, S.B., Rao, B.N.: Heat transfer in a viscoelastic boundary layer flow through a porous medium. Comput. Mech. **34**, 27–37 (2004)
21. Rajua, C.S.K., Sandeepb, N., Sugunammac, V., Jayachandra Babua, M., Ramana Reddy, J.V.: Heat and mass transfer in magnetohydrodynamic Casson fluid over an exponentially permeable stretching surface. JESTECH. **19**, 45–52 (2016)
22. Riley, N.: Magnetohydrodynamic free convection. J. Fluid. Mech. **18**, 577–586 (1964)
23. Rohni, A.M., Ahmad, S., Ismail, AIMd, Pop, I.: Boundary layer flow and heat transfer over an exponentially shrinking vertical sheet with suction. Int. J. Ther. Sci. **64**, 264–272 (2013)
24. Sakiadis, B.C.: Boundary layer behaviour on continuous solid surfaces. AIChE. J. **7**, 26–28 (1961)
25. Sandeep, N., Sulochana, C,B,R.K.: Unsteady MHD radiative flow and heat transfer of a dusty nanofluid over an exponential stretching surface. JESTECH. **19**, 227–240 (2016)
26. Seddeek, M.A.: Effect of Non-Darcian on forced convection heat transfer over a flat plate in a porous medium with temperature dependent viscosity. Int. Commun. Heat. Mass. Transf. **32**, 258–265 (2005)
27. Sharma, R.R., Singh, G.: Effects of variable thermal conductivity, viscous dissipation on steady MHD natural convection flow of low Prandtl fluid on an inclined porous plate with Ohmic heating. Meccanica. **45**, 237–247 (2010)
28. Sing, K.R., Cowling, T.G.: Thermal conduction in magnetohydrodynamics. J. Mech. Appl. Math. **16**, 1–5 (1963)
29. Singh, P., Tewari, K.: Non-Darcy free convection from vertical surfaces in thermally stratified porous medium. Int. J. Eng. Sci. **31**, 1233–1242 (1993)
30. Sparrow, M.E., Cess, R.Z.D.: Effect of magnetic field on free convection heat transfer. Int. J. Heat. Mass. Transf. **3**, 267–274 (1961)

Chapter 13
Heat and Mass Transfer in MHD Boundary Layer Flow over a Nonlinear Stretching Sheet in a Nanofluid with Convective Boundary Condition and Viscous Dissipation

Prashant G. Metri, M. Subhas Abel and Sergei Silvestrov

Abstract We analyzed the boundary layer flow and heat transfer over a stretching sheet due to nanofluids with the effects of magnetic field, Brownian motion, thermophoresis, viscous dissipation and convective boundary conditions. The transport equations used in the analysis took into account the effect of Brownian motion and thermophoresis parameters. The highly nonlinear partial differential equations governing flow and heat transport are simplified using similarity transformation. Resultant ordinary differential equations are solved numerically using the Runge–Kutta–Fehlberg and Newton–Raphson schemes based on the shooting method. The solutions velocity temperature and nanoparticle concentration depend on parameters such as Brownian motion, thermophoresis parameter, magnetic field and viscous dissipation, which have a significant influence on controlling the dynamics of the considered problem. Comparison with known results for certain particular cases shows an excellent agreement.

Keywords Brownian motion · Convective boundary conditions · Magnetohydrodynamics (MHD) · Nanoliquid · Thermophoresis

13.1 Introduction

Modern nanotechnology provides new opportunities to process and produce materials with average crystallite sizes below 50 nm. Nanofluids can be considered to be the

P.G. Metri (✉) · S. Silvestrov
Division of Applied Mathematics, The School of Education, Culture and Communication, Mälardalen University, Box 883, 721 23 Västerås, Sweden
e-mail: prashant.g.metri@mdh.se

S. Silvestrov
e-mail: sergei.silvestrov@mdh.se

M.S. Abel
Department of Mathematics, Gulbarga University, Gulbarga, Karnataka, India
e-mail: msabel2001@yahoo.co.uk

S. Silvestrov and M. Rančić (eds.), *Engineering Mathematics I*,
Springer Proceedings in Mathematics & Statistics 178,
DOI 10.1007/978-3-319-42082-0_13

next generation heat transfer fluids because they offer exciting new possibilities to enhance heat transfer performance compared to pure liquids. They are expected to have superior properties compared to conventional heat transfer fluids, as well as fluids containing micro-sized metallic particles. Also, nanofluids can improve abrasion-related properties as compared to the conventional solid/fluid mixtures. The development of nanofluids is still hindered by several factors such as the lack of agreement between results, poor characterization of suspensions, and the lack of theoretical understanding of the mechanisms.

A nanofluid is a fluid containing nanometer sized particles called nanoparticles. These fluids are engineered colloidal suspension of nanoparticles in a base fluid. The nanoparticles used in nanofluids are typically made of metals, oxides, carbides, or carbon nanotubes. Common base fluids include water, ethylene glycol and oil. Nanofluids have novel properties that make them potentially useful in many applications in heat transfer, including microelectronics, fuel cells, pharmaceutical processes, and hybrid-powered engine, engine cooling/vehicle thermal management, domestic refrigerator, chiller, heat exchanger, in grinding, machining and in boiler gas temperature reduction. They demonstrate enhanced thermal conductivity and the convective heat transfer coefficient compared to the base fluid. Knowledge of the rheological behavior of nanofluids is found to be very vital in deciding their suitability for convective heat transfer applications.

In the present world of fast technology, the cooling of electronic devices is one of the prominent industrial requirements, but the low thermal conductivity of classical heat transfer fluid such as water, oil and ethylene glycol, is the primary limitation. This leads to the creation of innovative technique in which the nanoscale size (1–100 nm) solid particles are suspended into classical heat transfer fluid in order to change the thermo-physical properties of host fluid, which enhance the heat transfer significantly. This colloidal suspension was first identified as nanofluid by Stephen U.S. Choi in 1995 at the Argonne National Laboratory [4]. The recent development of heat transfer nanofluids and their mathematical modeling [1] play a significant role in various industries. These fluids have numerous applications like cooling of electronics, transportation (engine cooling/vehicle thermal management), manufacturing, heat exchanger, nuclear systems cooling, biomedicine etc. [38, 42]. Several other studies have addressed various aspects of regular/nanofluids with stretching sheet [5, 6, 8, 9, 22, 24, 37, 43]. After the pioneering work by Sakiadis [39], a large amount of literature is available on boundary layer flow of Newtonian and non-Newtonian fluids over linear and nonlinear stretching surface. The problem of natural convection in a regular fluid past a vertical plate is a classical problem first studied theoretically by E. Pohlhausen in contribution to an experimental study by Schmidt and Beckmann [40].

In the past few years, convective heat transfer in nanofluids has become a topic of major current interest. Recently Khan and Pop [17] used the model of Kuznetsov and Nield [19] to study the boundary layer flow of a nanofluid past a stretching sheet with a constant surface temperature. Makinde and Aziz [22] considered to study the effect of a convective boundary condition on boundary layer flow, Wang [41] and Gorla et al. [7] to study the free convection on a vertical stretching surface, heat and mass transfer and nanoparticle fraction over a stretching surface in a nanofluid. The transformed non-linear ordinary differential equations governing the flow are solved numerically by the Runge-Kutta fourth order method.

The solution of boundary layer equation for a power law fluid in MHD was obtained by Helmy [12]. Chiam [3] investigated hydromagnetic flow over a surface stretching with power law velocity using shooting method. Ishak et al. [15] investigated MHD flow and heat transfer adjacent to a stretching vertical sheet. Nourazar et al. [33] investigated MHD forced convective flow of nanofluid over a horizontal stretching sheet with variable magnetic field with the effect of viscous dissipation. Hamad [10] obtained analytical solution by considering the effect of magnetic field for electrical conducting nanofluid flow over a linearly stretching sheet. Rana et al. [36] investigated the numerical solution of unsteady MHD flow of nanofluid on the rotating stretching sheet.

The effects of nanofluids could be considering in different ways such as dynamic effects which include the effects of Brownian motion and thermophoresis diffusion [13, 25, 34], and the static effects of Maxwell's theory [20, 21, 23, 28]. Recently, many researchers, using similarity solution, have examined the boundary layer flow, heat and mass transfer of nanofluids over stretching sheets. Khan and Pop [17] have analyzed the boundary-layer flow of a nanofluid past a stretching sheet using a model in which the Brownian motion and thermophoresis effects were taken into account. They reduced the whole governing partial differential equations into a set of non-linear ordinary differential equations and solved them numerically. In addition, the set of ordinary differential equations which was obtained by Khan and Pop [11] has been solved by Hassani et al. [16] using homotopy analysis method. After that many researchers using similarity solution approach, have extended the heat transfer of nanofluids over stretching sheets and examined the other effects such as the chemical reaction and heat radiation [22], convective boundary condition [35], nonlinear stretching velocity [29], partial slip boundary condition [30], magnetic nanofluid [31], partial slip and convective boundary condition [32], heat generation/absorption [14], thermal and solutal slip [26], nano non-Newtonian fluid [27], and Oldroyd-B nanofluid [2]. At the present time, it is not clear when the boundary layer approximations are adequate for analysis of flow and heat transfer of nanofluids over a stretching sheet in the case of flow and heat transfer of nanofluids. As mentioned, the enhancement of the thermal conductivity of nanofluids is the most outstanding thermo-physical property of nanofluids. In all of the previous studies [2, 11, 14, 16–18, 22, 26, 27, 29–32, 35], the effect of local volume fraction of nano particles on the thermal conductivity of the nanofluid was neglected. However, in the work of Buongiorno [1], it has been reported that the local concentration of nanoparticles may significantly affect the local thermal conductivity of the nanofluids.

In this paper, our main objective is to investigate the effect of a convective boundary condition boundary layer flow, heat transfer and nanoparticle fraction profiles over a stretching surface in nanofluid, with viscous dissipation. The governing boundary layer equations have been transformed to a two-point boundary value problem in similarity variables, and these have been solved numerically. The effects of embedded parameters on fluid velocity, temperature and particle concentration have been shown graphically. It is hoped that the results obtained will not only provide useful information for applications, but also serve as a balance to the previous studies.

13.2 Convective Transport Equations

Consider steady two-dimensional $(x,\ y)$ boundary layer flow of a nanofluid past a stretching sheet with a linear velocity variation with the distance x i.e. $u_w = cx^n$, where c is a real positive number with respect to a stretching rate, and n is a nonlinear stretching parameter, and x is the coordinate measured from the location where the sheet velocity is zero. The governing equations are:

$$\frac{\partial u}{\partial x} + \frac{\partial u}{\partial y} = 0, \tag{13.1}$$

$$u\frac{\partial u}{\partial x} + v\frac{\partial u}{\partial y} = \nu\left(\frac{\partial^2 u}{\partial y^2}\right) - \frac{\sigma B_0^2 u}{\rho}, \tag{13.2}$$

$$u\frac{\partial T}{\partial x} + v\frac{\partial T}{\partial y} = \alpha\left(\frac{\partial^2 T}{\partial y^2}\right) + \tau\left\{D_B\left(\frac{\partial C}{\partial y}\frac{\partial T}{\partial y}\right) + \frac{D_T}{T_\infty}\left[\left(\frac{\partial T}{\partial y}\right)^2\right]\right\} + \frac{\nu}{c_p}\left(\frac{\partial u}{\partial y}\right)^2, \tag{13.3}$$

$$u\frac{\partial C}{\partial x} + v\frac{\partial C}{\partial y} = D_B\left(\frac{\partial^2 C}{\partial y^2}\right) + \left(\frac{D_T}{T_\infty}\right)\left(\frac{\partial^2 T}{\partial y^2}\right), \tag{13.4}$$

where u and v are the velocity components along the x and y directions respectively, p is the fluid pressure, ρ_f is the density of base fluid, ν is the kinematic viscosity of the base fluid, α is the thermal diffusivity of the base fluid, $\tau = \frac{\rho c_p}{\rho c_f}$ is the ratio of nanoparticle heat capacity and the base fluid heat capacity, D_B is the Brownian diffusion coefficient, D_T is the thermophoretic diffusion coefficient, and T is the local temperature (Fig. 13.1).

The associated boundary conditions are:

$$y = 0,\ u = ax^n,\ v = 0,\ -k\frac{\partial T}{\partial y} = h(T_f - T),\ C = C_w, \tag{13.5}$$

$$y \to \infty,\ u = 0\ ,\ v = 0,\ T = T_\infty,\ C = C_\infty. \tag{13.6}$$

We introduce the following dimensionless quantities

$$\eta = y\sqrt{\frac{a(n+1)}{2\nu}}x^{\frac{n-1}{2}},\ \phi = \frac{C - C_\infty}{C_w - C_\infty},\ u = ax^n f'(\eta),$$
$$v = -\sqrt{\frac{a\nu(n+1)}{2}}x^{\frac{n-1}{2}}\left\{f + \frac{(n-1)}{n+1}\right\}\eta f',\ \theta = \frac{T - T_\infty}{T_f - T_\infty}. \tag{13.7}$$

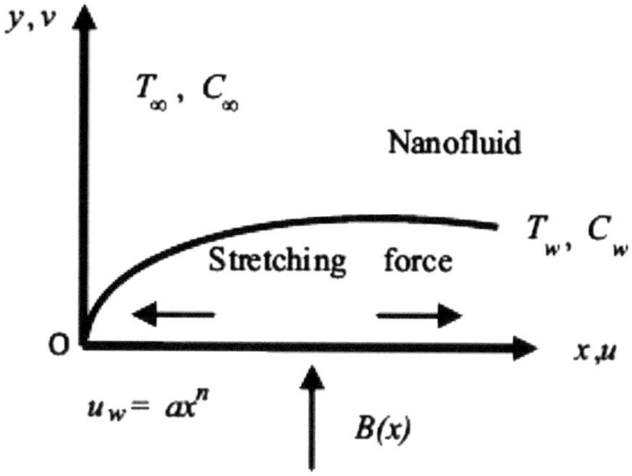

Fig. 13.1 Nano boundary layer flow over a nonlinear stretching sheet

Substituting (13.7) in (13.2)–(13.6), we obtain the following set of equations,

$$f''' + ff'' - f'^2 - \left(\frac{2n}{n+1}\right) f'^2 = 0, \tag{13.8}$$

$$\theta'' + Prf\theta' + PrNb\phi'\theta' + PrNt\theta'^2 + PrEcf''^2 = 0, \tag{13.9}$$

$$\phi'' + Lef\phi' + \frac{Nt}{Nb}\theta'' = 0, \tag{13.10}$$

subject to the following boundary conditions

$$f(0) = 0,\ f'(0) = 1,\ \theta'(0) = Bi[1 - \theta(0)],\ \phi(0) = 1, \tag{13.11}$$

$$f'(\infty) = 0,\ \theta(\infty) = 0,\ \phi(\infty) = 0, \tag{13.12}$$

where primes denote differentiation with respect to η and the five parameters appearing in Eqs. (13.9)–(13.12) are defined as follows

$$Pr = \frac{\nu}{\alpha},\ Le = \frac{\nu}{D_B},$$

$$Nb = \frac{(\rho c)_p D_B (C_w - C_\infty)}{(\rho c)_f^V},\ Nt = \frac{(\rho c)_p D_T (T_f - T_\infty)}{(\rho c)_f^V T_\infty},$$

$$Bi = \frac{h(\frac{\nu}{a})^{\frac{1}{2}}}{k},\ M = \frac{2\sigma B_0^2}{a\rho_f(n+1)}, \tag{13.13}$$

with $Nb = 0$ there is no thermal transport due to buoyancy effects created as a result of nanoparticle concentration gradients.

Here, we note that (13.8) with the corresponding boundary conditions on f provided by (13.11) has a closed form solution which is given by

$$f(\eta) = 1 - e^{-\eta}. \tag{13.14}$$

In (13.14), Pr, Le, Nb, Nt and Bi denote the Prandtl number, the Lewis number, the Brownian motion parameter, the thermophoresis parameter and the Biot number, respectively. The reduced Nusselt number Nur and the reduced Sherwood number Shr are obtained in terms of the dimensionless temperature at the surface $\theta'(0)$ and the dimensionless concentration at the sheet surface $\phi'(0)$ respectively, i.e.

$$Nur = Re_x^{\frac{-1}{2}}\ Nu = -\theta'(0), \tag{13.15}$$

$$Shr = Re_x^{\frac{-1}{2}}\ Nu = -\phi'(0), \tag{13.16}$$

where

$$Nu = \frac{q_w x}{k(T_w - T_\infty)},\quad Sh = \frac{q_m x}{D_B(\phi_w - \phi_\infty)},\quad Re_x = \frac{u_w(x)x}{\nu}. \tag{13.17}$$

13.3 Results and Discussion

Equations (13.8)–(13.10) subject to the boundary conditions, (13.11) and (13.12), were solved numerically using the Runge-Kutta-Fehlberg method. As a further check of the accuracy of our numerical computations, Table 13.1 contains a comparison of our results for the reduced Nusselt number and the reduced Sherwood number with those reported by Khan and Pop [17] for $Le = 10,\ Pr = 10,\ Bi = \infty,\ M = 10$. The infinitely large Biot number simulates the isothermal stretching model used in [17] as noted earlier. The results for all combination values of Brownian motion parameter Nb and the thermophoresis parameter Nt used in our computations, showed an exact match between our results and the ones reported in [17]. The first five entries show that for a fixed thermophoresis parameter $Nt = 0.1$, the reduced Nusselt number decreases sharply with the increasing in Brownian motion, that as Nb is increased from 0.1 to 0.5. However, the reduced Sherwood number increases substantially as Nb is increased from 0.1 to 0.2 but tends to plateau beyond $Nb = 0.2$. These observations are consistent with the initial slopes of the temperature and concentration profiles to be discussed later. As the Brownian motion intensifies, it impacts a larger extent of the fluid, causing the thermal boundary layer to thicken, which in turn decreases the reduced Nusselt number. The thickening of the boundary layer due to stronger Brownian motion will be highlighted again when the temperature profiles are discussed. It will be seen from the concentration profiles appearing later in the discussion that the initial slope of the curve and the extend of the concentra-

Table 13.1 Comparison of results for the reduced Nusselt number $-\theta'(0)$ and the reduced Sherwood number $-\phi'(0)$ with Khan and Pop [17]

Nb	Nt	Nur	Shr	Nur present	Shr present
0.1	0.1	0.9524	2.1294	0.5230	2.0507
0.2	0.1	0.5056	2.3819	0.3561	2.2346
0.3	0.1	0.2522	2.4100	0.2082	2.2797
0.4	0.1	0.1194	2.3997	0.1077	2.2846
0.5	0.1	0.0543	2.3836	0.0513	2.2767
0.1	0.2	0.6932	2.2740	0.4761	1.9851
0.1	0.3	0.5201	2.5286	0.44235	2.0231

tion boundary layer are not affected significantly beyond $Nb = 0.2$ and hence the plateau in the Sherwood number behavior. The last four entries in Table 13.1 show that the reduced Nusselt number decreases as the thermophoresis diffusion penetrates deeper into the fluid and causes the thermal boundary layer to thicken. However, the increase in the thermophoresis parameter enhances the Sherwood number, conclusion that is consistent with the results of Khan and Pop [17]. In Table 13.2, we listed some particular parameter and physical quantities name.

We now turn our attention to the discussion of graphical results that provide additional insights into the problem under investigation.

Figure 13.2 shows the temperature distribution in the thermal boundary layer for different values of Brownian motion and the thermophoresis parameters. As both Nb and Nt increase in the boundary layer thickness, as noted earlier in discussing the tabular data, the surface temperature increases and the curves become less steep

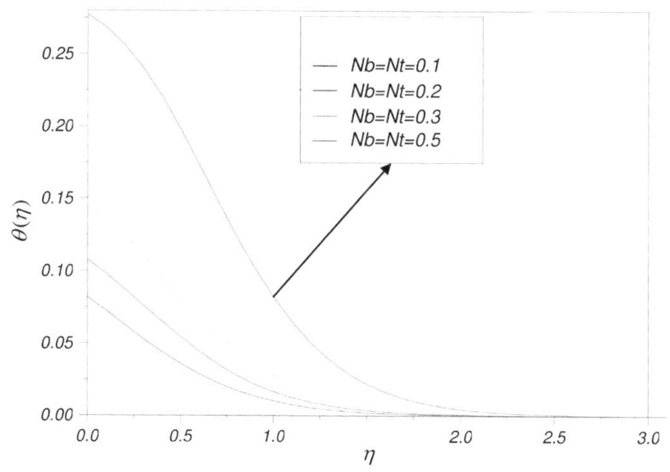

Fig. 13.2 Effect of Nt and Nb on temperature profiles when $M = 2$, $Ec = Le = 5$, $Pr = 5$, $Bi = 0.1$

indicating a diminution of the reduced Nusselt number. As seen in Fig. 13.3, the effect of Lewis number on the temperature profiles is noticeable only in the region close to the sheet as the curves tend to merge at larger distances from the sheet. The Lewis number expresses the relative contribution of thermal diffusion rate to species diffusion rate in the boundary layer regime. An increase of Lewis number will reduce thermal boundary layer thickness and will be accompanied by a increase in the temperature. It also reveals that the temperature gradient at surface of sheet increases. There will be much greater reduction in concentration boundary layer thickness than the thermal boundary layer thickness over an increment in Lewis number.

Figure 13.4 illustrates the effect of Biot number on the thermal boundary layer. As expected, the stronger convection results in higher surface temperatures, causing the thermal effect to penetrate deeper into the quiescent fluid. The temperature profiles depicted in Fig. 13.5 show that as the Prandtl number increases, the thickness of the thermal boundary layer decreases as the curve become increasingly steeper. As a consequence, the reduced Nusselt number, being proportional to the initial slope increases. This pattern is reminiscent of the convective boundary layer flow in a

Table 13.2 Nomenclature

B_i	Biot number	a	Positive constant associated with linear stretching
D_B	Brownian diffusion coefficient	D_T	Thermophoretic diffusion coefficient
f	Dimensionless steam function	g	Gravitational acceleration
h	Convective heat transfer coefficient	k	Thermal conductivity of the nanofluid
Le	Lewis number	Nb	Brownian motion parameter
Nt	Thermophoresis parameter	Nu	Nusselt number
Nur	Reduced Nusselt number	Pr	Prandtl number
p	Pressure	Re_x	Local Reynolds number
Sh	Sherwood number	Shr	Reduced Sherwood number
M	Magnetic number	Ec	Eckert number
T_f	Temperature of the hot fluid	T_w	Sheet surface (wall) temperature
T_∞	Ambient temperature	u, v	Velocity components in x and y directions
C	Nanoparticle volume fraction	c_w	Nanoparticle volume fraction at the wall
Greek symbols			
α	Thermal diffusivity of the base fluid	η	Similarity variable
θ	Dimensionless temperature	ϕ	Dimensionless volume fraction
μ	Absolute viscosity of the base fluid	ν	Kinematic viscosity of the base fluid
ρ_f	Density of the base fluid	ρ_p	Nanoparticle mass density
$(\rho c)_f$	Heat capacity of the base fluid	$(\rho c)_p$	Heat capacity of the nanoparticle material
ψ	Stream function		

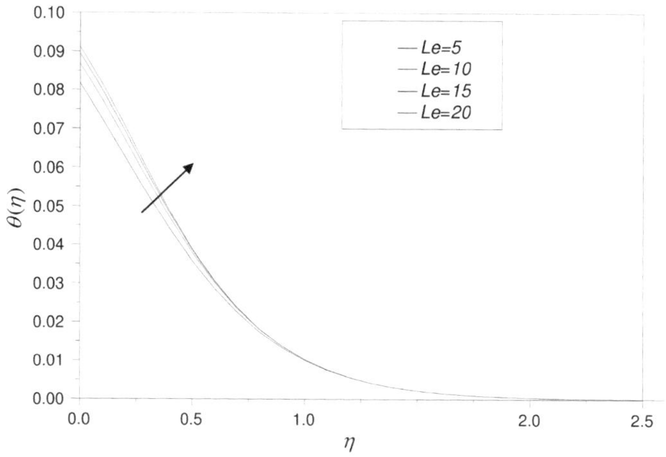

Fig. 13.3 Effect of Le on temperature profiles when $n = 2$, $M = 2$, $Nt = Nb = 0.1$, $Pr = 5$, $Ec = 5$, $Bi = 0.1$

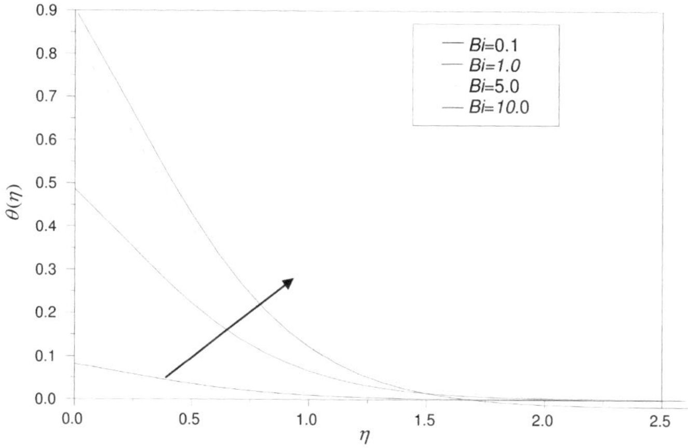

Fig. 13.4 Effect of Bi on temperature profiles when $n = 1.5$, $M = 2$, $Nt = Nb = 0.1$, $Pr = Le = 5$, $Ec = 5$, $Bi = 0.1$

regular fluid [12]. Figure 13.6 shows that the effect of Magnetic number on the temperature profiles is noticeable only in the region close to the sheet as the curves tend to merge at larger distances from the sheet.

Figure 13.7 reveals the effect made by the Viscous dissipation on temperature profile. On observing the temperature graph, the wall temperature of the sheet increases as the values of Ec increases. Moreover, when values of Ec increases the thermal boundary layer thickness increases. This is due to fact that the heat transfer rate at the surface decreases as Ec increases.

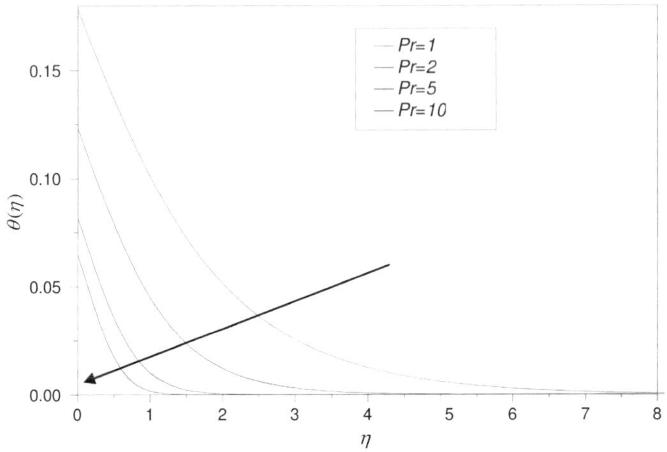

Fig. 13.5 Effect of Pr on temperature profiles when $n = 1$, $M = 2$, $Nt = Nb = Bi = 0.1$, $Le = 5$, $Ec = 5$

Figure 13.8 shows the influence of the change of Brownian motion parameter Nb and thermophoresis parameter Nt on concentration profile when $Nb = Nt$. It is noticed that as thermophoresis parameter increases in the concentration boundary layer thickness and the surface decreases as both Nb and Nt increase, which represents the mass transfer rate. Consequently, concentration on the surface of sheet increases. This is due to fact that the thermophoresis parameter Nt is directly proportional to the mass transfer coefficient associated with fluid.

Figure 13.9 illustrates the effect of Lewis number on concentration profile. When the Lewis number increases the concentration profile decreases and concentration boundary layer thickness decreases. This is probably due to fact that mass transfer rate increases as Lewis number increases. It also reveals that the concentration gradient at surface sheet increases. Moreover, the concentration at the surface of sheet decreases as values of Le increase.

Figure 13.10 reveals the effect of Biot number on nanoparticle concentration profile. It is concluded that concentration distribution as well as concentration boundary layer thickness increase for higher values of the Biot number.

Figure 13.11 shows the influence of magnetic field parameter M on the concentration profile. Magnetic field is increased in the concentration boundary layer thickness. However, an increment in concentration boundary layer is not significant. Similar to other common fluids, the nanofluids show similar characteristics regarding the influence of the magnetic field.

It is observed from Fig. 13.12 that the effect of nonlinear stretching parameter n on the dimensionless velocity profile is to decrease velocity slightly with increase of nonlinear stretching parameter n. In Fig. 13.13 it is noticed that nonlinear stretching parameter n enhances temperature negligibly in the boundary layer region.

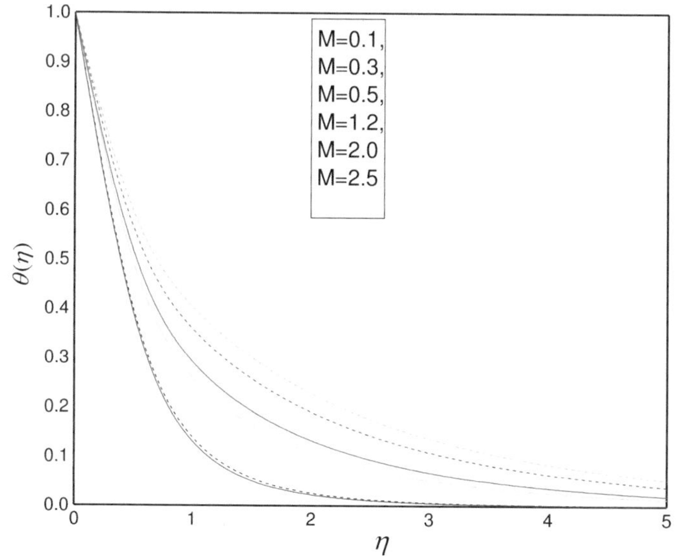

Fig. 13.6 Effect of M on temperature profiles when $n = 2$, $Nt = Nb = Bi = 0.1$, $Pr = Le = 5$, $Ec = 5$

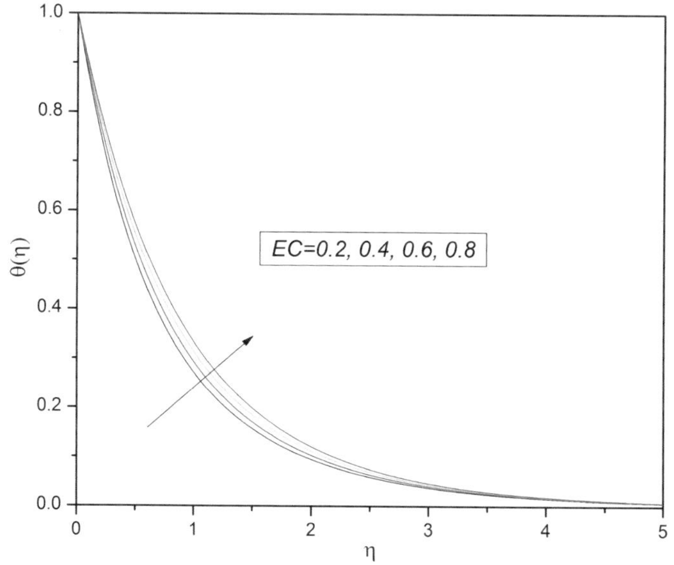

Fig. 13.7 Effect of Eckert number on temperature profiles for different values of $n = 1.5$, $Nt = Nb = 5$, $Pr = 1.0$, $Bi = 0.5$, $M = 2$

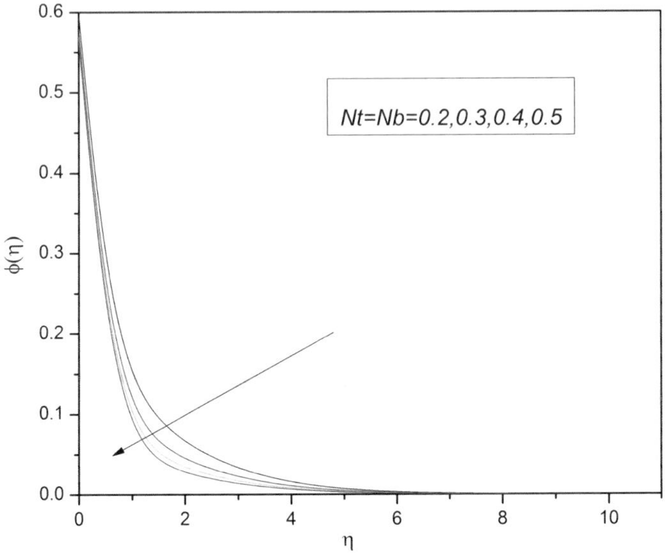

Fig. 13.8 Effect of Nt and Nb on concentration profiles when $n = 2$, $Bi = 0.1$, $Le = Pr = 5$, $M = 1.5$

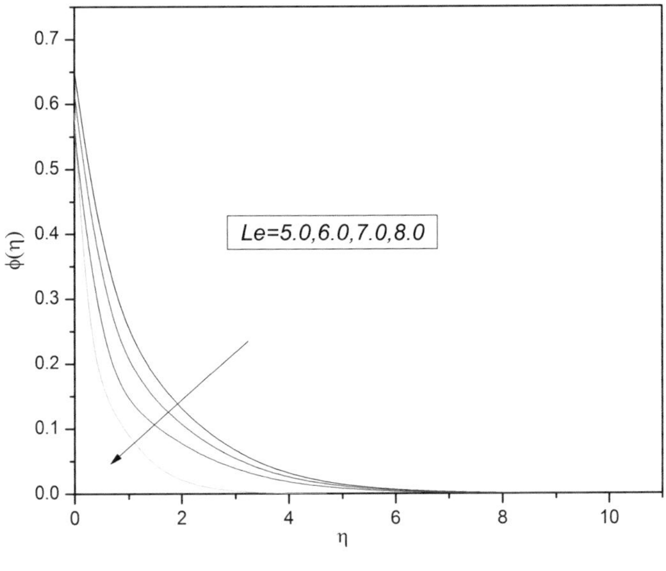

Fig. 13.9 Effect of Le on concentration profiles for different values of $n = 2$, $Nt = Nb = 0.1$, $Pr = 5$, $Bi = 0.1$

13.4 Conclusion

We have studied the boundary layer flow and heat transfer within a nanofluid on a stretching sheet using a shooting method that involves the Runge-Kutta-Fehlberg and Newton–Raphson schemes. The effects of some governing parameters like Lewis number, Brownian motion parameter, thermophoresis parameter, convective Biot number, Magnetic parameter, nonlinear stretching parameter were analysed. The obtained numerical results are excellent agreement for some limiting cases with

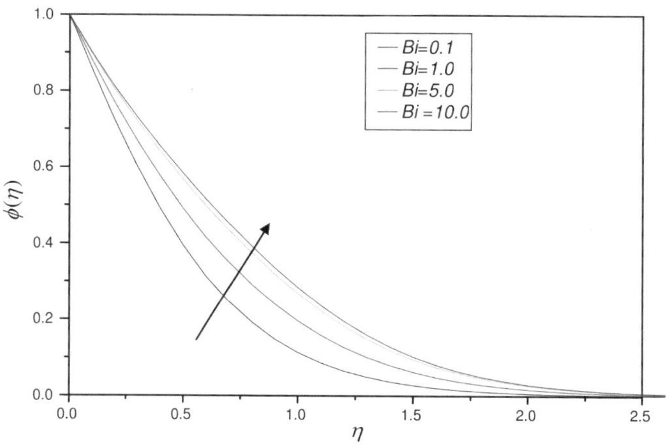

Fig. 13.10 Effect of Bi on concentration profiles for different values of $n = 2$, $Nt = Nb = 0.1$, $Pr = Le = 5$

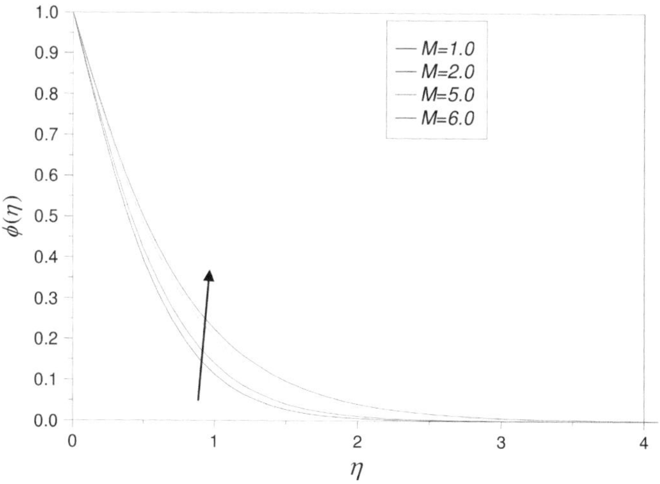

Fig. 13.11 Effect of M on concentration profiles for different values of $n = 2$, $Nt = Nb = Bi = 0.1$, $Pr = Le = 5$

reference ones (Khan and Pop [17]). Some of the important findings of our analysis obtained by the graphical representation are listed below:

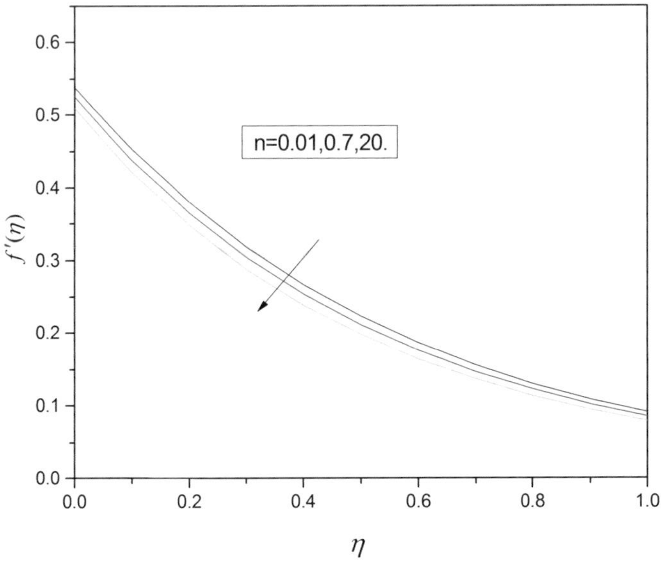

Fig. 13.12 Effect of nonlinear stretching parameter n on velocity profile for various values of $M = 2$, $Nt = Nb = Bi = 0.1$, $Pr = Le = 5$

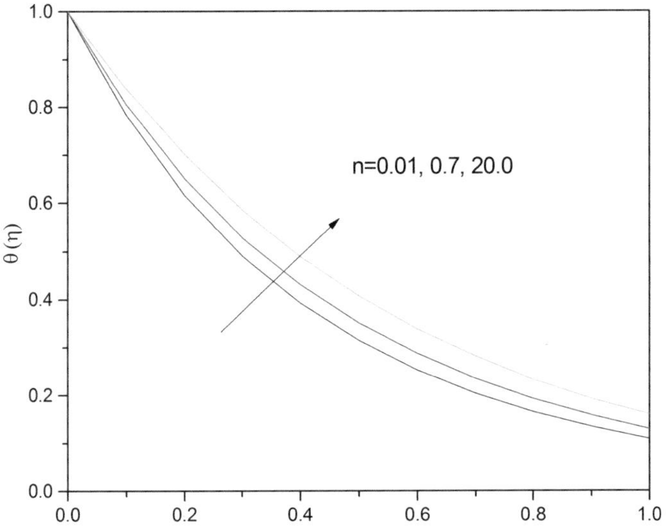

Fig. 13.13 Effect of nonlinear stretching parameter n on temperature profile for various values of $M = 1$, $Nt = Nb = Bi = 0.1$, $Pr = Le = 5$

1. For infinitely large Biot number characterizing the convective heating, which corresponds to the constant temperature boundary condition, the present results and those reported by Khan and Pop [17] match up to four decimal places.
2. The increasing of thermophoresis parameter Nt and the Brownian motion parameter Nb is to increase the temperature in the boundary layer which consequently reduces the heat transfer rate at the surface.
3. Velocity profile decreases with an increase in nonlinear stretching sheet parameter.
4. Concentration boundary layer thickness increases with an increase in the Biot number and the magnetic field parameter.
5. A rising value in Nb and the decreasing in Nt produce a decrease in the nanoparticle concentration, as a result the local Sherwood number increases.
6. An increase in parameter Nb decreases the local Nusselt number $-\theta'(0)$, but the opposite is true in local Sherhood number $-\phi'(0)$.

Acknowledgements The first author is grateful to Erasmus Mundus project FUSION (Featured Europe and south-east Asia mobility Network) for support and to the Division of Applied Mathematics, School of Education, Culture and Communication at Mälardalen University for creating excellent research environment during his visit and work on this paper.

References

1. Buongiorno, J.: Convective transport in nanofluids. ASME. J. Heat Transf. **128**, 240–250 (2006)
2. Chandrasekar, M., Suresh, S.: A review on the mechanisms of heat transport in nanofluids. Heat Transf. Eng. **30**, 1136–1150 (2009)
3. Chiam, T.C.: Hydromagnetic flow over a surface stretching with a power law velocity. Int. J. Eng. Sci. **33**, 429–435 (1995)
4. Choi, U.S., Eastman, J.A.: Enhancing thermal conductivity of fluids with nanoparticles. In: ASME International Mechanical Engineering Congress and Exposition, San Francisco, USA (1995)
5. Cortell, R.: Viscous flow and heat transfer over a nonlinear stretching sheet. Appl. Math. Comput. **184**, 864–873 (2007)
6. Gbadeyan, J.A., Olanrewaju, M.A., Olanrewaju, P.O.: Boundary layer flow of a nanofluid past a stretching sheet with a convective boundary condition in the presence of magnetic field and thermal radiation. Aust. J. Basic Appl. Sci. **5**, 1323–1334 (2011)
7. Gorla, R.S.R., Sidawi, I.: Free convection on a vertical stretching surface with suction and blowing. Appl. Sci. Res. **52**, 247–257 (1994)
8. Goyal, M., Bhargava, R.: Boundary layer flow and heat transfer of viscoelastic nanofluids past a stretching sheet with partial slip conditions. Appl. Nanosci., 1–7 (2013)
9. Goyal, M., Bhargava, R.: Numerical study of thermo diffusion effects on boundary layer flow of nanofluids over a power law stretching sheet. Microfluid. Nanofluid., 1322–1326 (2014)
10. Hamad, M.A.A.: Analytical solution of natural convection flow of a nanofluid over a linearly stretching sheet in the presence of magnetic field. Int. Commun. Heat Mass Transf. **38**, 487–492 (2011)
11. Hassani, M., Tabar, M.M., Nemati, H., Domairry, G., Noori, F.: An analytical solution for boundary layer flow of a nanofluid past a stretching sheet. Int. J. Therm. Sci. **50**, 2256–2263 (2011)

12. Helmy, K.A.: Solution of the boundary layer equation for a power law fluid in magneto hydrodynamics. Acta Mech. **102**, 25–37 (1994)
13. Ibrahim, W., Makinde, O.D.: The effect of double stratification on boundary-layer flow and heat transfer of nanofluid over a vertical plate. Comput. Fluids **86**, 433–441 (2013)
14. Ibrahim, W., Shankar, B.: MHD boundary layer flow and heat transfer of a nanofluid past a permeable stretching sheet with velocity, thermal and solutal slip boundary conditions. Comput. Fluids **75**, 1–10 (2013)
15. Ishak, A., Nazar, R., Pop, I.: Hydromagnetic flow and heat transfer adjacent to a stretching vertical sheet. Heat Mass Transf. **44**, 921–927 (2008)
16. Kahar, R.A., Kandasamy, R., Muhaimin, I.: Scaling group transformation for boundary-layer flow of a nanofluid past a porous vertical stretching surface in the presence of chemical reaction with heat radiation. Comput. Fluids **52**, 15–21 (2011)
17. Khan, W.A., Pop, I.: Boundary-layer flow of a nanofluid past a stretching sheet. Int. J. Heat Mass Transf. **53**, 2477–2483 (2010)
18. Khanafer, K., Vafai, K.: A critical synthesis of thermophysical characteristics of nanofluids. Int. J. Heat Mass Transf. **54**, 4410–4428 (2011)
19. Kuznetsov, A.V., Nield, D.A.: Natural convective boundary-layer flow of a nanofluid past a vertical plate. Int. J. Therm. Sci. **49**, 243–247 (2010)
20. Makinde, O.D.: Computational modelling of nanofluids flow over a convectively heated unsteady stretching sheet. Curr. Nanosci. **9**, 673–678 (2013)
21. Makinde, O.D.: Effects of viscous dissipation and Newtonian heating on boundary-layer flow of nanofluids over a flat plate. Int. J. Numer. Methods Heat Fluid Flow **23**, 1291–1303 (2013)
22. Makinde, O.D., Aziz, A.: Boundary layer flow of a nanofluid past a stretching sheet with a convective boundary condition. Int. J. Therm. Sci. **50**, 1326–1332 (2011)
23. Makinde, O.D., Khan, W.A., Aziz, A.: On inherent irreversibility in Sakiadis flow of nanofluids. Int. J. Exergy **13**, 159–174 (2013)
24. Mansur, S., Ishak, A.: The flow and heat transfer of a nanofluid past a stretching/shrinking sheet with convective boundary condition. Abstr. Appl. Anal., Article ID 350647 (2013)
25. Mutuku, W.N., Makinde, O.D.: Hydromagnetic bioconvection of nanofluid over a permeable vertical plate due to gyrotactic microorganisms. Comput, Fluids (2014)
26. Nadeem, S., Mehmood, R., Akbar, N.S.: Non-orthogonal stagnation point flow of a nano non-Newtonian fluid towards a stretching surface with heat transfer. Int. J. Heat Mass Transf. **57**, 679–689 (2013)
27. Nadeem, S., Haq, R.U., Akbar, N.S., Lee, C., Khan, Z.H.: Numerical study of boundary layer flow and heat transfer of Oldroyd-B nanofluid towards a stretching sheet. **8**, (2013)
28. Njane, M., Mutuku, W.N., Makinde, O.D.: Combined effect of Buoyancy force and Navier slip on MHD flow of a nanofluid over a convectively heated vertical porous plate. Sci. World. J. (2013)
29. Noghrehabadi, A., Pourrajab, R., Ghalambaz, M.: Effect of partial slip boundary condition on the flow and heat transfer of nanofluids past stretching sheet prescribed constant wall temperature. Int. J. Therm. Sci. **54**, 253–261 (2012)
30. Noghrehabadadi, A., Ghalambaz, M., Ghanbarzadeh, A.: Heat transfer of magnetohydrodynamic viscous nanofluids over an isothermal stretching sheet. J. Thermophys. Heat Transf. **26**, 686–699 (2012)
31. Noghrehabadi, A., Pourrajab, R., Ghalambaz, M.: Flow and heat transfer of nanofluids over stretching sheet taking into account partial slip and thermal convective boundary conditions. Heat Mass Transf. **49**, 1357–1366 (2013)
32. Noghrehabadi, A., Saffarian, M., Pourrajab, M., Ghalambaz, M.: Entropy analysis for nanofluid flow over a stretching sheet in the presence of heat generation/absorption and partial slip. J. Mech. Sci. Tech. **27**, 927–937 (2013)
33. Nourazar, S.S., Matin, M.H., Simiari, M.: The HPM applied to MHD nanofluid flow over a horizontal stretching plate. J. Appl. Math., 810–827 (2011)
34. Olanrewaju, A.M., Makinde, O.D.: On boundary layer stagnation point flow of a nanofluid over a permeable flat surface with Newtonian heating. Chem. Eng. Commun. **200**, 836–852 (2013)

35. Rana, P., Bhargava, R.: Flow and heat transfer of a nanofluid over a nonlinearly stretching sheet: a numerical study. Commun. Nonlinear Sci. Numer. Simul. **17**, 212–226 (2012)
36. Rana, P., Bhargava, R., Beg, O.A.: Finite element simulation of unsteady magnetohydrodynamic transport phenomena on a stretching sheet in a rotating nanofluid. J. Nanoeng. Nanosyst. **227**, 77–99 (2011)
37. Rashidi, M.M., Friedoonimehr, N., Hosseini, A., Beg, A.O., Hung, T.K.: Homotopy simulation of nanofluid dynamics from an nonlinear stretching isothermal sheet with transpiration. Meccanica **49**, 469–482 (2014)
38. Saidur, R., Leong, K.Y., Mohammad, H.A.: A review on applications and challenges of nanofluids. Renew. Sustain. Energy Rev. **15**, 1646–1668 (2011)
39. Sakiadas, B.C.: Boundary layer behavior on continuous solid surfaces: I boundary layer equations for two dimensional and flow. AIChE. J. **7**, 26–28 (1961)
40. Schmidt, E., Beckmann, W.: Das Temperature-und Geschwindikeitsfeld voneiner warme abgebenden senkrechten platte bei naturlicher konvection, II. Die Versuche und ihre Ergibnisse Forcsh Ingenieurwes **1**, 391–406 (1930)
41. Wang, C.Y.: Free convection on a vertical stretching surface. J. Appl. Math. Mech. (ZAMM) **69**, 418–420 (1989)
42. Wong, K.V., Leon, O.D.: Application of nanofluids: current and future. Adv. Mech. Eng. Article ID **51965**, 1–11 (2009)
43. Zaimi, K., Ishak, A., Pop, I.: Boundary layer flow and heat transfer over a nonlinearly permeable stretching/shrinking sheet in a nanofluid. Sci. Rep. **4** (2014)

Chapter 14
Effect of Time-Periodic Boundary Temperature Modulations on the Onset of Convection in a Maxwell Fluid–Nanofluid Saturated Porous Layer

Jawali C. Umavathi, Kuppalapalle Vajravelu, Prashant G. Metri and Sergei Silvestrov

Abstract The linear stability of Maxwell fluid–nanofluid flow in a saturated porous layer is examined theoretically when the walls of the porous layers are subjected to time-periodic temperature modulations. A modified Darcy–Maxwell model is used to describe the fluid motion, and the nanofluid model used includes the effects of the Brownian motion. The thermal conductivity and viscosity are considered to be dependent on the nanoparticle volume fraction. A perturbation method based on a small amplitude of an applied temperature field is used to compute the critical value of the Rayleigh number and the wave number. The stability of the system characterized by a critical Rayleigh number is calculated as a function of the relaxation parameter, the concentration Rayleigh number, the porosity parameter, the Lewis number, the heat capacity ratio, the Vadász number, the viscosity parameter, the conductivity variation parameter, and the frequency of modulation. Three types of temperature modulations are considered, and the effects of all three types of modulations are found to destabilize the system as compared to the unmodulated system.

J.C. Umavathi (✉)
Department of Mathematics, Gulbarga University, Gulbarga, Karnataka, India
e-mail: drumavathi@rediffmail.com

J.C. Umavathi
Department of Engineering, University of Sannio, Benevento, Italy

K. Vajravelu
Department of Mathematics, University of Central Florida, Orlando, FL 32816, USA
e-mail: kuppalapalle.vajravelu@ucf.edu

K. Vajravelu
Department of Mechanical, Material and Aerospace Engineering, University of Central Florida, Orlando, FL 32816, USA

P.G. Metri · S. Silvestrov
Division of Applied Mathematics, The School of Education, Culture and Communication, Mälardalen University, Box 883, 721 23 Västerås, Sweden
e-mail: prashant.g.metri@mdh.se

S. Silvestrov
e-mail: sergei.silvestrov@mdh.se

S. Silvestrov and M. Rančić (eds.), *Engineering Mathematics I*,
Springer Proceedings in Mathematics & Statistics 178,
DOI 10.1007/978-3-319-42082-0_14

Keywords Thermal modulation · Nanofluid · Galerkin method · Stability analysis

14.1 Introduction

Heat transfer enhancement in the base flow of fluid dispersion of nanoscale particles was reported by Masuda et al. [16]. The presence of nanoparticles in the fluid significantly increases the effective thermal conductivity of the mixture. The term nanofluid was coined by Choi [5] to refer to a fluid containing a dispersion of nanoparticles. These enhanced properties and behavior imply an enormous potential of nanofluids for device miniaturization and process intensification which could have impacts on many industrial sectors including chemical processing, transportation, electronics, medicine, energy, and the environment (see for details Chen et al. [4]). Several attempts were made to explain abnormal increases in the thermal conductivity and viscosity of nanofluids (Buongiorno [3], Vadász [34, 35]). However, a satisfactory explanation has yet to be found as emphasized by Eastman et al. [7] in their recent comprehensive review of the nanofluid literature. On the other hand, Buongiorno [3] focused on heat transfer enhancement of nanofluids in convective situations. He focused on the further heat transfer enhancement observed in convective situations: Buongiorno noted that the observation of convective heat transfer enhancement by several researchers could be due to the dispersion of the suspended nanoparticles, but he argued that this effect is too small to explain the observed enhancement. Also, Buongiorno noted that the absolute velocity of a nanoparticle could be viewed as the sum of the base fluid velocity and a relative velocity (that he called the slip velocity). He considered, in turn, seven slip mechanisms: inertia, Brownian diffusion, thermophoresis, diffusiophoresis, Magnus effect, fluid drainage, and gravity settling. After examining each of these effects, he concluded that in the absence of turbulence, the effects of the Brownian diffusion and the thermophoresis are important. Based on these two effects, Buongiorno formulated the conservation equations.

The Bénard problem (the onset of convection in a horizontal layer uniformly heated from below) for a nanofluid was studied by Tzou [32] on the basis of the transport equations of Buongiorno [3]. The corresponding problem for flow in a porous medium (the Horton–Rogers–Lapwood problem) was studied by Nield and Kuznetsov [21] using the Darcy model.

An alternative approach is to ignore special phenomena such as Brownian motion and thermophoresis but instead examine the effect of the variation of thermal conductivity and viscosity with the nanofluid particle fraction, using expressions used in the theory of mixtures. This approach was employed by Tiwari and Das [31] to study the cross-diffusion effects. It is assumed that the nanofluid is diluted so that the nanofluid volume fraction is small compared with unity. Then they assumed that the volume fraction is a linear function of the vertical coordinate. The vertical heterogeneity (especially the case of horizontal layers) was studied by McKibbin and O'Sullivan [18] and Leong and Lai [13]; and horizontal heterogeneity was studied by Nield [19], and Gounat and Caltagirone [10]. More general aspects of conductivity heterogeneity were discussed by Braester and Vadász [2], and Rees and Riley [23]. Simmons et al. [28] have pointed out that in many heterogeneous geological systems, hydraulic properties such as the hydraulic conductivity of the system under consideration can vary by many orders of magnitude and sometimes rapidly over small spatial scales. They also pointed out that the onset of instability is controlled by very local conditions in the vicinity of the evolving boundary layer and not by the global layer properties or indeed some average property of that macroscopic layer. They also pointed out that any averaging process would remove the very structural controls and physics that are expected to be important in controlling the onset, growth, and/or decay of instabilities in a highly heterogeneous system for the general case involving both vertical heterogeneity and horizontal heterogeneity. For this complicated situation no exact analytical solution can be expected to exist, but it is reasonable to seek an approximate analytical solution, based on the expectation that for weak heterogeneity, the solution would not differ dramatically from the solution for the homogeneous case. Following this approach, an extension of the Galerkin approximate method has been widely employed (see, for example, Finlayson [9]). In the context of the onset of convection, the commonly used Galerkin method involves trial functions of the vertical coordinate only. Thus, to a first approximation, the thermal conductivity and the viscosity can be taken as weak functions of the vertical coordinate. This means that we can treat the problem as one involving a weakly heterogeneous porous medium (Nield [20]).

Many working fluids of practical interest are viscoelastic rather than Newtonian. For this reason, current interest in this area is concerned with studies of the various viscoelastic models such as Maxwell fluids (Sokolov and Tanner [29]), Oldroyd type models (Khayat [12], Siddheshwar et al. [27]), Rivlin–Ericksen fluids (Siddheshwar and Srikrishna [26]), and Walters-B liquids (El-Sayed [8]). Analogous studies on viscoelastic fluid convection in porous media are those by Shekar and Jayalatha [24], Tan and Masuoka [30], and Shivakumara et al. [25].

Recently, Wang and Tan [36] have made a stability analysis of double diffusive convection of Maxwell fluid in a porous medium. It is worthwhile to point out that the first viscoelastic rate type model, which is still used widely, is due to Maxwell [17]. While Maxwell did not develop this model for polymeric liquids, he recognized that such fluid has a means for storing energy characterizing its viscous nature. Recently, Malashetty et al. [15] have studied double diffusive convection in a viscoelastic fluid saturated porous layer using the Oldroyd model. Very recently, Awad et al. [1] used the Darcy–Brinkman–Maxwell model to study linear stability analysis of a Maxwell fluid with cross-diffusion and double-diffusive convection.

Nonetheless, the studies related to the effects of thermal modulation on the onset of convection in a viscoelastic fluid-saturated porous medium have not received much attention. Chung Liu [6] has examined the stability of a horizontally extended second-grade fluid layer heated from below subject to temperature modulation at walls.

Motivated by the above studies, in the present paper, we study the effect of thermal modulation on the onset of convection in a Maxwell fluid and nanofluid saturated porous medium. The boundary temperature modulation alters the basic temperature distribution from linear to nonlinear which helps in effective control of convective instability. The difficulty in dealing with such instability problems is that one has to solve time-dependent stability equations with variable coefficients, and to our knowledge no work has been initiated for such fluids in this direction. The resulting eigenvalue problem is solved by a perturbation technique with amplitude of the temperature modulation as a perturbation parameter. In particular, it is shown that the onset of convection can be advanced by a proper tuning of the frequency of the boundary temperature modulation.

14.2 Mathematical Formulation

We consider an infinite horizontal porous layer saturated with a nanofluid, confined between the planes $z^* = 0$ and $z^* = H$, with the vertically downward gravity force acting on it. A Cartesian frame of reference is chosen with the origin in the lower boundary and the z-axis vertically upwards. The Boussinesq approximation, which states that the variation in density is negligible everywhere in the conservation except in the buoyancy term, is assumed to hold. The conservation equations take the form

$$\nabla^* \cdot v_D^* = 0. \tag{14.1}$$

Here v_D^* is the nanofluid Darcy velocity and $v_D^* = (u^*, v^*, w^*)$.

The conservation equation for the nanoparticles, in the absence of thermophoresis and chemical reactions, takes the form

$$\frac{\partial \phi^*}{\partial t^*} + \frac{1}{\varepsilon} v_D^* \cdot \nabla \phi^* = \nabla^* \cdot [D_B \nabla^* \phi^*], \tag{14.2}$$

where ϕ^* is the nanoparticle volume fraction, ε is the porosity, and D_B is the Brownian diffusion coefficient. We use the Darcy model for a porous medium. Hence, the momentum equation can be written as

$$\left(1 + \tilde{\lambda}\frac{\partial}{\partial t^*}\right)\frac{\rho}{\varepsilon}\frac{\partial v_D^*}{\partial t^*} = \left(1 + \tilde{\lambda}\frac{\partial}{\partial t^*}\right)(-\nabla^* p^* + \rho g) - \frac{\mu_{eff}}{K} v_D^*. \tag{14.3}$$

Here ρ is the overall density of the nanofluid, which we assume to be given by

$$\rho = \phi^* \rho_p + (1 - \phi^*)\rho_0[1 - \beta_T(T^* - T_0^*)], \tag{14.4}$$

where ρ_p is the particle density, ρ_0 is a reference density for the fluid, and β_T is the thermal volumetric expansion. The thermal energy equation for a nanofluid can be written as

$$(\rho c)_m \frac{\partial T^*}{\partial t^*} + (\rho c)_f v_D^* \cdot \nabla^* T^* = k_m \nabla^{*2} T^* + \varepsilon(\rho c)_p [D_B \nabla^* \phi^* \cdot \nabla T^*]. \tag{14.5}$$

The conservation of nanoparticle mass requires that

$$\frac{\partial \phi^*}{\partial t^*} + \frac{1}{\varepsilon} v_D^* \cdot \nabla^* \phi^* = D_p \nabla^{*2} \phi^*. \tag{14.6}$$

Here c is the fluid specific heat (at constant pressure), k_m is the overall thermal conductivity of the porous medium saturated by the nanofluid, and c_p is the nanoparticle specific heat of the material constituting the nanoparticles (following Nield and Kuznetsov [22]). Thus,

$$k_m = \varepsilon k_{eff} + (1-\varepsilon)k_s, \tag{14.7}$$

where k_{eff} is the effective conductivity of the nanofluid (fluid plus nanoparticles) and k_s is the conductivity of the solid material forming the matrix of the porous medium.

We now introduce the viscosity and the conductivity dependence on nanoparticle fraction. Following Tiwari and Das [31], we adopt the formulas, based on a theory of mixtures,

$$\frac{\mu_{eff}}{\mu_f} = \frac{1}{(1-\phi^*)^{2.5}}, \tag{14.8}$$

$$\frac{k_{eff}}{k_f} = \frac{(k_p + 2k_f) - 2\phi^*(k_f - k_p)}{(k_p + 2k_f) + \phi^*(k_f - k_p)}. \tag{14.9}$$

Here k_f and k_p are the thermal conductivities of the fluid and the nanoparticles, respectively. In the case where ϕ^* is small compared with unity, we can approximate these formulas by

$$\frac{\mu_{eff}}{\mu_f} = 1 + 2.5\phi^*, \tag{14.10}$$

$$\frac{k_{eff}}{k_f} = \frac{(k_p + 2k_f) - 2\phi^*(k_f - k_p)}{(k_p + 2k_f) + \phi^*(k_f - k_p)} = 1 + 3\phi^* \frac{(k_p - k_f)}{(k_p + 2k_f)}. \tag{14.11}$$

We assume that the volumetric fractions of the nanoparticles are constant on the boundaries. Thus, the boundary conditions are

$$w^* = 0, \quad \phi^* = \phi_0^* \quad \text{at} \quad z^* = 0, \tag{14.12}$$

$$w^* = 0, \quad \phi^* = \phi_1^* \quad \text{at} \quad z^* = H. \tag{14.13}$$

For thermal modulation, the external driving force is modulated harmonically in time by varying the temperature of the lower and upper horizontal boundary. Accordingly, we take

$$T(z,t) = T_0 + \frac{\triangle T}{2}[1 + \varepsilon_1 cos(\Omega t)] \quad \text{at} \quad z^* = 0, \tag{14.14}$$

$$T(z,t) = T_0 - \frac{\Delta T}{2}[1 - \varepsilon_1 cos(\Omega t + \phi)] \quad \text{at} \quad z^* = H, \tag{14.15}$$

where ε_1 represents a small amplitude of modulation (which is used as a perturbation parameter to solve the problem), Ω the frequency of modulation, and ϕ the phase angle. We consider three types of modulation, viz.,
Case (a): Symmetric (in phase, $\phi = 0$),
Case (b): Asymmetric (out of phase, $\phi = \pi$), and
Case (c): Only lower wall temperature is modulated while the upper one is held at constant temperature ($\phi = -i\infty$).

14.3 Basic State Problem

The basic state of the fluid is quiescent and is given by

$$\rho_b \vec{g} + \nabla p_b = 0, \tag{14.16}$$

$$(\rho c)_m \frac{\partial T_b^*}{\partial t^*} = k_m \nabla^2 T^*, \tag{14.17}$$

$$\frac{d^2 \phi_b^*}{dz^2} = 0. \tag{14.18}$$

The solution of (14.17) satisfying the thermal conditions as given in (14.14) and (14.15) is $T_b = T_1(z) + \varepsilon_t T_2(z,t)$ where

$$T_1(z) = T_R + \frac{\Delta T}{2}\left(1 - \frac{2z}{H}\right), \tag{14.19}$$

$$T_2(z,t) = Re[\{b(\lambda)e^{\frac{\lambda z}{H}} + b(-\lambda)e^{\frac{-\lambda z}{H}}\}e^{-i\omega t}], \tag{14.20}$$

with

$$\lambda = (1-i)\left(\frac{(\rho c)_m \omega H^2}{2k_m}\right), \quad b(\lambda) = \frac{\Delta T}{2}\left(\frac{e^{-i\phi} - e^{-\lambda}}{e^{\lambda} - e^{-\lambda}}\right), \tag{14.21}$$

and *Re* stands for real part. We do not record the expressions of p_b and ρ_b as these are not explicitly required in the remaining part of the paper.

14.4 Linear Stability Analysis

Let the basic state be distributed by an infinitesimal perturbation. We now have,

$$v = v', \quad p = p_b + p', \quad T = T_b + T', \quad \phi = \phi_b + \phi', \tag{14.22}$$

where a prime indicates that the quantities are infinitesimal perturbations. Substituting (14.22) into (14.1)–(14.7) and linearizing by neglecting products of primed quantities, we have,

$$(1 + \lambda_1 s)(\nabla p - RT\widehat{e_z} + Rn\phi\widehat{e_z} + \gamma_a s v) + \tilde{\mu} v = 0, \tag{14.23}$$

$$\frac{\partial T'}{\partial t} + w' \frac{\partial T_b}{\partial z} = \tilde{k}\frac{\partial^2 T}{\partial z^2} + \frac{N_B}{Le}\left(\frac{\partial T_b}{\partial z} + \frac{\partial T'}{\partial z} + \frac{\partial \phi'}{\partial z}\frac{\partial T_b}{\partial z}\right), \tag{14.24}$$

$$\frac{1}{\sigma}\frac{\partial \phi'}{\partial t} + \frac{1}{\varepsilon} w' = \frac{1}{Le}\nabla^2 \phi', \tag{14.25}$$

$$w' = 0, \quad T' = 0, \quad \phi' = 0 \quad \text{at} \quad z = 0, 1. \tag{14.26}$$

We introduce the following transformations:

$$(x, y, z) = \frac{(x^*, y^*, z^*)}{H}, \quad t = \frac{t^* \alpha_m}{\sigma H^2}, \quad (u, v, w) = \frac{(u^*, v^*, w^*)H}{\alpha_m}, \quad p = \frac{p^* K}{\mu_f \alpha_m},$$
$$\phi = \frac{\phi^* - \phi_0^*}{\phi_1^* - \phi_0^*}, \quad T = \frac{T^* - T_c^*}{T_h^* - T_c^*}, \quad \omega = \frac{\sigma \Omega H^2}{\alpha_m}, \quad s = \frac{\partial}{\partial t},$$

with

$$\alpha_m = \frac{k_m}{(\rho c_p)_f}, \quad \sigma = \frac{(\sigma c_p)_m}{(\rho c_p)_f}, \quad \tilde{\mu} = \frac{\mu_{eff}}{\mu_f}, \quad \tilde{k_p} = \frac{k_p}{k_f}, \quad \tilde{k_s} = \frac{k_s}{k_f}, \quad \tilde{k} = \frac{k_m}{k_s}.$$

The dimensionless parameters that appear are these:

- $Pr = \frac{\mu_f}{\rho \alpha_m}$ - the Prandtl number,
- $Da = \frac{K}{H^2}$ - the Darcy number,
- $Va = \frac{\varepsilon^2 Pr}{Da}$ - the Vadász number,
- $\lambda_1 = \frac{\tilde{\lambda} \alpha_m}{\sigma H^2}$ - the relaxation parameter (also known as the Deborah number),
- $\gamma_a = \frac{\varepsilon}{\sigma Va}$ - the acceleration coefficient,
- $Le = \frac{\alpha_m}{D_m}$ - the nanofluid Lewis number,
- $R = \frac{R_0 g K (1 - \phi_0^*) \beta_T \Delta T^* H}{\mu_f \alpha_m}$ - the nanoparticle Rayleigh number, and
- $N_B = \frac{(\rho c_p)_p}{(\rho c)_f}(\phi_1^* - \phi_0^*)$ - modified particle-density increment.

In deriving (14.23), the term proportional to the product of ϕ and T (Oberbeck–Boussinesq approximation) is neglected. This assumption is likely to be valid in the case of small temperature gradients in a dilute suspension of nanoparticles: For regular fluid the parameters Rn and N_B are zeros.

We eliminate pressure by operating on (14.23) with $\widehat{e_z}$ curl curl and using the identity *curl curl* $\equiv$ *grad div* $-\nabla^2$ results in

$$[(1+\lambda_1 s)s\gamma_a+\tilde{\mu}]\,\nabla^2 w' = (1+\lambda_1 s)[R\,\nabla_H^2 - Rn\,\nabla_H^2\,\phi']. \tag{14.27}$$

Here ∇_H^2 is the two-dimensional Laplacian operator on the horizontal plane. By combining the (14.24)–(14.26), we obtain the equations for the vertical component of velocity w in the form (dropping prime)

$$\begin{aligned}&\left[\frac{\partial}{\partial t}-\nabla^2\gamma\right]\left[\frac{1}{\sigma}\frac{\partial}{\partial t}-\frac{\nabla^2}{Le}\right][\nu+s\gamma_a(1+\lambda_1 s)]\,\nabla^2 w -\\ &\qquad -\frac{(1+\lambda_1 s)Rn}{\varepsilon}\left[\frac{\partial}{\partial t}-\nabla^2\gamma\right]\nabla_1^2 w +\\ &\qquad +(1+\lambda_1 s)R\frac{\partial T_b}{\partial z}\left[\frac{1}{\sigma}\frac{\partial}{\partial t}-\frac{\nabla^2}{Le}\right]\nabla_1^2 w = 0,\end{aligned} \tag{14.28}$$

where, $\nu = 1+1.25(\phi_1^*+\phi_0^*)$, and $\eta = \varepsilon+(1-\varepsilon)\tilde{k}_s+\frac{3(\phi_1^*+\phi_0^*)\varepsilon}{2}\left(\frac{\tilde{k}_p-1}{\tilde{k}_p+2}\right)$.

It is worth noting that the factor ν comes from the mean value of $\tilde{\mu}(z)$ over the range [0, 1], and the factor η is the mean value of $\tilde{k}(z)$ over the same range. That means that when evaluating the critical Rayleigh number, it is a good approximation to base that number on the mean values of the viscosity and conductivity based in turn on the basic solution for the nanofluid fraction (following Nield and Kuznetsov [22]).

The boundary condition (14.26) is applied to (14.27) resulting in the following boundary condition for w:

$$w = \frac{d^2 w}{dz^2} = 0 \quad \text{at} \quad z = 0, 1. \tag{14.29}$$

Using (14.19), the dimensionless temperature gradient appearing in (14.24) may be written as

$$\frac{\partial T_b}{\partial z} = -1+\varepsilon f, \tag{14.30}$$

where

$$f = Re\left[A(\lambda)e^{\lambda z}+A(-\lambda)e^{-\lambda z}e^{-i\omega}t\right], \text{ for} \tag{14.31}$$

$$A(\lambda) = \frac{\lambda}{2}\left(\frac{e^{-i\varphi}-e^{-\lambda}}{e^{\lambda}-e^{-\lambda}}\right), \text{ and } \lambda = (1-i)\left(\frac{\sigma\omega}{2}\right)^{\frac{1}{2}}.$$

14.5 Method of Solution

We seek the eigenfunctions w and eigenvalues Ra of (14.28) for the basic temperature gradient given by (14.30) that departs from the linear profile $\frac{\partial T_b}{\partial z} = -1$ by quantities of order ε_1. We therefore assume the solution of (14.28) is in the form

$$(w, R) = (w_0, R_0) + \varepsilon_1(w_1, R_1) + \varepsilon_1^2(w_2, R_2) + \dots . \tag{14.32}$$

Substituting (14.32) into (14.28) and equating the coefficients of various powers of ε_t on either side of the resulting equation, we obtain the following system of equations up to the order of ε_t^2:

$$Lw_0 = 0, \tag{14.33}$$

$$Lw_1 = (1 + \lambda_1 s)\left[\left(\frac{R_0 \omega G}{\sigma}\nabla_1^2 + \frac{R_0 f}{Le}\right)\nabla_1^2 - \frac{R_1}{Le}\nabla^2\nabla_1^2\right] w_0, \tag{14.34}$$

$$Lw_2 = (1 + \lambda_1 s)\left[R_0\left(\frac{\omega G}{\sigma} + \frac{f}{Le}\nabla^2\right) - \frac{R_1}{Le}\nabla^2\right]\nabla_1^2 w_1 + \\ +(1 + \lambda_1 s)R_1\left(\frac{\omega G}{\sigma} + \frac{f}{Le}\nabla^2 + \frac{R_2}{Le}\nabla^2\right)\nabla_1^2 w_0, \tag{14.35}$$

where

$$L = \left(1 + \lambda\frac{\partial}{\partial t}\right)\left(\frac{\partial}{\partial t} - \nabla^2\gamma\right)\left(\frac{1}{\sigma}\frac{\partial}{\partial t} - \frac{\nabla^2}{Le}\right)\left(\nu + \gamma_a\frac{\partial}{\partial t}\right)\nabla^2 - \\ -\frac{Rn}{\varepsilon}\left(\frac{\partial}{\partial t} - \nabla^2\gamma\right)\nabla_1^2 + \frac{R_0}{Le}\nabla^2\nabla_1^2,$$

and w_0, w_1, w_2 are required to satisfy the boundary condition in (14.29).

We now assume the solutions for (14.33) are of the form $w_0 = w_0(z)exp[i(lx + my)]$ where $w_0(z) = w_0^n(z) = sin(n\pi z), n = 1, 2, 3 \dots$ and l, m are the wave numbers in the xy plane such that $l^2 + m^2 = \alpha^2$. The corresponding eigenvalues are given by

$$R_0 = \frac{(n^2\pi^2 + \alpha^2)^2\nu\gamma}{\alpha^2} - \frac{RnLe\gamma}{\varepsilon}. \tag{14.36}$$

For a fixed value of the wave number α, the least eigenvalue occurs at $n = 1$ and is given by

$$R_0 = \frac{(\pi^2 + \alpha^2)^2\nu\gamma}{\alpha^2} - \frac{RnLe\gamma}{\varepsilon}, \tag{14.37}$$

and R_{0c} assumes the minimum value

$$R_{0c} = 4\pi^2 \nu\gamma - \frac{RnLe\gamma}{\varepsilon}. \tag{14.38}$$

These are the values reported by Horton and Rogers [11] in the absence of concentration Rayleigh number Rn.

The equation for w_1 then takes the form

$$Lw_1 = R_0\alpha^2(1 - \lambda_1 i\omega)\left(\frac{\omega}{\sigma}G + \frac{(D^2 - \alpha^2)f}{Le}\right)\sin\pi z, \tag{14.39}$$

where $D = \frac{d}{dz}$ and $G = I.P.[\{A(\lambda)e^{\lambda z}\} + \{A(-\lambda)e^{-\lambda z}\}e^{-i\omega t}]$. Thus,

$$D^2 f \sin\pi z = (\lambda^2 - \pi^2)f\sin\pi z + 2\lambda\pi f' \cos\pi z \tag{14.40}$$

with $f' = R.P.[\{A(\lambda)e^{\lambda z}\} + \{A(-\lambda)e^{-\lambda z}\}e^{-i\omega t}]$.

Using (14.40), (14.39) becomes

$$Lw_1 = R_0\alpha^2(-1 + \lambda_1 i\omega)\left(\frac{\omega}{\sigma}G\sin\pi z - L_1 f\sin\pi z + \frac{2\lambda\pi f'}{Le}\cos\pi z\right), \tag{14.41}$$

where $L_1 = \frac{i\omega+\pi^2+\alpha^2}{Le}$.

We solve (14.41) for w_1 by expanding the right hand side of it in Fourier series expansion and inverting the operator L for this we need the following Fourier series expansions

$$g_{nm}(\lambda) = 2\int_0^1 e^{\lambda z}\sin(m\pi z)\sin(n\pi z)dz = \frac{-4nm\pi^2\lambda[1 + (-1)^{n+m+1}e^z]}{[\lambda^2 + (n+m)^2\pi^2][\lambda^2 + (n-m)^2\pi^2]}, \tag{14.42}$$

$$f_{nm}(\lambda) = 2\int_0^1 e^{\lambda z}\cos(m\pi z)\cos(n\pi z)dz = \frac{2\lambda[\lambda^2 + (n+m)^2\pi^2][1 + (-1)^{n+m+1}e^z]}{[\lambda^2 + (n+m)^2\pi^2][\lambda^2 + (n-m)^2\pi^2]}, \tag{14.43}$$

where

$$e^{\lambda z}\sin(m\pi z) = \sum_{n=1}^{\infty} g_{nm}(\lambda)\sin(n\pi z), \tag{14.44}$$

$$e^{\lambda z}\cos(m\pi z) = \sum_{n=1}^{\infty} f_{nm}(\lambda)\cos(n\pi z). \tag{14.45}$$

Now,

$$L(\omega, n) = A + i\omega B, \tag{14.46}$$

where

$$A = \left[\omega^2\gamma_a(n^2\pi^2+\alpha^2)^2\left(\frac{1}{Le}+\frac{\gamma}{\sigma}\right)(1+\lambda_1\nu) + \frac{\omega^2}{\sigma}(n^2\pi^2+\alpha^2)(\nu-\lambda_1\omega^2\gamma_a)+\right.$$
$$+(n^2\pi^2+\alpha^2)^3\frac{\gamma}{Le}(-\nu+\lambda_1\omega^2\gamma_a) + \frac{Rn}{\varepsilon}\alpha^2(\gamma(n^2\pi^2+\alpha^2)-\lambda_1\omega^2)+$$
$$\left.+\left(4\pi^2\nu\gamma - \frac{RnLe\gamma}{\varepsilon}\right)\frac{\alpha^2}{Le}(n^2\pi^2+\alpha^2)\right],$$

$$B = \left[(n^2\pi^2+\alpha^2)^2\left(\frac{1}{Le}+\frac{\gamma}{\sigma}\right)(\nu-\lambda_1\gamma_a\omega^2) + \frac{\omega^2}{\sigma}(n^2\pi^2+\alpha^2)(-\gamma_a-\lambda_1\nu)+\right.$$
$$+(n^2\pi^2+\alpha^2)^3\frac{\gamma}{Le}(\gamma_a+\lambda_1\nu) + \frac{Rn}{\varepsilon}\alpha^2(-1+\gamma\lambda_1(n^2\pi^2+\alpha^2)) -$$
$$\left.-\left(4\pi^2\nu\gamma - \frac{RnLe\gamma}{\varepsilon}\right)\frac{\alpha^2\lambda_1}{Le}(n^2\pi^2+\alpha^2)\right].$$

It is easily seen that:

$$L\left[\sin(n\pi z)e^{-i\omega t}\right] = L(\omega, n)\sin(n\pi z)e^{i\omega t},$$

and

$$L\left[\cos(n\pi z)e^{-i\omega t}\right] = L(\omega, n)\cos(n\pi z)e^{i\omega t},$$

and (14.41) now becomes

$$Lw_1 = (-1+\lambda_1 i\omega)\alpha^2 R_0\left[\frac{\omega}{\sigma}I.P.\sum_{n=1}^{\infty}A_n(\lambda)\sin n\pi z e^{i\omega t} - \right. \tag{14.47}$$
$$\left.-L_1 R.P.\sum_{n=1}^{\infty}A_n(\lambda)\sin n\pi z e^{i\omega t} + \frac{2\lambda\pi}{Le}R.P.\sum_{n=1}^{\infty}B_n(\lambda)\cos n\pi z e^{i\omega t}\right],$$

$$Lw_1 = (-1+\lambda_1 i\omega)\alpha^2 R_0\left[\frac{\omega}{\sigma}I.P.\sum_{n=1}^{\infty}\frac{A_n(\lambda)}{L(\omega, n)}\sin n\pi z e^{i\omega t} - \right. \tag{14.48}$$
$$-L_1 R.P.\sum_{n=1}^{\infty}\frac{A_n(\lambda)}{L(\omega, n)}\sin n\pi z e^{i\omega t} +$$

$$+ \frac{2\lambda\pi}{Le} R.P. \sum_{n=1}^{\infty} \frac{B_n(\lambda)}{L(\omega, n)} \cos n\pi z e^{i\omega t} \Bigg],$$

where $A_n = A(\lambda)g_{n1}(\lambda) + A(-\lambda)g_{n1}(-\lambda)$, and $B_n = A(\lambda)f_{n1}(\lambda) + A(-\lambda)$ $f_{n1}(-\lambda)$.

To simplify (14.34) for w_2 we need

$$Lw_2 = (1 + \lambda_1 s)\left[R_0\left(\frac{\omega G}{\sigma} + \frac{\nabla^2 f}{Le}\right)\nabla_1^2 w_1 - R_2\frac{\nabla^2}{Le} \cdot \nabla_1^2 w_1\right]. \tag{14.49}$$

The equation for then can be written as

$$Lw_2 = (1 - \lambda_1 i\omega)\left[R_0\left(\frac{\omega G}{\sigma} - L_n f\right)w_1 + \frac{2DfDw_1}{Le}\right] - R_2\frac{\alpha^2}{Le}(\pi^2 + \alpha^2), \tag{14.50}$$

where $L_n = \frac{i\omega + n^2\pi^2 + \alpha^2}{Le}$.

We shall not require the solution of this equation but merely use it to determine R_2. The solvability condition requires that the time-independent part of the right hand side of (14.50) must be orthogonal to $\sin(\pi z)$. Multiplying equation (14.50) by $\sin(\pi z)$ and integrating between 0 and 1 we obtain

$$R_2 = \frac{2LeR_0(1 - 2i\lambda\omega)}{\nabla^2} \int_0^1 \left(\overline{\frac{\nabla^2 f}{Le} \frac{\omega G}{\sigma}}\right) w_1 \sin(\pi z) dz, \tag{14.51}$$

where an upper bar denotes the time average.

We have the Fourier series expansions

$$f \sin \pi z = R.P. \sum A_n(\lambda) \sin n\pi z e^{i\omega t}, \tag{14.52}$$
$$Df \sin \pi z = R.P. \sum \lambda C_n(\lambda) sinn\pi z e^{i\omega t},$$

where $C_n(\lambda) = A(\lambda)g_{n1}(\lambda) - A(-\lambda)g_{n1}(-\lambda)$.

Using (14.52) in (14.51) we obtain

$$R_2 = \frac{LeR_0^2\alpha^2}{2(\pi^2 + \alpha^2)} \cdot \tag{14.53}$$

$$\left[\left(-\frac{\omega^2}{\sigma^2} - \overline{L_n}L_1\right) R.P. \sum \frac{|A_n|^2}{|L(\omega, n)|^2} L^*(\omega, n)(1 - 2i\lambda_1\omega)(-1 + i\lambda_1\omega) + \right]$$
$$+ \left[\frac{4n\pi^2\lambda_1}{Le^2} R.P. \sum \overline{\lambda_1 C_n} \frac{B_n}{|L(\omega, n)|^2} L^*(\omega, n)(1 - 2i\lambda_1\omega)(-1 + i\lambda_1\omega)\right],$$

where $L^*(\omega, n)$ is the complex conjugate of $L(\omega, n)$, and

$$| A_n(\lambda) |^2 = \frac{16n^2\pi^4\omega^2}{(\omega^2 + (n+1)^4\pi^4)(\omega^2 + (n-1)^4\pi^4)}.$$

The critical value of R_2, denoted by R_{2c}, is obtained at the wave number given by equation $\alpha_c = \pi$ for the following three different cases:

1. When the oscillating temperature field is symmetric so that the wall temperatures are modulated in phase (with $\phi = 0$).
2. When the wall temperature field is antisymmetric corresponding to out-of-phase modulation (with $\phi = \pi$).
3. When only the temperature of the bottom wall is modulated, the upper wall being held at a constant temperature (with $\phi = -i\infty$).

14.6 Results and Discussion

The effect of thermal modulation on the onset of convection in a layer of Maxwell fluid and nanofluid saturated porous medium is investigated using linear stability analysis. A perturbation technique with amplitude of the modulating temperature as a perturbation parameter is used to find the critical thermal Rayleigh number as a function of frequency of the modulation, relaxation parameter, concentration Rayleigh number, porosity parameter, Lewis number, heat capacity ratio, Vadász number, conductivity, and viscosity variation parameters. The sign of R_{2c} characterizes the stabilizing or destabilizing effects of modulation. A positive R_{2c} indicates that the modulation effect is to stabilize the flow: while a negative R_{2c} indicates the effect is to destabilize, compared to the system in which modulation is absent. We present below the results for three different wall temperature oscillating mechanisms: They are, symmetric, asymmetric, and lower wall temperature modulation only.

In Figs. 14.1, 14.2, 14.3, 14.4, 14.5, 14.6, 14.7 and 14.8, the variations of critical Rayleigh number R_{2c} with frequency ω for different governing parameters are presented for the case of symmetric temperature modulation. It can be seen from these figures that for small frequencies the critical Rayleigh number R_{2c} is negative indicating the destabilized flow. For moderate and high frequencies, the critical Rayleigh number R_{2c} is positive indicating that the effect of symmetric modulation is to stabilize the system. It can also be seen that as R_{2c} decreases to its minimum value (thus producing maximum destabilization), and then increases to its maximum stabilizing value, and finally decreases to zero as the frequency increases from zero to infinity. That is, in the presence of thermal modulation, convection occurs at lower values of the Rayleigh number compared to the unmodulated system.

Figure 14.1 shows the effect of the relaxation parameter λ_1 on the critical Rayleigh number R_{2c} for fixing the other governing parameters in the case of symmetric modulation. It is seen that an increase in the value of the relaxation parameter increases the

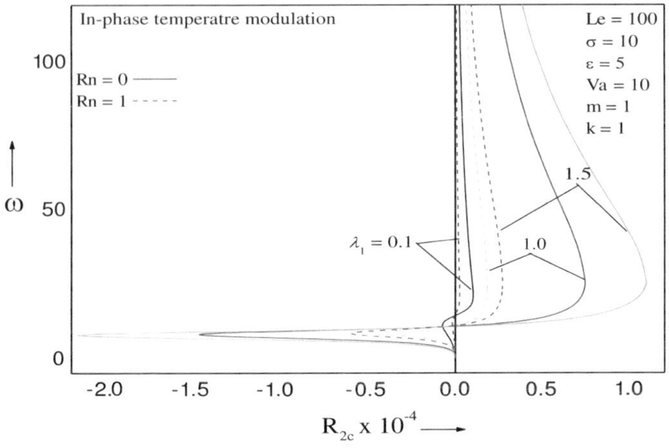

Fig. 14.1 Variation of R_{2c} with ω for different values of λ_1 and Rn

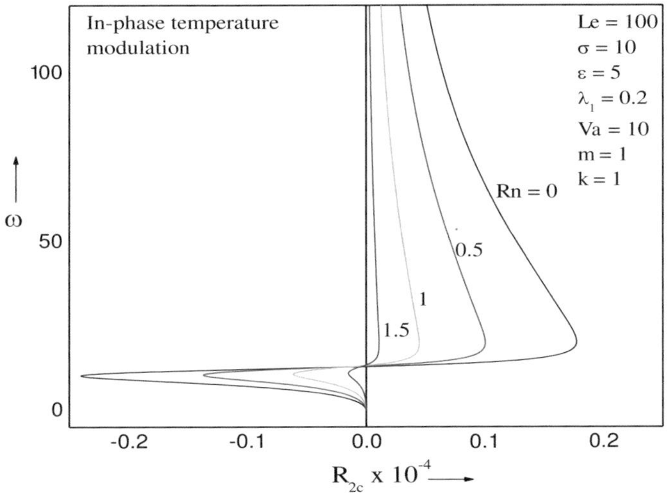

Fig. 14.2 Variation of R_{2c} with ω for different values of Rn

magnitude of R_{2c}. At small frequencies, R_{2c} increases negatively, while R_{2c} increases positively with the relaxation parameter at moderate and high frequencies for both regular and nanofluids. Hence the effect of the relaxation parameter is to destabilize the system for small frequencies while its effect is to stabilize the system for moderate and high frequencies. This agrees well with the results obtained by Malashetty and Begum [14] for a clear fluid. Figure 14.1 also indicates that the peak negative value of R_{2c} increases with an increase in the value of λ_1 which is the result obtained by Shivakumara et al. [25] for a viscoelastic fluid.

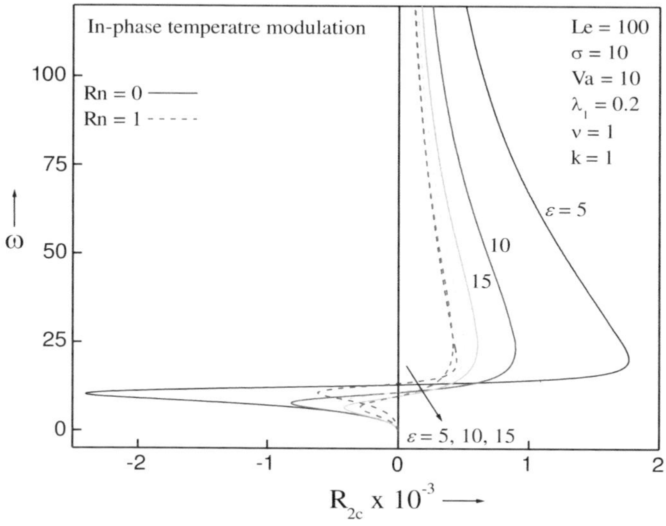

Fig. 14.3 Variation of R_{2c} with ω for different values of ε and Le

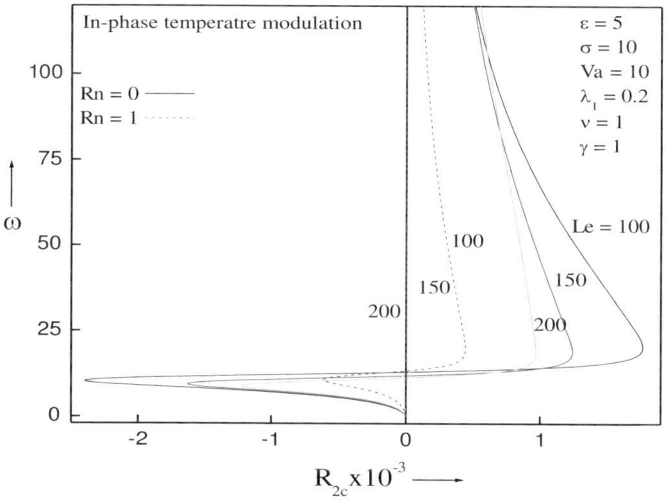

Fig. 14.4 Variation of R_{2c} with ω for different values of Rn and Le

Figure 14.2 shows the variation of R_{2c} with ω for different values of the concentration Rayleigh number Rn : $Rn > 0$ indicates top heavy nanoparticles and $Rn < 0$ indicates bottom heavy nanoparticles. Here also it is observed that for small frequencies, R_{2c} is negative indicating that the symmetric modulation has destabilizing effect while for moderate and large values of frequencies its effect is stabilizing for both regular and nanofluids. This is similar to the observed results of Umavathi [33]. The effect of porosity parameter ε for symmetric modulation is shown in Fig. 14.3. It is

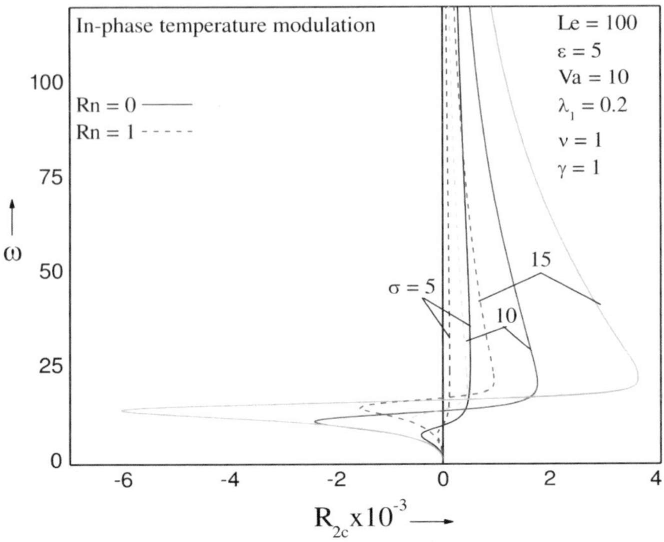

Fig. 14.5 Variation of R_{2c} with ω for different values of Rn and σ

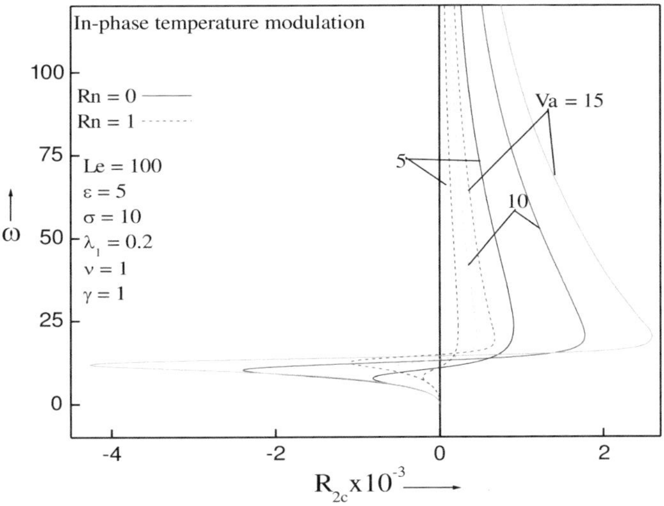

Fig. 14.6 Variation of R_{2c} with ω for different values of Rn and Vadász number Va

observed that as ε increases, the value of $| R_{2c} |$ becomes small indicating that the larger values of ε decrease the effect of modulation. Here also it is observed that as ω increases, R_{2c} increases to its maximum value initially and then starts decreasing with further increase in ω. When ω is very large, all the curves for different porosity ε coalesce and $| R_{2c} |$ approaches to zero. Figure 14.4 depicts the variation of R_{2c} with frequency ω for different values of Lewis number Le. An increase in the value of the

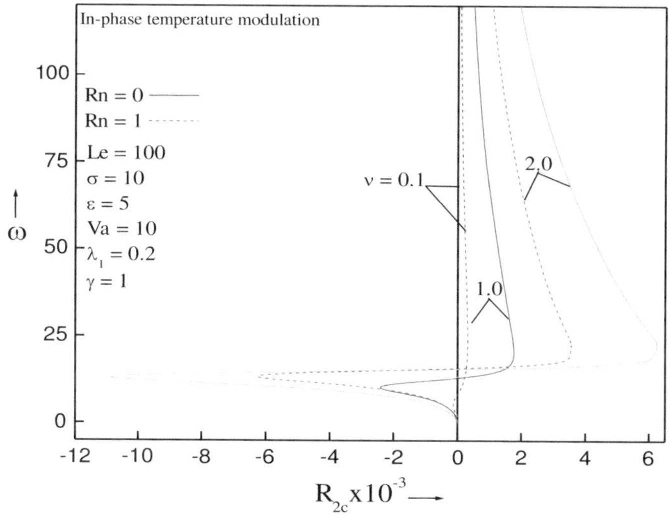

Fig. 14.7 Variation of R_{2c} with ω for different values of Rn and ν

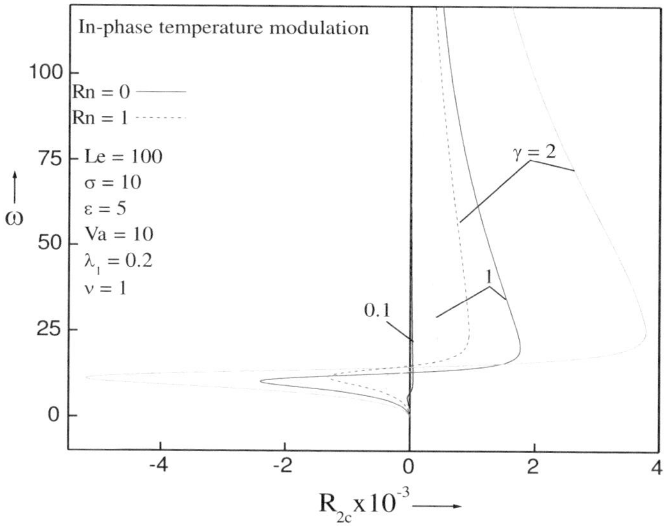

Fig. 14.8 Variation of R_{2c} with ω for different values of Rn and γ

Lewis number decreases the value of $| R_{2c} |$ indicating that the effect of increasing Le is to reduce the effect of thermal modulation for regular and nanofluids. As ω increases, $| R_{2c} |$ increases to its maximum value initially and then decreases with further increase in ω. For large, ω all the curves for different Lewis number coincide, and $| R_{2c} |$ approaches to zero for both regular and nanofluids. The effect of thermal capacity ratio σ and ω is shown in Fig. 14.5. As σ increases, $| R_{2c} |$ decreases for

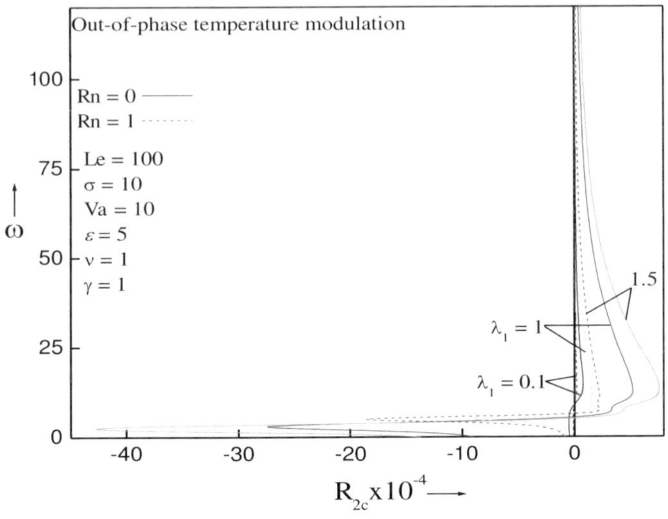

Fig. 14.9 Variation of R_{2c} with ω for different values of Rn and λ_1

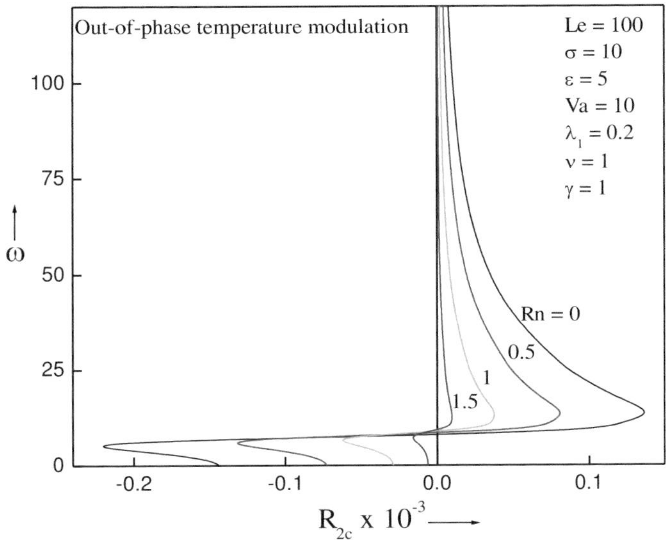

Fig. 14.10 Variation of R_{2c} with ω for different values of Rn

both regular and nanofluids. Here also $| R_{2c} |$ increases to its maximum value initially as ω increases and then starts decreasing with further increase in ω. The effect of Vadász number Va shows a similar nature as that of heat capacity ratio σ as seen in Fig. 14.6. The effects of viscosity variation parameter υ and conductivity variation parameter γ are shown in Figs. 14.7 and 14.8, respectively. As υ and γ increase, $| R_{2c} |$ decreases indicating that the viscosity and conductivity ratio stabilizes the

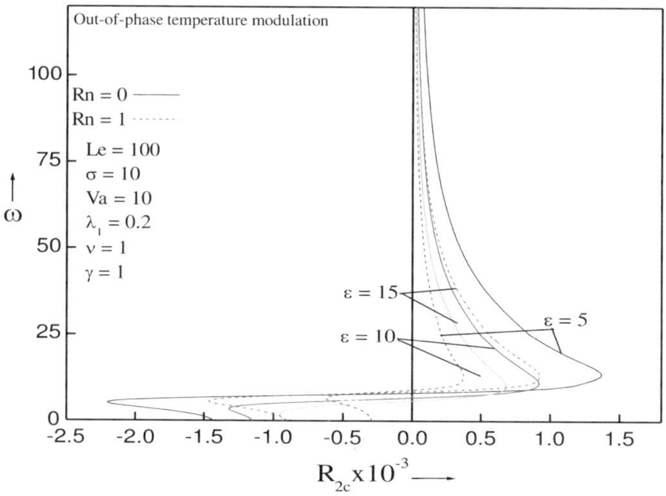

Fig. 14.11 Variation of R_{2c} with ω for different values of Rn and ε

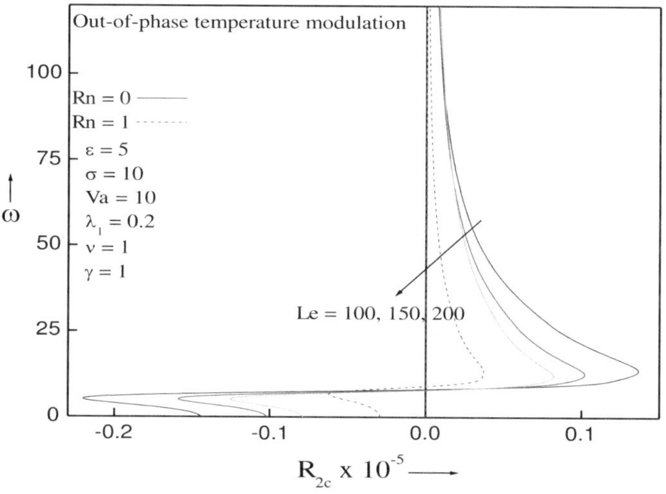

Fig. 14.12 Variation of R_{2c} with ω for different values of Rn and Le

system. As ω increases, $|\ R_{2c}\ |$ increases to its maximum value initially and then starts decreasing with further increase in ω.

The results obtained for the case of asymmetric modulation are presented in Figs. 14.9, 14.10, 14.11, 14.12, 14.13, 14.14, 14.15 and 14.16. All these figures show that for all parameters, small frequencies have destabilizing effects while for moderate and large values of the frequency, their effects are to stabilize the system. It is seen from Fig. 14.9 that an increase in the value of λ_1 increases the magnitude of R_{2c}. The effect of the concentration Rayleigh number Rn, porosity parameter ε, Lewis

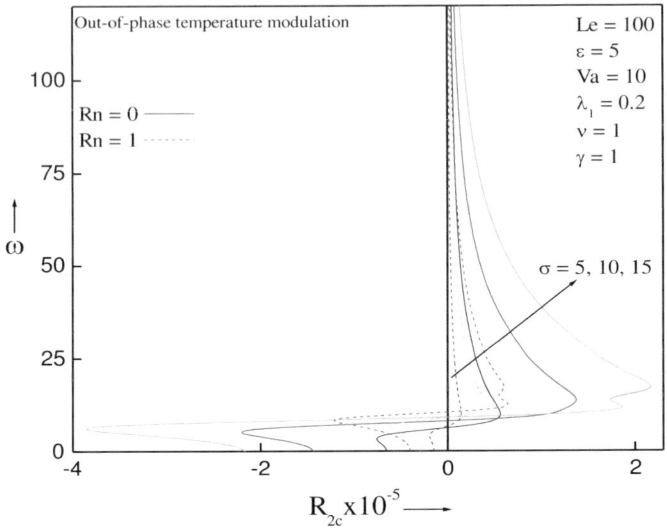

Fig. 14.13 Variation of R_{2c} with ω for different values of Rn and σ

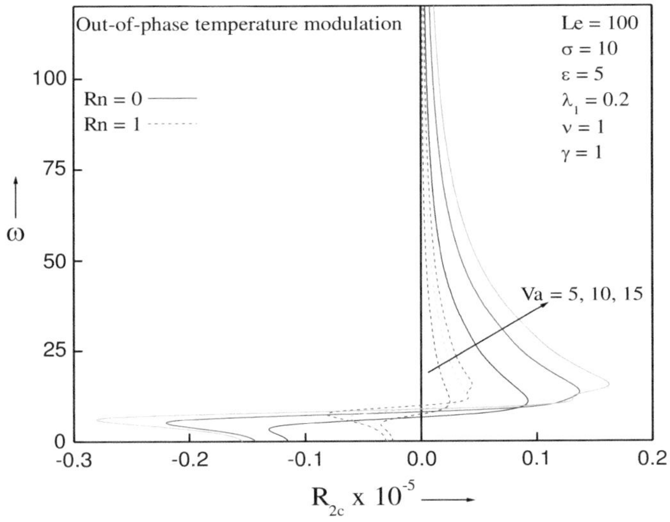

Fig. 14.14 Variation of R_{2c} with ω for different values of Rn and Va

number Le, thermal capacity ratio σ, Vadász number Va, viscosity and conductivity variation parameters υ and γ is the same as in the case of symmetric modulation, and hence a detailed explanation is not presented. The variation of all the governing parameters for the case of only lower wall temperature modulation produce similar effects as for asymmetric modulation and hence not shown pictorially.

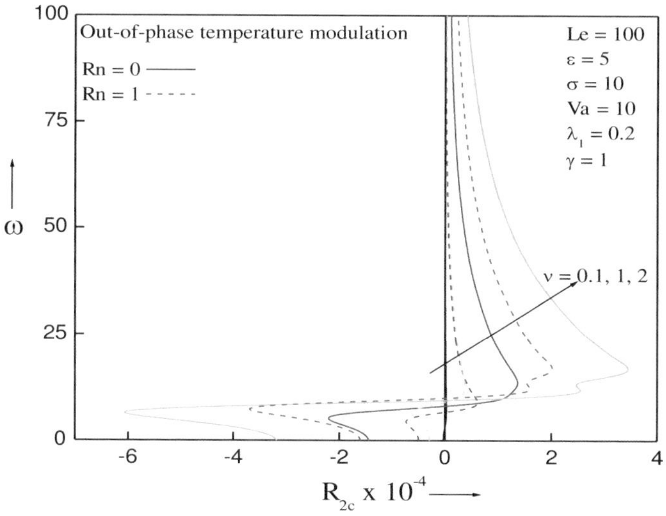

Fig. 14.15 Variation of R_{2c} with ω for different values of Rn and ν

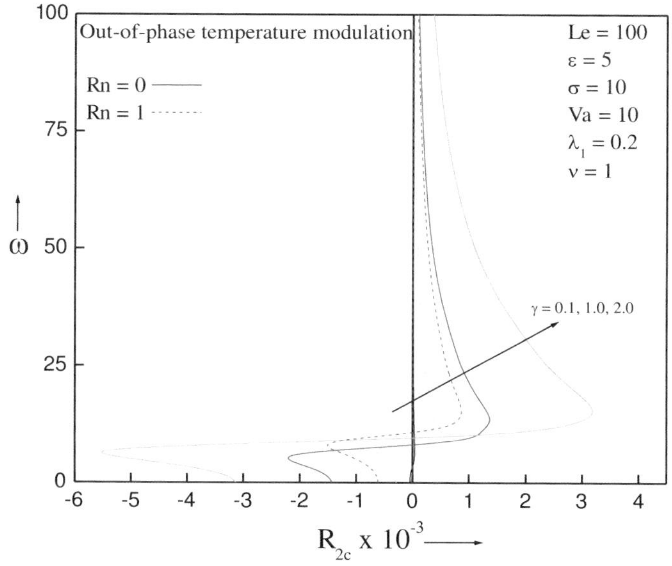

Fig. 14.16 Variation of R_{2c} with ω for different values of Rn and γ

From Figs. 14.1, 14.2, 14.3, 14.4, 14.5, 14.6, 14.7, 14.8, 14.9, 14.10, 14.11, 14.12, 14.13, 14.14, 14.15 and 14.16, one can observe that the peak values of for a regular fluid compared to a nanofluid for all the governing parameters. A nanofluid has a more stabilizing effect compared to a regular fluid.

Table 14.1 Nomenclature

c	Nanofluid specific heat at constant pressure	c_p	Specific heat of the nanoparticle material
$(\rho c)_m$	Effective heat capacity of the porous medium	d_p	Nanoparticle diameter
g	Gravitational acceleration	D_B	Brownian diffusion coefficient $\frac{m^2}{s}$
h_p	Specific enthalpy of the nanoparticle Specific enthalpy of the nanoparticle materialmaterial	H	Dimensional layer depth (m)
j_p	Diffusion mass flux for the nanoparticles	$j_{p,T}$	Thermophoretic diffusion
k	Thermal conductivity of the nanofluid	k_B	Boltzman constant
k_m	Effective thermal conductivity of the porous medium	k_p	Thermal conductivity of the particle material
Le	Lewis parameter	N_A	Modified diffusivity ratio
N_B	Modified particle-density increment	p^*	Pressure
p	Dimensionless pressure, $\frac{p^*K}{\mu\alpha_m}$	q	Energy flux relative to a frame moving with the nanofluid velocity
R	Thermal Rayleigh–Darcy number	Rn	Concentration Rayleigh number
t^*	time	t	Dimensionless time, $t^*\alpha_m/\sigma H^2$
T^*	Nanofluid temperature	T	Dimensionless temperature, $\frac{T^*-T_c^*}{T_h^*-T_c^*}$
T_c^*	Temperature at the upper wall	T_h^*	Temperature at the lower wall
T_R	Reference temperature	(u, v, w)	Dimensionless Darcy velocity components $\frac{(u^*,v^*,w^*)H}{\alpha_m}$
v	Nanofluid velocity	v_D	Darcy velocity εv
v_D^*	Dimensionless Darcy velocity (u^*, v^*, w^*)	γ_a	Non dimensional acceleration coefficient
Va	Vadász number	(x, y, z)	Dimensionless Cartesian coordinate
$\frac{(x^*,y^*,z^*)}{H}$	Vertically upward coordinate	(x^*, y^*, z^*)	Cartesian coordinates

Greek symbols

α_m	Thermal diffusivity of the porous medium $\frac{k_m}{(\rho c)_f}$	$\tilde{\beta}$	Proportionality factor
γ	Conductivity variation parameter	λ_1	Relaxation parameter
ε	Porosity of the medium	ε_t	Amplitude of the modulation
μ	Viscosity of the fluid	ν	Viscosity variation parameter
ρ	Fluid density	ρ_p	Nanoparticle mass density
σ	Parameter	ϕ^*	Nanoparticle volume fraction

(continued)

Table 14.1 (continued)

ϕ	Relative nanoparticle volume fraction, $\frac{\phi^*-\phi_c^*}{\phi_h^*-\phi_c^*}$	Ω	Dimensional frequency
ω	Dimensionless frequency $\left(=\frac{\Omega H^2}{K}\right)$		
ψ	Phase angles		
$\psi=0$	Symmetric modulation	$\psi=\pi$	Antisymmetric modulation
$\psi=-i\infty$	Only lower wall temperature modulation		

14.7 Conclusion

The effect of thermal modulation on the onset of convection in a Maxwell fluid and nanofluid saturated porous layer was studied using a linear stability analysis and the following conclusions were drawn (Table 14.1):

1. The effect of all three types of modulations namely, symmetric, asymmetric, and only with lower wall temperature modulations is found to be destabilizing compared to the unmodulated system.
2. Low frequency symmetric modulation is destabilizing while high frequency symmetric modulation is always stabilizing for both regular and nanofluids.
3. Large values of the concentration Rayleigh number are found to stabilize the system for all types of modulations.
4. The viscosity and conductivity variation parameters produce more stability for the system.
5. The nanofluid is found to be more stabilizing compared to regular fluid in all three types of temperature modulations.

Acknowledgements One of the authors, J.C. Umavathi, is thankful for the financial support under the UGC-MRP F.43-66/2014 (SR) Project, and also to Prof. Maurizio Sasso, supervisor and Prof. Matteo Savino co-coordinator of the ERUSMUS MUNDUS "Featured eUrope and South/south-east Asia mobility Network FUSION" for their support to do Post-Doctoral Research.

References

1. Awad, F.G., Sibanda, P., Motsa, S.S.: On the linear stability analysis of a Maxwell fluid with double-diffusive convection. Appl. Math. Model. **34**, 3509–3517 (2010)
2. Braester, C., Vadász, P.: The effect of a weak heterogeneity of a porous medium on natural convection. J. Fluid Mech. **254**, 345–362 (1993)
3. Buongiorno, J.: Convective transport in nanofluids. ASME J. Heat Transf. **128**, 240–250 (2006)
4. Chen, H.S., Ding, Y.L., Tan, C.Q.: Rheological behavior of nanofluids. New. J. Phys. **9**, 1–25 (2007)

5. Choi, S.: Enhancing thermal conductivity of fluids with nanoparticles. In: Siginer, D.A., Wang, H.P. (eds.) Developments and Applications of Non-Newtonian Flows) 231/MD - 66, pp. 99–105. ASME, New York (1995)
6. Chung Liu, I.: Effect of modulation on onset of thermal convection of a second grade fluid layer. Int. J. Non-Linear Mech. **39**, 1647–1657 (2004)
7. Eastman, J.A., Choi, S.: LI, S., Thompson, L.J.: Anomalously increased effective thermal conductivities of ethylene-glycol-based nanofluids containing copper nanoparticles. Appl. Phys. Lett. **78**, 718–720 (2001)
8. El-Sayed, M.F.: Electro hydrodynamic instability of two superposed Walters B viscoelastic fluids in relative motion through porous medium. Arch. Appl. Mech. **71**, 717–732 (2001)
9. Finlayson, B.A.: The Method of Weighted Residuals and Variation Principles. Academic Press, New York (1972)
10. Gounot, J., Caltagirone, J.P.: Stabilite et convection naturelle au sein d'une couche poreuse non homogene. Int. J. Heat Mass Transf. **32**, 1131–1140 (1989)
11. Horton, W., Rogers, F.T.: Convection currents in a porous medium. J. Appl. Phys. **16**, 367–370 (1945)
12. Khayat, R.W.: Chaos and over stability in the thermal convection of viscoelastic fluids. J. Non-Newton. Fluid Mech. **53**, 227–255 (1994)
13. Leong, J.C., Lai, F.C.: Natural convection in rectangular layers porous cavities. J. Thermophys. Heat Transf. **18**, 457–463 (2004)
14. Malashetty, M.S., Begum, I.: Effect of thermal/gravity modulation on the onset of convection in a Maxwell fluid saturated porous layer. Transp. Porous Media **90**, 889–909 (2011)
15. Malashetty, M.S., Swamy, M., Heera, R.: The onset of convection in a binary viscoelastic fluid saturated porous layer. Z. Angew. Math. Mech. **89**, 356–369 (2009)
16. Masuda, H., Ebata, A., Teramae, K., Hishinuma, N.: Alteration of thermal conductivity and viscosity of liquid by dispersing ultra-fine particles. Netsu Bussei/Jpn. J. Thermophys. Prop. **7**, 227–233 (1993)
17. Maxwell, J.C.: On the dynamical theory of gases. Philos. Trans. R. Soc. Lond. Ser. A. **157**, 26–78 (1866)
18. McKibbin, R., O'Sullivan, M.J.: Heat transfer in a layered porous medium heated from below. J. Facil. Manag. **111**, 141–173 (1981)
19. Nield, D.A.: Convective heat transfer in porous media with columnar structures. Transp. Porous Media **2**, 177–185 (1987)
20. Nield, D.A.: General heterogeneity effects on the onset of convection in a porous medium. In: Vadász, P. (ed.) Emerging Topics in Heat and Mass Transfer in Porous Media, pp. 63–84. Springer, New York (2008)
21. Nield, D.A., Kuznetsov, A.V.: Thermal instability in a porous medium layer saturated by a nanofluid. Int. J. Heat Mass Transf. **52**, 5796–5801 (2009)
22. Nield, D.A., Kuznetsov, A.V.: The onset of convection in a layer of a porous medium saturated by a nanofluid: effects of conductivity and viscosity variation and cross diffusion. Transp. Porous Media **92**, 837–846 (2012)
23. Rees, D.A.S., Riley, D.S.: The three-dimensionality of finite-amplitude convection in a layered porous medium heated from below. J. Fluid Mech. **211**, 437–461 (1990)
24. Sekhar, G.N., Jayalatha, G.: Elastic effects on Rayleigh-Bénard convection in liquids with temperature-dependent viscosity. Int. J. Therm. Sci. **49**, 67–75 (2010)
25. Shivakumara, I.S., Lee, J., Malashetty, M.S., Sureshkumar, S.: Effect of thermal modulation on the onset of convection in Walters B viscoelastic fluid-saturated porous medium. Transp. Porous Media **87**, 291–307 (2011)
26. Siddheshwar, P.G., Srikrishna, C.V.: Unsteady nonlinear convection in a second-order fluid. Int. J. Nonlinear Mech. **37**, 321–330 (2002)
27. Siddheshwar, P.G., Sekhar, G.N., Jayalatha, G.: Effect of time-periodic vertical oscillations of the Rayleigh-Bénard system on nonlinear convection in viscoelastic liquids. J. Non-Newton. Fluid Mech. **165**, 1412–1418 (2010)

28. Simmons, C.T., Fenstemaker, T.R., Sharp, J.M.: Variable-density flow and solute transport in heterogeneous porous media: Approaches, resolutions and future challenges. J. Contam. Hydrol. **52**, 245–275 (2001)
29. Sokolov, M., Tanner, R.I.: Convective stability of a general viscoelastic fluid heated from below. Phys. Fluid **15**, 534–539 (1972)
30. Tan, W.C., Masouka, T.: Stability analysis of Maxwell fluid in a porous medium heated from below. Phys. Lett. A. **360**, 454–460 (2007)
31. Tiwari, R.K., Das, M.K.: Heat transfer augmentation in a two-sided lid-driven differentially heated square cavity utilizing nanofluids. Int. J. Heat Mass Transf. **50**, 2002–2018 (2007)
32. Tzou, D.Y.: Thermal instability of nanofluids in natural convection. Int. J. Heat Mass Transf. **51**, 2967–2979 (2008)
33. Umavathi, J.C.: Effect of thermal modulation on the onset of convection in a porous medium layer saturated by a nanofluid. Transp. Porous Media **98**, 59–79 (2013)
34. Vadász, P.: Heat transfer enhancement in nanofluids suspensions: possible mechanisms and explanations. Int. J. Heat Mass Transf. **48**, 2673–2683 (2005)
35. Vadász, P.: Heat conduction in nanofluid suspensions. ASME. J. Heat Transf. **128**, 465–477 (2006)
36. Wang, S., Tan, W.C.: Stability analysis of double-diffusive convection of Maxwell fluid in a porous medium heated from below. Phys. Lett. A. **372**, 3046–3050 (2008)

Chapter 15
Effect of First Order Chemical Reaction on Magneto Convection in a Vertical Double Passage Channel

J. Pratap Kumar, Jawali C. Umavathi, Prashant G. Metri and Sergei Silvestrov

Abstract The objective of this paper is to study magneto-hydrodynamic flow in a vertical double passage channel taking into account the presence of the first order chemical reaction. The channel is divided into two passages by means of a thin, perfectly conducting plane baffle and hence the velocity will be individual in each stream. The governing equations are solved by using regular perturbation technique valid for small values of the Brinkman number and differential transform method valid for all values of the Brinkman number. The results are obtained for velocity, temperature and concentration. The effects of various dimensionless parameters such as thermal Grashof number, mass Grashof number, Brinkman number, first order chemical reaction parameter, and Hartman number on the flow variables are discussed and presented graphically for open and short circuits. The validity of solutions obtained by differential transform method and regular perturbation method are in good agreement for small values of the Brinkman number. Further the effects of governing parameters on the volumetric flow rate, species concentration, total heat rate, skin friction and Nusselt number are also observed and tabulated.

Keywords Chemical reaction · Double passage channel · Differential transform method · Regular perturbation method

---

J. Pratap Kumar (✉) · J.C. Umavathi
Department of Mathematics Gulbarga University Gulbarga, Karnataka, India
e-mail: p_rathap@yahoo.com; p_rathap@rediffmail.com

J.C. Umavathi
Department of Engineering, University of Sannio, Benevento, Italy
e-mail: drumavathi@rediffmail.com

P.G. Metri · S. Silvestrov
Division of Applied Mathematics, The School of Education, Culture and Communication, Mälardalen University, 883, 721 23 Västerås, Sweden
e-mail: prashant.g.metri@mdh.se

S. Silvestrov
e-mail: sergei.silvestrov@mdh.se

S. Silvestrov and M. Rančić (eds.), *Engineering Mathematics I*,
Springer Proceedings in Mathematics & Statistics 178,
DOI 10.1007/978-3-319-42082-0_15

15.1 Introduction

Magneto-hydrodynamics (MHD) is the branch of continuum mechanics which deals with the flow of electrically conducting fluids in electric and magnetic fields. Many natural phenomena and engineering problems are worth being subjected to an MHD analysis. Magneto-hydrodynamic equations are ordinary electromagnetic and hydrodynamic equations modified to take into account the interaction between the motion of the fluid and the electromagnetic field. The formulation of the electromagnetic theory in a mathematical form is known as Maxwell's equations.

The flow and heat transfer of electrically conducting fluids in channels and circular pipes under the effect of a transverse magnetic field occurs in magneto-hydrodynamic (MHD) generators, pumps, accelerators and flow meters and have applications in nuclear reactors, filtration, geothermal systems and others. The interest in the outer magnetic field effect on heat-physical processes appeared seventy years ago. Research in magneto-hydrodynamics grew rapidly during the late 1950s as a result of extensive studies of ionized gases for a number of applications. Blum et al. [1] carried out one of the first works in the field of heat and mass transfer in the presence of a magnetic field. Many exciting innovations were put forth in the areas of MHD propulsion [5], remote energy deposition for drag reduction [32], plasma actuators, radiation driven hypersonic wind tunnel, MHD control of flow and heat transfer in the boundary layer [2, 23, 24, 39], enhanced plasma ignition [11] and combustion stability. Extensive research however has revealed that additional and refined fidelity of physics in modeling and analyzing the interdisciplinary endeavor are required to reach a conclusive assessment. In order to ensure a successful and effective use of electromagnetic phenomena in industrial processes and technical systems, a very good understanding of the effects of the application of a magnetic field on the flow of electrically conducting fluids in channels and various geometric elements is required.

The present trend in the field of chemical reaction analysis is to give a mathematical model for the system to predict the reactor performance. Much research was being carried out across the globe. The study of heat and mass transfer with chemical reaction is given primary importance in chemical and hydro-metallurgical industries. A study on chemical reaction on the flow past an impulsively started vertical plate with uniform heat and mass flux was made by Muthucumaraswamy and Ganesan [21]. The same type of problem with inclusion of constant wall suction was studied by Makinde and Sibanda [17]. Fan et al. [6] studied the same problem over a horizontal moving plate. Kandasamy and Anjalidevi [10] investigated the effect of chemical reaction of the flow over a wedge. Sattar [29] investigated the effect of free and forced convection boundary layer flow through a Porous medium with large suction. Atul Kumar Singh [30] analyzed the MHD free convection and mass transfer flow with heat source and thermal diffusion. Recently Prathap Kumar et al. [12–16] have studied Taylor dispersion of solute for immiscible fluids for viscous and for composite Porous media.

Umavathi [33] studied the combined effect of viscous and applied electrical field in a vertical channel. Later Umavathi and her group analyzed magneto-hydrodynamic flow and heat transfer for various geometries [18–20, 34–38]. Yao [41] studied the natural convection heat transfer from isothermal vertical wavy surfaces, such as sinusoidal surfaces. Rees and Pop [27] examined the natural convection flow over a vertical wavy surface with constant wall temperature in Porous media saturated with Newtonian fluids. Hossain and Rees [8] studied the heat and mass transfer in natural convection flow along a vertical wavy surface with constant wall temperature and concentration for Newtonian fluid. Cheng [3] presented the solution of heat and mass transfer in natural convection flow along a vertical wavy surface in Porous medium saturated with Newtonian fluid.

When the channel is divided into several passages by means of plane baffles, as usually occurs in heat exchangers or electronic equipment, it is quite possible to enhance the heat transfer performance between the walls and fluid by the adjustment of each baffle position and strength of separate flow streams. In such configurations, perfectly conductive and thin baffles may be used to avoid significant increase of the transverse thermal resistance. Chin-Hsiang et al. [4] studied the thermal characteristics of hydro dynamically and thermally fully developed flow in an asymmetrical heated horizontal channel, which is divided into two passages (by means of a baffle) for two separate flow streams. Salah El-Din [28] studied analytically, the laminar fully developed combined convection in a vertical double passage channel with different wall temperature and concluded that heat transfer in the channel is affected significantly by the baffle position.

The differential transformation method (DTM) is a numerical method based on a Taylor expansion. This method constructs an analytical solution in the form of a polynomial. The concept of differential transform method was first proposed and applied to solve linear and nonlinear initial value problems in electric circuit analysis by Zhou [42]. Unlike the traditional high order Taylor series method which requires a lot of symbolic computations, the differential transform method is an iterative procedure for obtaining Taylor series solutions. This method will not consume too much computer time when applying to nonlinear or parameter varying systems. But, it is different from Taylor series method that requires computation of the high order derivatives. The differential transform method is an iterative procedure that is described by the transformed equations of original functions for solution of differential equations. This method is well addressed in [9, 22, 25, 26, 40].

Keeping in view the practical applications where there is a requirement of enhancement of heat transfer by inserting a baffle and the effects of chemical reaction, it is the aim of this paper to understand the flow nature by inserting a baffle in a vertical channel filled with chemically reacting conducting fluid.

15.2 Mathematical Formulation

Consider a steady, two-dimensional laminar fully developed free convection flow in an open ended vertical channel filled with purely viscous electrically conducting fluid. The x-axis is taken vertically upward, and parallel to the direction of buoyancy, and the y-axis is normal to it. A uniform magnetic field is applied normal to the plates and uniform electric field is applied perpendicular to the plate. The thermal conductivity, dynamic viscosity, thermal and concentration expansion coefficients are considered as constant. The Oberbeck-Boussinesq approximation is assumed to hold and for the evaluation of the gravitational body force, the density ρ is assumed to depend on temperature according to the equation of state $(\rho = \rho_0(1 - \beta(T - T_0)))$. It is also assumed that the magnetic Reynolds number is sufficiently small so that the induced magnetic field can be neglected and the induced electric field is assumed to be negligible. Ohmic and viscous dissipations are included in the energy equation. The flow is assumed to be steady, laminar and fully developed. The walls are maintained at constant but different temperatures. The channel is divided into two passages by means of thin, perfectly conducting plane baffle and each stream will have its own pressure gradient and hence the velocity will be individual in each stream. After inserting the baffle, the fluid in Stream-I is concentrated.

The governing equations for velocity, temperature and concentration are
Stream-I

$$\rho g\beta_T(T_1 - Tw_2) + \rho g\beta_c(C_1 - C_0) - \frac{dP}{dX} + \mu\frac{d^2U_1}{dY^2} - \sigma_e(E_0 + B_0U_1)B_0 = 0, \tag{15.1}$$

$$\frac{d^2T_1}{dY^2} + \frac{\nu}{\alpha C_p}\left(\frac{dU_1}{dY}\right)^2 + \frac{\sigma_e}{\alpha\rho C_p}(E_0 + B_0U_1)^2 = 0, \tag{15.2}$$

$$D\frac{d^2C_1}{dY^2} - K_1C_1 = 0, \tag{15.3}$$

Stream-II

$$\rho g\beta_T(T_1 - Tw_2) - \frac{\partial P}{\partial X} + \mu\frac{d^2U_2}{dY^2} - \sigma_e(E_0 + B_0U_2)B_0 = 0, \tag{15.4}$$

$$\frac{d^2T_2}{dY^2} + \frac{\nu}{\alpha C_p}\left(\frac{dU_2}{dY}\right)^2 + \frac{\sigma_e}{\alpha\rho C_p}(E_0 + B_0U - 2)^2 = 0, \tag{15.5}$$

which are subject to the boundary conditions on velocity, temperature and concentration as

$$U_1 = 0,\ T_1 = T_{W_1} = 0,\ C = C_1 \quad \text{at} \quad Y = -h,$$
$$U_2 = 0,\ T_1 = T_{W_2} = 0, \quad \text{at} \quad Y = h,$$

$$U_1 = 0,\ U_2 = 0,\ T_1 = T_2,\ \frac{dT_1}{dY} = \frac{dT_2}{dY},\ C = C_2 \ \text{at}\ Y = h^*. \tag{15.6}$$

Introducing the following non-dimensional variables in the governing equations for velocity temperature and concentration as

$$u_i = \frac{U_i}{U_1},\ \theta = \frac{T_i - T_{w_2}}{T_{w_1} - T_{w_2}},\ Gr = \frac{g\beta_T \Delta T h^3}{\upsilon^2}, \tag{15.7}$$

$$Gc = \frac{g\beta_c \Delta C h^3}{\upsilon^2},\ \phi_1 = \frac{C - C_0}{C_1 - C_0},\ Re = \frac{\overline{U_1} h}{\upsilon},\ Br = \frac{\overline{U}_1^2 \mu}{k \,\Delta T},$$

$$Y^* = \frac{Y^*}{h},\ p = \frac{h^2}{\mu \overline{U}_1} \frac{dp}{dX},\ \Delta T = T_{w_2} - T_{w_1},$$

$$\Delta C = C_1 - C_0,\ Y = \frac{y}{h},\ M^2 = \frac{\sigma_e B_0^2 h^2}{\mu},$$

$$E = \frac{E_0}{B_0 \overline{u}_1},\ \alpha = \frac{k_1 h^2}{D},\ n = \frac{C_2 - C_0}{C_1 - C_0}.$$

one obtains the momentum, energy and concentration equations corresponding to Stream-I and Stream-II as

Stream-I

$$\frac{d^2 u_1}{dy^2} + GR_T \theta_1 + GR_C \phi_1 - p - M^2 (E + u_1) = 0, \tag{15.8}$$

$$\frac{d^2 \theta_1}{dy^2} + Br\left(\left(\frac{du_1}{dy}\right)^2 + M^2 (E + u_1)^2\right) = 0, \tag{15.9}$$

$$\frac{d^2 \phi_1}{dy^2} - \alpha^2 \phi_1 = 0, \tag{15.10}$$

Stream-II

$$\frac{d^2 u_2}{dy^2} + GR_T \theta_2 - p - M^2 (E + u_2) = 0, \tag{15.11}$$

$$\frac{d^2 \theta_2}{dy^2} + Br\left(\left(\frac{du_2}{dy}\right)^2 + M^2 (E + u_2)^2\right) = 0, \tag{15.12}$$

which are subject to the boundary conditions

$$u_1 = 0,\ \theta_1 = 1,\ \phi_1 = 1, \quad \text{at} \quad y = -1, \tag{15.13}$$
$$u_2 = 0,\ \theta_2 = 0 \quad \text{at} \quad y = 1,$$

$$u_1 = 0,\ u_2 = 0,\ \theta_1 = \theta_2,\ \frac{d\theta_1}{dy} = \frac{d\theta_2}{dy},\ \phi_1 = n \quad \text{at} \quad y = y^*,$$

where $GR_T = \frac{Gr}{Re}$ and $GR_C = \frac{Gc}{Re}$.

15.3 Solutions

Solution of (15.10) using boundary condition (15.13) becomes

$$\phi_1 = B_1 \cosh(\alpha y) + B_2 \sinh(\alpha y). \tag{15.14}$$

15.3.1 Perturbation Method

Equations (15.8), (15.9), (15.11) and (15.12) are coupled non-linear ordinary differential equations. Approximate solutions can be found by using the regular perturbation method. The perturbation parameter *Br* is usually small and hence regular perturbation method can be strongly justified. Adopting this technique, solutions for velocity and temperature are assumed in the form

$$u_i(y) = u_{i0}(y) + Bru_{i1}(y) + Br^2 u_{i2}(y) + \cdots, \tag{15.15}$$

$$\theta_i(y) = \theta_{i0}(y) + Br\theta_{i1}(y) + Br^2\theta_{i2}(y) + \cdots. \tag{15.16}$$

Substituting (15.15) and (15.16) in (15.8), (15.9), (15.11) and (15.12), and equating the coefficients of like power of *Br* to zero and one, we obtain the zero and first order equations as

Stream-I

Zeroth order equations

$$\frac{d^2u_{10}}{dy^2} + GR_T\theta_{10} + GR_C\phi_1 - p - M^2(E + u_{10}) = 0, \tag{15.17}$$

$$\frac{d^2\theta_{10}}{dy^2} = 0. \tag{15.18}$$

First order equations

$$\frac{d^2u_{11}}{dy^2} + GR_T\theta_{11} - M^2u_{11} = 0, \tag{15.19}$$

$$T_b = T_1(z) + \varepsilon_t T_2(z, t),$$

$$\frac{d^2\theta_{11}}{dy^2} + \left(\left(\frac{du_{10}}{dy}\right)^2 + M^2(E + u_{10})^2\right) = 0. \tag{15.20}$$

Stream-II

Zeroth order equations

$$\frac{d^2u_{20}}{dy^2} + GR_T\theta_{20} - p - M^2(E + u_{20}) = 0, \tag{15.21}$$

$$\frac{d^2\theta_{20}}{dy^2} = 0. \tag{15.22}$$

First order equations

$$\frac{d^2u_{21}}{dy^2} + GR_T\theta_{21} - p - M^2(E + u_{20}) = 0, \tag{15.23}$$

$$\frac{d^2\theta_{21}}{dy^2} + \left(\left(\frac{du_{20}}{dy}\right)^2 + M^2(E + u_{20})^2\right) = 0. \tag{15.24}$$

The corresponding boundary conditions reduces to

Zeroth order

$$\begin{aligned} u_{10} = 0,\ \theta_{10} = 1,\ \phi_1 = 1 \quad \text{at} \quad y = -1, \\ u_{20} = 0,\ \theta_{20} = 0 \quad \text{at} \quad y = 1, \\ u_{10} = 0,\ u_{20} = 0,\ \theta_{10} = \theta_{20},\ \frac{d\theta_{10}}{dy} = \frac{d\theta_{20}}{dy},\ \phi_1 = n \quad \text{at} \quad y = y^*. \end{aligned} \tag{15.25}$$

First order

$$\begin{aligned} u_{11} = 0,\ \theta_{11} = 0 \quad \text{at} \quad y = -1, \\ u_{21} = 0,\ \theta_{21} = 0 \quad \text{at} \quad y = 1, \\ u_{11} = 0,\ u_{21} = 0,\ \theta_{11} = \theta_{21},\ \frac{d\theta_{11}}{dy} = \frac{d\theta_{21}}{dy},\ \phi_1 = n \quad \text{at} \quad y = y^*. \end{aligned} \tag{15.26}$$

The solutions of the zeroth and first order Eqs. (15.17)-(15.24) using the boundary conditions as in (15.25) and (15.26) become

Zeroth-order solutions

Stream-I

$$\theta_{10} = z_1 y + z_2, \tag{15.27}$$

$$u_{10} = A_1 \cosh(My) + A_2 \sinh(My) + r_1 + r_2 y + r_3 \cosh(\alpha y) + r_4 \sinh(\alpha y). \tag{15.28}$$

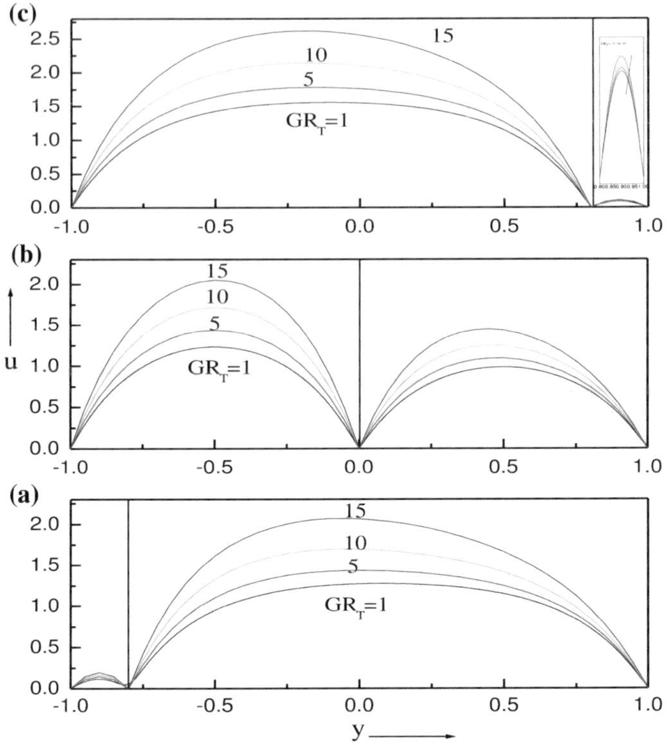

Fig. 15.1 Velocity profiles for different values of thermal Grashof number GR_T at **a** $y^* = 0.8$, **b** $y^* = 0$, **c** $y^* = 0.8$

Stream-II

$$\theta_{20} = z_3 y + z_4, \tag{15.29}$$

$$u_{20} = A_3 \cosh(My) + A_4 \sinh(My) + r_5 + r_6 y. \tag{15.30}$$

First order solution
Stream-I

$$\begin{aligned} \theta_{11} = E_1 + E_2 y + q_1 y^2 + q_2 y^3 + q_3 y^4 + q_4 \cosh(\alpha y) + \\ + q_5 \sinh(\alpha y) + q_6 \cosh(2\alpha y) + q_7 \sinh(2\alpha y) + q_8 \cosh(2My) + \\ + q_9 \sinh(2My) + q_{10} \cosh(My) + q_{11} \sinh(My) + q_{12} y \cosh(My) + \\ + q_{13} y \sinh(My) + q_{14} y \cosh(\alpha y) + q_{15} y \sinh(\alpha y) + q_{16} \cosh(\alpha + M)y + \\ + q_{17} \cosh(\alpha - M)y + q_{18} \sinh(\alpha + M)y + q_{19} \sinh(\alpha - M)y, \end{aligned} \tag{15.31}$$

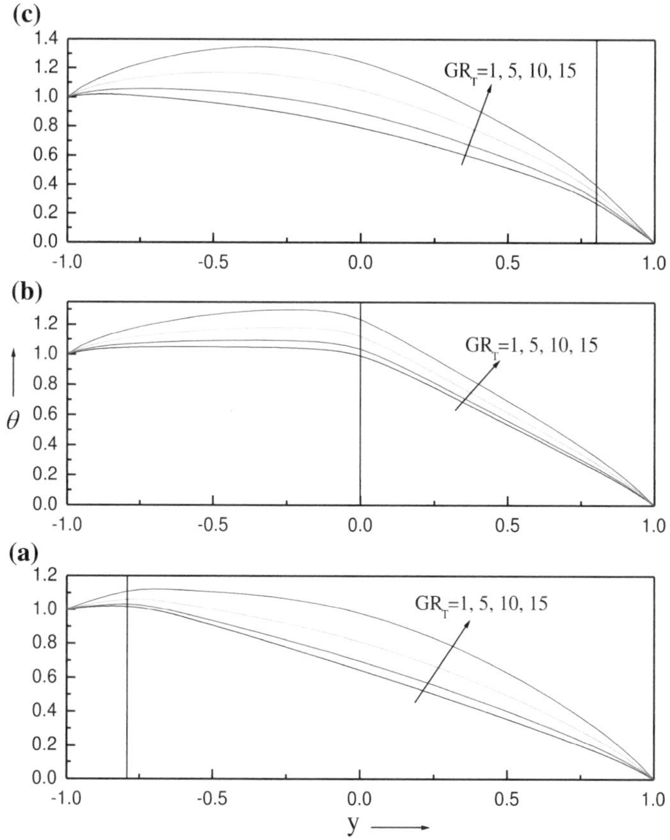

Fig. 15.2 Temperature profiles for different values of thermal Grashof number GR_T at **a** $y^* = 0.8$, **b** $y^* = 0$, **c** $y^* = 0.8$

$$
\begin{aligned}
u_{11} = {} & E_3 \cosh(My) + E_4 \sinh(My) + H_1 + H_2 y + H_3 y^2 + \\
& + H_4 y^3 + H_5 y^4 + H_6 \cosh(\alpha y) + H_7 \sinh(\alpha y) + \\
& + H_8 \cosh(2\alpha y) + H_9 \sinh(2\alpha y) + H_{10} \cosh(2My) + \\
& + H_{11} \sinh(2My) + H_{12} y \cosh(My) + H_{13} y \sinh(My) + \\
& + H_{14} y \cosh(\alpha y) + H_{15} y \sinh(\alpha y) + H_{16} \cosh(\alpha + M)y + \\
& + H_{17} \cosh(\alpha - M)y + H_{18} \sinh(\alpha + M)y + H_{19} \sinh(\alpha - M)y + \\
& + H_{20} y^2 \cosh(My) + H_{21} y^2 \sinh(My).
\end{aligned}
\tag{15.32}
$$

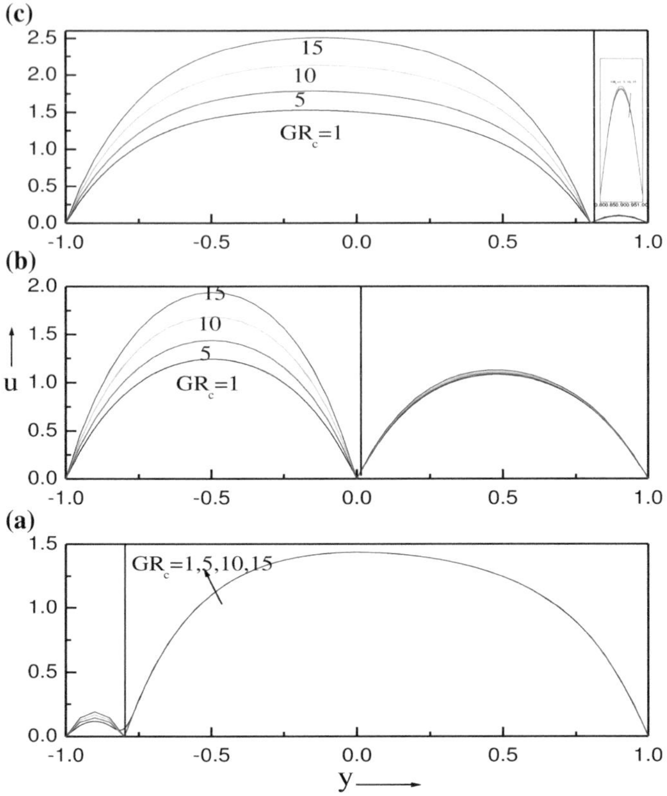

Fig. 15.3 Velocity profiles for different values of concentration Grashof number GR_C at **a** $y^* = -0.8$, **b** $y^* = 0$, **c** $y^* = 0.8$

Stream-II

$$\begin{aligned}\theta_{21} &= E_5 + E_6 y + F_1 y^2 + F_2 y^3 + F_3 y^4 + \\ &+ F_4 \cosh(2My) + F_5 \sinh(2My) + F_6 \cosh(My) + \\ &+ F_7 \sinh(My) + F_8 y \cosh(My) + F_9 y \sinh(My),\end{aligned} \tag{15.33}$$

$$\begin{aligned}u_{21} &= E_7 \cosh(My) + E_8 \sinh(My) + \\ &+ H_{22} + H_{23} y + H_{24} y^2 + H_{25} y^3 + H_{26} y^4 + \\ &+ H_{27} \cosh(2My) + H_{28} \sinh(2My) + H_{29} y \cosh(My) + \\ &+ H_{30} y \sinh(My) + H_{31} y^2 \cosh(My) + H_{32} y^2 \sinh(My).\end{aligned} \tag{15.34}$$

The dimensionless total volume flow rate is given by

$$Qv = Qv_1 + Qv_2, \tag{15.35}$$

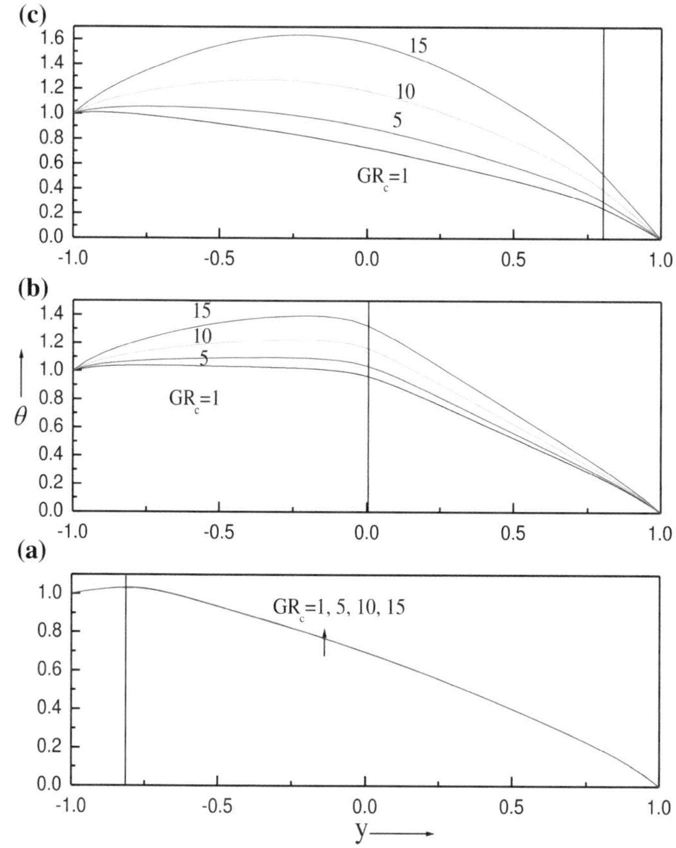

Fig. 15.4 Temperature profiles for different values of concentration Grashof number GR_C at **a** $y^* = -0.8$, **b** $y^* = 0$, **c** $y^* = 0.8$

where

$$Qv_1 = \int_{-1}^{y^*} u_1 dy, \quad Qv_2 = \int_{y^*}^{1} u_2 dy. \tag{15.36}$$

The dimensionless total heat rate added to the fluid is given by

$$E = H_{E_1} + H_{E_2}, \tag{15.37}$$

$$H_{E_1} = \int_{-1}^{y^*} u_1 \theta_1 dy, \quad H_{E_2} = \int_{y^*}^{1} u_2 \theta_2 dy. \tag{15.38}$$

The dimensionless total species rate added to the fluid is given by

$$Cs = Cs_1 + Cs_2, \tag{15.39}$$

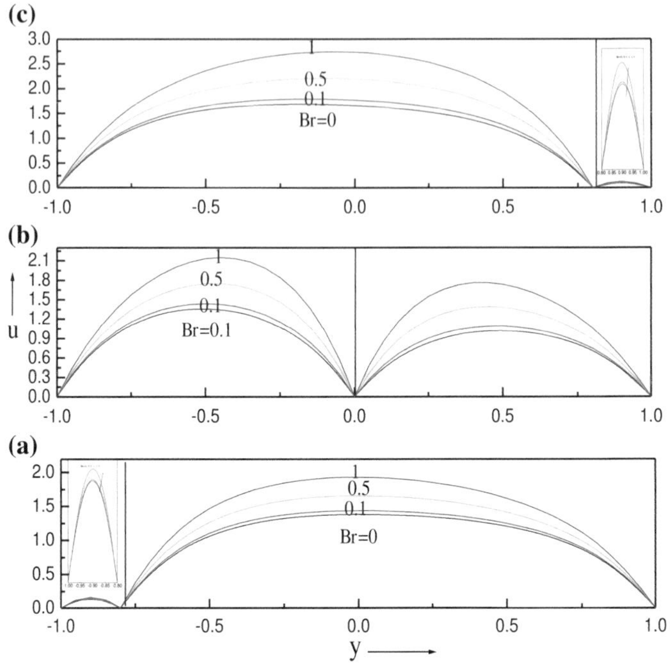

Fig. 15.5 Velocity profiles for different values of Brikman number Br at **a** $y^* = -0.8$, **b** $y^* = 0$, **c** $y^* = 0.8$

where

$$Cs_1 = \int_{-1}^{y^*} u_1 \phi_1 dy, \quad Cs_2 = \int_{y^*}^{1} u_2 \phi_2 dy. \tag{15.40}$$

15.3.2 Basic Concept of Differential Transform Method

The analytical solutions obtained in Sect. 15.3.1 are valid only for small values of Brinkman number Br. In many practical problems mentioned earlier, the values of Br are usually large. In that case analytical solutions are difficult, and hence we resort to semi-numerical-analytical method known as Differential Transform Method (DTM).

The general concept of DTM is explained here: The kth differential transformation of an analytical function $F(k)$ is defined as (Zhou [42])

$$F(k) = \frac{1}{k!}\left[\frac{d^k f(\eta)}{d\eta^k}\right]_{\eta=\eta_0}, \tag{15.41}$$

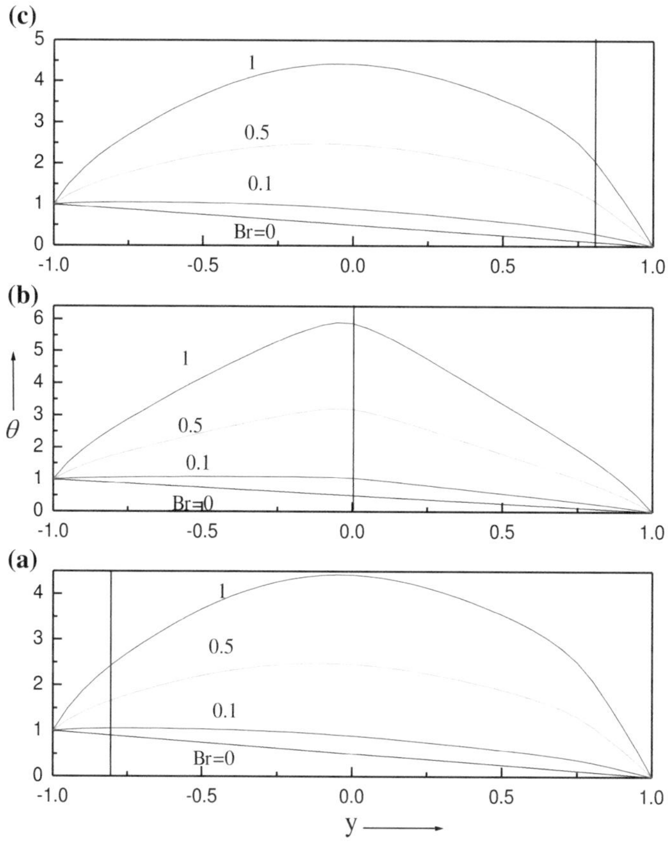

Fig. 15.6 Temperature profiles for different values of Brikman number Br at **a** $y^* = -0.8$, **b** $y^* = 0$, **c** $y^* = 0.8$

and the inverse differential transformation is given by

$$f(\eta) = \sum_{k=0}^{\infty} F(k)(\eta - \eta_0)^k. \tag{15.42}$$

Combining (15.41) and (15.42)

$$f(\eta) = \sum_{k=0}^{\infty} F(k)\frac{(\eta - \eta_0)^k}{k!}\frac{d^k f(\eta)}{d\eta^k}|_{\eta=\eta_0}. \tag{15.43}$$

From (15.42), it can be seen that the differential transformation method is derived from Taylor's series expansion. In real applications the sum $\sum_{k=0}^{\infty} F(k)(\eta - \eta_0)^k$ is very small and can be neglected when k is sufficiently large. So $f(\eta)$ can be expressed by a finite series, and (15.42) may be written as

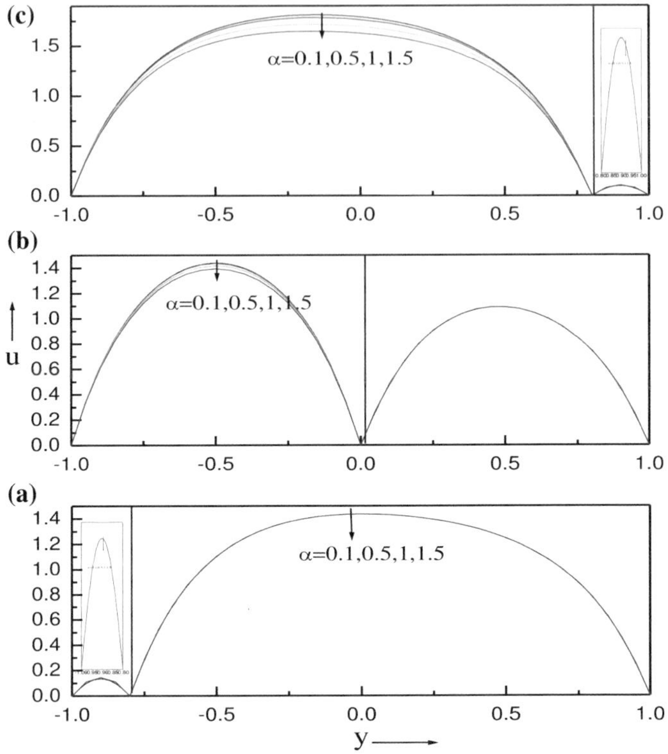

Fig. 15.7 Velocity profiles for different values of chemical reaction parameter α at **a** $y^* = -0.8$, **b** $y^* = 0$, **c** $y^* = 0.8$

$$f(\eta) = \sum_{k=0}^{n} F(k)(\eta - \eta_0)^k, \tag{15.44}$$

where the value of k depends on the convergence requirement in real applications and $F(k)$ is the differential transform of $f(\eta)$.

15.4 Results and Discussion

The velocity, temperature and concentration fields for an electrically conducting fluid in a vertical double passage are presented in Figs. 15.1, 15.2, 15.3, 15.4, 15.5, 15.6, 15.7, 15.8, 15.9, 15.10, 15.11, 15.12 and 15.13. The channel is divided into two passages by inserting a thin perfectly conducting baffle. The fluid is concentrated in passage one only after inserting the baffle. The basic equations are solved by using

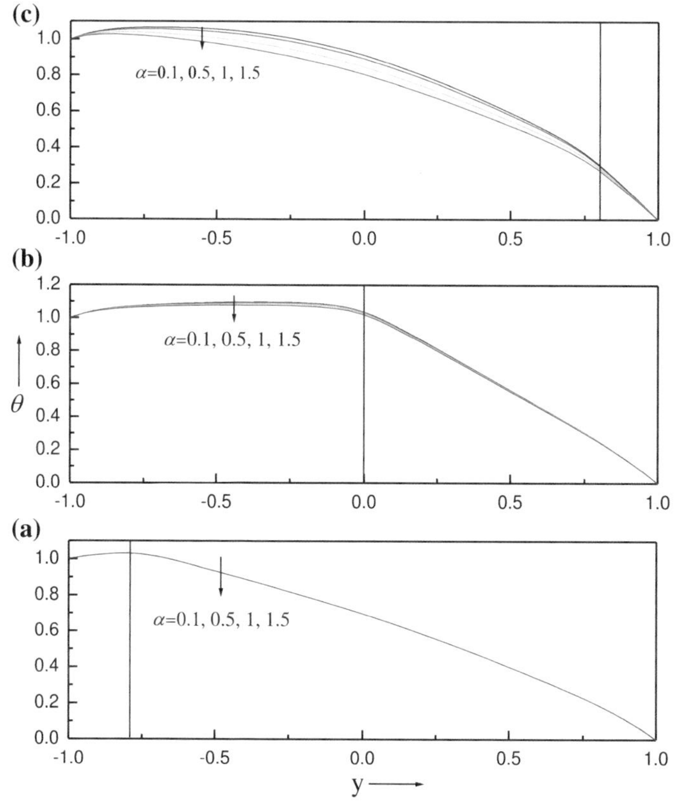

Fig. 15.8 Temperature profiles for different values of chemical reaction parameter α at **a** $y^* = -0.8$, **b** $y^* = 0$, **c** $y^* = 0.8$

regular perturbation method valid for small values of Br. The Brinkman number is exploited as a perturbation parameter. To understand the fluid nature for large values of Brinkman number, the coupled nonlinear ordinary differential equations are solved by DTM which is a semi-numerical-analytical method. The case $E = 0$ corresponds to short circuit and $E \neq 0$ corresponds to the open circuit case. The values of thermal Grashof number, mass Grashof number, Brinkman number, pressure gradient, first order chemical reaction parameter, wall concentration ratio, and Hartman number are fixed as 5, 5, –5, 0.5, 1, 4 for open circuit ($E = -1$) for all the graphs and tables except the varying parameter.

The effect of thermal Grashof number GR_T (ratio of thermal Grashof number to Reynolds number) on the velocity and temperature fields is shown in Figs. 15.1 and 15.2 at three different baffle positions ($y^* = -0.8,\ 0,\ 0.8$) keeping the left wall at higher temperature. It is observed from Figs. 15.1 and 15.2 that the velocity and temperature increases at all the baffle positions as the thermal Grashof number increases. However when the baffle is placed near left wall the maximum point of

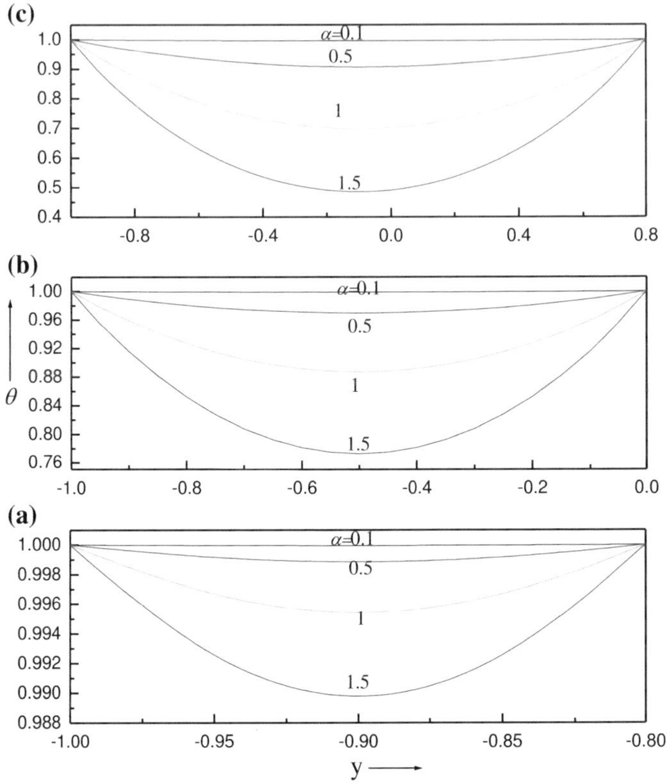

Fig. 15.9 Concentration profiles for different values of chemical reaction parameter α at **a** $y^* = -0.8$, **b** $y^* = 0$, **c** $y^* = 0.8$

velocity is in Stream-II, when the baffle is placed near the centre of the channel and near the right wall, the maximum point of velocity is in Stream-I. It is also seen from Fig. 15.1 that the velocity profiles in both the passages are not equal though there is a no-slip condition at the walls and near the baffle. One can infer this result as, the thermal Grashof number increases as the buoyancy force increases and the wall conditions on temperature are not equal (left wall is at higher temperature when compared to right wall). It is seen from Fig. 15.2 that the temperature profiles look similar at all the baffle positions. This is due to the reason that the temperature and heat flux are considered as continuous at the baffle positions. The optimum value of temperature is seen in Stream-II for the baffle position near the left wall and in Stream-I when the baffle position is near the right wall.

The effect of mass Grashof number GR_C (ratio of mass Grashof number to Reynolds number) on the flow is shown in Figs. 15.3 and 15.4 at all baffle positions. The increase in mass Grashof number increases the velocity and temperature fields in both the streams at all the baffle positions. The enhancement of velocity is

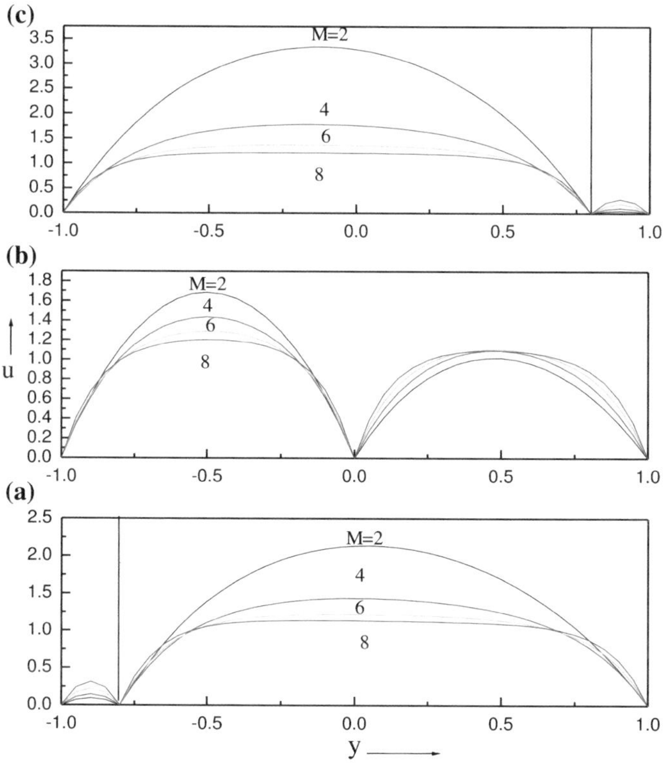

Fig. 15.10 Velocity profiles for different values of Hartmann number M at **a** $y^* = -0.8$, **b** $y^* = 0$, **c** $y^* = 0.8$

significant in Stream-I when compared to Stream-II. The reason for this nature is that fluid is concentrated only in Stream-I. It is viewed from Fig. 15.4b that though the fluid is not concentrated in Stream-II there is a slight increase in the velocity field when the baffle is positioned near the centre of the channel. This is because of the conditions imposed on temperature and heat flux. That is to say that, there is heat transfer from Stream-I to Stream-II and there is no mass transfer from Stream-I to Stream-II. However due to transfer of heat from Stream-I to Stream-II results in increase in thermal and concentration buoyancy forces and hence enhancement of velocity in small magnitude is observed in Stream-II as GR_C increases. There is no effect of GR_C in Stream-II when the baffle is positioned near the left and right walls. Similar results were also observed by Fasogbon [7] for regular channel in the absence of baffle. From Fig. 15.4a it is seen that the temperature does not vary significantly in both the streams as GR_C increases when the baffle is positioned near the left wall. However its influence is dominated in Stream-I when the baffle is placed in the centre of the channel and near the right wall.

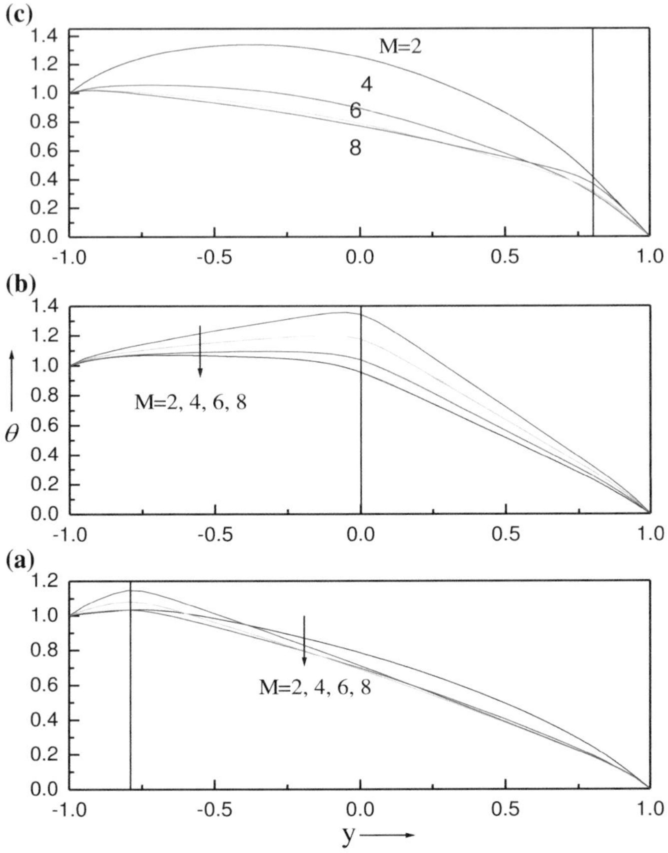

Fig. 15.11 Temperature profiles for different values of Hartmann number M at **a** $y^* = -0.8$, **b** $y^* = 0$, **c** $y^* = 0.8$

As the Brinkman number increases, the velocity and temperature increases in both streams at all baffle positions as seen in Figs. 15.5 and 15.6, respectively. One can infer this nature is due to the fact that increase in Brinkman number increases the viscous dissipation and hence increases the temperature, which in turn influences the velocity field. Here also temperature profiles looks similar at all baffle positions.

The effect of the first order chemical reaction parameter α, on the velocity, temperature and concentration fields is observed in Figs. 15.7, 15.8 and 15.9, respectively. As α increases the velocity and temperature decreases in Stream-I and does not influence in Stream-II. This is due to the fact that the fluid in Stream-I is concentrated. At any position of the baffle the effect of α is to minimize the concentration field. Figure 15.9 shows the concentration profiles when the baffle is positioned at $y^* = -0.8,\ 0,\ 0.8$ and respectively. Similar results were also observed by Srinivas

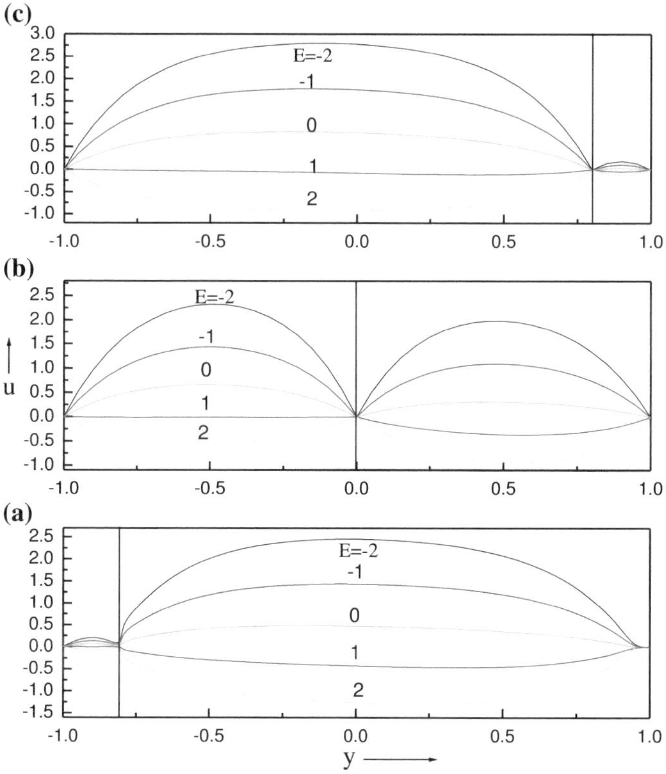

Fig. 15.12 Velocity profiles for different values of electric field load parameter E at **a** $y^* = -0.8$, **b** $y^* = 0$, **c** $y^* = 0.8$

and Muturaj [31] for mixed convective flow in a vertical channel in the absence of baffle.

The effect of Hartman number M on velocity and temperature field is shown in Figs. 15.10 and 15.11, respectively at three positions of the baffle. As the Hartman number M increases velocity decreases in both streams at different positions of the baffle for open circuits. The Hartman number M represents the ratio of the Lorentz force to the viscous force, implying that the larger the Hartman number, the stronger the retarding effect on the velocity field. Hence as Hartman number M increases the velocity decreases at all baffle positions. Further, as the width of the passage increases, the velocity profiles are flattened. Hence the velocity profiles are wider in Stream-II when the baffle is placed near the left wall, in Stream-I when the baffle is placed near the right wall and narrow when the baffle is placed in the centre of the channel. The temperature profiles are also reduced as the Hartman number increases in both the streams when the baffle is positioned at the centre and at the right wall. The temperature decreases as M increases at $M = -0.5$ (approximately) and onwards, where as it increases in Stream-I. This nature can be inferred as when the baffle is near

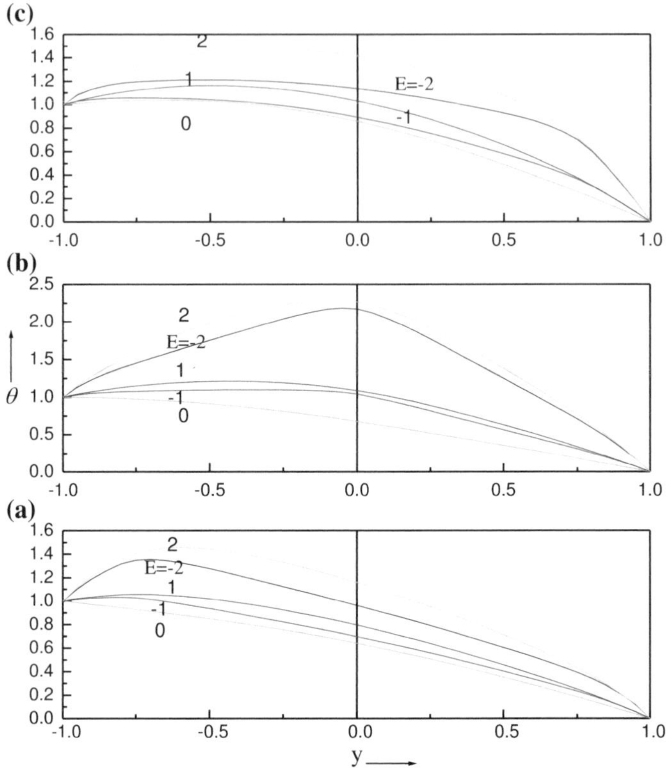

Fig. 15.13 Temperature profiles for different values of electric field load parameter E at **a** $y^* = -0.8$, **b** $y^* = 0$, **c** $y^* = 0.8$

the right wall, which is at higher temperature, is not much influenced by the retarding effect of the Lorenz force. Further one can also observe that the temperature profile is wider in Stream-I when the baffle is placed near the right wall when compared to the baffle positioned in the centre of the channel.

The effect of the applied electric field E on the velocity and temperature is displaced in Figs. 15.12 and 15.13, respectively for both open and short circuits. The effect of negative E is to add the flow while the effect of positive E is to oppose the flow as compared to the case for the short circuit. Since the direction of flow is reversed by changing the values of E, the results can be applied to the practical problem where there is a requirement of reversal flow. The temperature increases as the electric field load parameter increases in both passages at all baffle positions. However, the magnitude of temperature is large for positive E when compared to negative E. Here also the temperature profiles are flat in wider passage. The effect of Hartman number and electric field load parameter on the flow show similar results as observed by Umavathi [33] in the absence of baffle and the first order chemical reaction.

Table 15.1 Comparison of velocity and temperature with $Br = 0$ and $y^* = 0.0$

y	Velocity			Temperature		
	DTM	PM	% of Error (%)	DTM	PM	% of Error (%)
−1	0	0	0.00	1.000000	1.000000	0.00
−0.75	1.204365	1.204365	0.00	0.875000	0.875000	0.00
−0.5	1.550479	1.550479	0.00	0.750000	0.750000	0.00
−0.25	1.168996	1.168996	0.00	0.625000	0.625000	0.00
0	0	0	0.00	0.500000	0.500000	0.00
0	0	0	0.00	0.500000	0.500000	0.00
0.25	0.707605	0.707605	0.00	0.375000	0.375000	0.00
0.5	0.901861	0.901861	0.00	0.250000	0.250000	0.00
0.75	0.672236	0.672236	0.00	0.125000	0.125000	0.00
1	0	0	0.00	0	0	0.00

Table 15.2 Comparison of velocity and temperature with $Br = 0.05$ and $y^* = 0.0$

y	Velocity			Temperature		
	DTM	PM	% of Error (%)	DTM	PM	% of Error (%)
−1	0	0	0.00	1.000000	1.000000	0.00
−0.75	1.277758	1.247721	3.00	1.036972	0.971676	6.53
−0.5	1.665603	1.618226	4.74	1.019641	0.907179	11.25
−0.25	1.269718	1.228412	4.13	0.980143	0.835212	14.49
0	0	0	0.00	0.886478	0.727643	15.88
0	0	0	0.00	0.886478	0.727643	15.88
0.25	0.798468	0.759023	3.94	0.700614	0.558826	14.18
0.5	1.003566	0.959327	4.42	0.486179	0.382006	10.42
0.75	0.735737	0.708361	2.74	0.257282	0.201896	5.54
1	0	0	0.00	0	0	0.00

Since the regular perturbation method (PM) is valid only for small values of Brinkman number, this restriction in relaxed by finding the solution of the governing equation using the DTM. Tables 15.1, 15.2, 15.3, 15.4, 15.5, 15.6, 15.7, 15.8 and 15.9 are the values of velocity and temperature when the baffle is positioned near the centre of the channel, near the left wall and near the right wall respectively. The validity of the DTM is justified by comparing the DTM solutions with the PM method in the absence of the Brinkman number. It is seen from Tables 15.1, 15.4, and 15.7 that the DTM and PM values are exact in the absence of the Brinkman number at all baffle positions in both streams. When the Brinkman number is 0.05, the percentage of error of the DTM and the PM is less when compared with the value of $Br = 0.15$ as seen in Tables 15.2, 15.3, 15.5, 15.6, 15.8, and 15.9, respectively at all baffle positions in both streams. Hence, one can conclude that regular perturbation

Table 15.3 Comparison of velocity and temperature with $Br = 0.15$ and $y^* = 0.0$

y	Velocity			Temperature		
	DTM	PM	% of Error (%)	DTM	PM	% of Error (%)
–1	0	0	0.00	1.000000	1.000000	0.00
–0.75	1.536883	1.334433	20.24	1.604460	1.165029	43.94
–0.5	2.072972	1.753718	31.93	1.972888	1.221536	75.14
–0.25	1.626963	1.347243	27.97	2.242187	1.255635	98.66
0	0	0	0.00	2.268726	1.182928	108.58
0	0	0	0.00	2.268726	1.182928	108.58
0.25	1.117535	0.861860	25.57	1.849663	0.926477	92.32
0.5	1.358661	1.074260	28.44	1.306977	0.646019	66.10
0.75	0.956506	0.780613	17.59	0.711477	0.355687	35.58
1	0	0	0.00	0	0	0.0000

Table 15.4 Comparison of velocity and temperature with $Br = 0$ and $y^* = -0.8$

y	Velocity			Temperature		
	DTM	PM	% of Error (%)	DTM	PM	% of Error (%)
–1	0	0	0.00	1.000000	1.000000	0.00
–0.95	0.069354	0.069354	0.00	0.975000	0.975000	0.00
–0.9	0.092187	0.092187	0.00	0.950000	0.950000	0.00
–0.85	0.069043	0.069043	0.00	0.925000	0.925000	0.00
–0.8	0	0	0.00	0.900000	0.900000	0.00
–0.8	0	0	0.00	0.900000	0.900000	0.00
–0.5	1.260132	1.260132	0.00	0.750000	0.750000	0.00
–0.2	1.805345	1.805345	0.00	0.600000	0.600000	0.00
0.1	1.907424	1.907424	0.00	0.450000	0.450000	0.00
0.4	1.673782	1.673782	0.00	0.300000	0.300000	0.00
0.7	1.087305	1.087305	0.00	0.150000	0.150000	0.00
1	0	0	0.00	0	0	0.00

method cannot be applied for large values of the Brinkman number as the error is greater. Further, one can infer from these tables that the percentage of error between the DTM and the PM is greater in the narrow passage $y^* = 0$ when compared with wider passage $y^* = -0.8,\ 0.8$, for different values of the Brinkman number.

The effect of thermal Grashof number GR_T, mass Grashof number GR_C, Brinkman number Br, chemical reaction parameter α and Hartmann number M on the volumetric flow rate in Stream-I and Stream-II at different positions of the baffle is tabulated in Table 15.10. It is seen that increase in the thermal Grashof number, mass Grashof number and the Brinkman number increases the volumetric flow rate of the fluid

Table 15.5 Comparison of velocity and temperature with $Br = 0.05$ and $y^* = -0.8$

y	Velocity			Temperature		
	DTM	PM	% of Error (%)	DTM	PM	% of Error (%)
–1	0	0	0.00	1.000000	1.000000	0.00
–0.95	0.070556	0.069878	0.07	1.015203	0.992367	2.28
–0.9	0.094068	0.093024	0.10	1.026025	0.984173	4.19
–0.85	0.070662	0.069774	0.09	1.034695	0.975545	5.91
–0.8	0	0	0.00	1.041877	0.966357	7.55
–0.8	0	0	0.00	1.041877	0.966357	7.55
–0.5	1.397079	1.324997	7.21	1.006091	0.866824	13.93
–0.2	2.013926	1.905942	10.80	0.894776	0.736583	15.82
0.1	2.129951	2.018219	11.17	0.739775	0.593012	14.68
0.4	1.860009	1.770253	8.98	0.543458	0.431243	11.22
0.7	1.195027	1.145360	4.97	0.304197	0.242983	6.12
1	0	0	0.00	0	0	0.00

Table 15.6 Comparison of velocity and temperature with $Br = 0.1$ and $y^* = -0.8$

y	Velocity			Temperature		
	DTM	PM	% of Error (%)	DTM	PM	% of Error (%)
–1	0	0	0.00	1.000000	1.000000	0.00
–0.95	0.066662	0.070402	0.37	0.966262	1.009733	4.35
–0.9	0.090302	0.093860	0.36	0.993598	1.018345	2.47
–0.85	0.068630	0.070504	0.19	1.042718	1.026091	1.66
–0.8	0	0	0.00	1.098087	1.032715	6.54
–0.8	0	0	0.00	1.098087	1.032715	6.54
–0.5	1.547555	1.389862	15.77	1.259219	0.983648	27.56
–0.2	2.257138	2.006539	25.06	1.236455	0.873166	36.33
0.1	2.400241	2.129013	27.12	1.101593	0.736025	36.56
0.4	2.092376	1.866724	22.57	0.859604	0.562487	29.71
0.7	1.331559	1.203414	12.81	0.508352	0.335965	17.24
1	0	0	0.00	0	0	0.00

flowing through the vertical channel. This is due to the reasons that increase in the thermal Grashof number and mass Grashof number increases the buoyancy which tends to accelerate the fluid flow, thus raising the volumetric flow rate. Increase in the Brinkman number increases the viscous dissipation and hence accelerates the fluid flow. The chemical reaction parameter and the Hartmann number reduces the volumetric flow rates in both streams at all baffle positions, which is an expected result. The effect of governing parameters on species concentration is shown in Table 15.11.

Table 15.7 Comparison of velocity and temperature with $Br = 0$ and $y^* = 0.8$

y	Velocity			Temperature		
	DTM	PM	% of Error (%)	DTM	PM	% of Error (%)
–1	0	0	0.00	1.000000	1.000000	0.00
–0.7	1.799059	1.799059	0.00	0.850000	0.850000	0.00
–0.4	2.596490	2.596490	0.00	0.700000	0.700000	0.00
–0.1	2.771982	2.771982	0.00	0.550000	0.550000	0.00
0.2	2.464927	2.464927	0.00	0.400000	0.400000	0.00
0.5	1.626232	1.626232	0.00	0.250000	0.250000	0.00
0.8	0	0	0.00	0.100000	0.100000	0.00
0.8	0	0	0.00	0.100000	0.100000	0.00
0.85	0.034303	0.034303	0.00	0.075000	0.075000	0.00
0.9	0.045492	0.045492	0.00	0.050000	0.050000	0.00
0.95	0.033991	0.033991	0.00	0.025000	0.025000	0.00
1	0	0	0.00	0	0	0.00

Table 15.8 Comparison of velocity and temperature with $Br = 0.01$ and $y^* = 0.8$

y	Velocity			Temperature		
	DTM	PM	% of Error (%)	DTM	PM	% of Error (%)
–1	0	0	0.00	1.000000	1.000000	0.00
–0.7	1.836544	1.827308	0.92	0.905152	0.892072	1.31
–0.4	2.660660	2.644651	1.60	0.785382	0.763931	2.15
–0.1	2.847653	2.828706	1.89	0.649513	0.624232	2.53
0.2	2.534629	2.517264	1.74	0.497826	0.473285	2.45
0.5	1.671046	1.660029	1.10	0.330520	0.311253	1.93
0.8	0	0	0.00	0.141015	0.131755	0.93
0.8	0	0	0.00	0.141015	0.131755	0.93
0.85	0.034751	0.034648	0.01	0.106057	0.098970	0.71
0.9	0.046004	0.045886	0.01	0.070900	0.066089	0.48
0.95	0.034312	0.034238	0.01	0.035550	0.033107	0.24
1	0	0	0.00	0	0	0.00

Increase in thermal Grashof number, mass Grashof number and Brinkman number accelerates the fluid flow, thus enhancing the mass transfer rate between the wall and the fluid flowing through the vertical channel. As the chemical reaction parameter α and Hartmann number M increase, the dimensionless total species rate added to the fluid decreases.

The dimensionless total heat rate added to the fluid is tabulated in Table 15.12 as functions of thermal Grashof number GR_T, mass Grashof number GR_C, Brinkman

Table 15.9 Comparison of velocity and temperature with $Br = 0.08$ and $y^* = 0.8$

y	Velocity			Temperature		
	DTM	PM	% of Error (%)	DTM	PM	% of Error (%)
–1	0	0	0.00	1.000000	1.000000	0.00
–0.7	2.503122	2.025055	47.81	1.858967	1.186577	67.24
–0.4	3.812342	2.981778	83.06	2.315717	1.211447	110.43
–0.1	4.213343	3.225770	98.76	2.456803	1.143857	131.29
0.2	3.794654	2.883624	91.10	2.275009	0.986280	128.87
0.5	2.480021	1.896608	58.34	1.778652	0.740023	103.86
0.8	0	0	0.00	0.884947	0.354042	53.09
0.8	0	0	0.00	0.884947	0.354042	53.09
0.85	0.043026	0.037067	0.60	0.681299	0.266759	41.45
0.9	0.055493	0.048650	0.68	0.472397	0.178715	29.37
0.95	0.040219	0.035966	0.43	0.251419	0.089852	16.16
1	0	0	0.00	0	0	0.00

Table 15.10 Volumetric flow rate

	$y^* = -0.8$		$y^* = 0$		$y^* = 0.8$	
GR_c	Q_{v1}	Q_{v2}	Q_{v1}	Q_{v2}	Q_{v1}	Q_{v2}
1	0.01692	1.75421	0.86099	0.69703	2.08058	0.01324
5	0.01948	1.95914	1.00169	0.76976	2.35915	0.01363
10	0.02277	2.28458	1.19766	0.87605	2.79541	0.01423
15	0.02623	2.73524	1.42699	1.00924	3.36282	0.015
GR_T						
1	0.01698	1.95903	0.87573	0.76438	2.07089	0.01356
5	0.01948	1.95914	1.00169	0.76976	2.35915	0.01363
10	0.0226	1.95934	1.16339	0.77893	2.73876	0.01374
15	0.02572	1.95959	1.32983	0.79081	3.13977	0.01389
α						
0.1	0.01949	1.95914	1.01724	0.77053	2.47398	0.01366
0.5	0.01949	1.95914	1.01318	0.77033	2.4396	0.01365
1	0.01948	1.95914	1.00169	0.76976	2.35915	0.01363
1.5	0.01946	1.95914	0.98573	0.76899	2.27668	0.01361
Br						
0	0.01927	1.89015	0.96297	0.72032	2.33602	0.01332
0.1	0.01949	1.95914	1.01318	0.77033	2.4396	0.01365
0.5	0.02037	2.23513	1.21378	0.9675	2.82813	0.01488
1	0.02148	2.58011	1.45188	1.20699	3.20953	0.01624
4	0.01948	1.95914	1.00169	0.76976	2.35915	0.01363

(continued)

Table 15.10 (continued)

	$y^* = -0.8$		$y^* = 0$		$y^* = 0.8$	
M	Q_{v1}	Q_{v2}	Q_{v1}	Q_{v2}	Q_{v1}	Q_{v2}
6	0.02986	1.79587	0.9518	0.81959	1.98524	0.02438
8	0.04219	1.74833	0.93402	0.85075	1.85845	0.03718
10	0.05516	1.73296	0.92916	0.87238	1.80541	0.05062
E						
–2	0.03001	3.36255	1.62606	1.3969	3.75639	0.02417
–1	0.01948	1.95914	1.00169	0.76976	2.35915	0.01363
1	0	–0.61266	–0.01995	–0.25741	–0.2003	–0.00642
2	–0.01002	–1.78105	–0.41721	–0.65744	–1.36252	–0.01592

Table 15.11 Species concentration

	$y^* = -0.8$	$y^* = 0$	$y^* = 0.8$		$y^* = -0.8$	$y^* = 0$	$y^* = 0.8$
GR_T	c_{s1}	c_{s1}	c_{s1}	Br	c_{s1}	c_{s1}	c_{s1}
1	0.01685	0.78409	1.59338	0	0.01918	0.86393	1.71696
5	0.0194	0.9122	1.80591	0.1	0.0194	0.9122	1.80591
10	0.02268	1.09063	2.1381	0.5	0.02029	1.10529	2.16172
15	0.02613	1.29942	2.56938	1	0.02141	1.34664	2.60648
GR_c				M			
1	0.01691	0.79748	1.5848	4	0.0194	0.9122	1.80591
5	0.0194	0.9122	1.80591	6	0.02975	0.86803	1.53088
10	0.02251	1.05948	2.09686	8	0.04202	0.85303	1.44083
15	0.02562	1.21107	2.404	10	0.05492	0.84967	1.40521
α				E			
0.1	0.01949	1.01624	2.4663	–2	0.02989	1.48077	2.87534
0.5	0.01947	0.98888	2.26397	–1	0.0194	0.9122	1.80591
1	0.0194	0.9122	1.80591	1	–5.33277E–4	–0.01818	–0.15355
1.5	0.01929	0.80804	1.35854	2	–0.00998	–0.37999	–1.04358

number Br, chemical reaction parameter α and Hartmann number M. As GR_T, GR_C and Br increases, buoyancy tends to accelerate the fluid flow raising the heat transfer rate between wall and fluid and thus increases the total heat rate added to the fluid in the vertical channel. The first order chemical reaction parameter and Porous parameter reduces the total heat rate added to the fluid in both streams at all baffle positions.

The magnitude of skin friction at the left and right wall increase as GR_T and GR_C increase, and decreases with α and M at all baffle positions in both streams as shown in Table 15.13. Similar nature is also observed on the Nusselt number at the left and

Table 15.12 Total energy flow

	$y^* = -0.8$		$y^* = 0$		$y^* = 0.8$	
GR_T	E_1	E_2	E_1	E_2	E_1	E_2
1	0.01718	1.01764	0.89702	0.37076	1.59443	0.00181
5	0.0199	1.23304	1.07937	0.43271	1.98767	0.00205
10	0.02359	1.66123	1.37037	0.54261	2.69272	0.00244
15	0.02777	2.36949	1.76313	0.7053	3.74289	0.00297
GR_c						
1	0.01734	1.23239	0.90022	0.40367	1.54667	0.00175
5	0.0199	1.23304	1.07937	0.43271	1.98767	0.00205
10	0.0231	1.23414	1.35645	0.48313	2.765	0.00257
15	0.02631	1.23557	1.70711	0.55024	3.8602	0.00325
α						
0.1	0.01991	1.23305	1.10395	0.43691	2.2107	0.00219
0.5	0.01991	1.23304	1.09749	0.4358	2.14137	0.00214
1	0.0199	1.23304	1.07937	0.43271	1.98767	0.00205
1.5	0.01988	1.23304	1.05468	0.42854	1.84223	0.00195
Br						
0	0.01829	0.86564	0.71276	0.18134	1.24812	6.66304E-4
0.1	0.0199	1.23304	1.07937	0.43271	1.98767	0.00205
0.5	0.02665	2.94332	2.92158	1.76675	5.64494	0.00823
1	0.03586	5.62265	6.06982	4.17362	1.7894	0.01745
M						
4	0.0199	1.23304	1.07937	0.43271	1.98767	0.00205
6	0.02979	1.11922	1.08716	0.52123	1.5434	0.00368
8	–2.80383	1.12487	1.1434	0.61442	1.43794	–1113.94
10	–1.90679 × 109	1.15488	1.2216	0.71078	1.42704	–372321
E						
–2	0.03547	2.96291	2.82766	1.74077	4.04434	0.00763
–1	0.0199	1.23304	1.07937	0.43271	1.98767	0.00205
1	–0.0005	–0.40204	–0.02321	–0.15097	–0.17827	–0.0010
2	–0.01207	–1.77489	–0.78954	–0.85354	–1.74273	–0.00507

right walls as it can be viewed in Table 15.14. That is to say that magnitude of Nusselt number increases at both plates as GR_T, GR_C and Br increases and decreases as *alpha* and M increase at all baffle positions in both streams.

Table 15.13 Skin friction

	σ_1			σ_2		
GR_T	$y^* = -0.8$	$y^* = 0$	$y^* = 0.8$	$y^* = -0.8$	$y^* = 0$	$y^* = 0.8$
1	2.56415	6.44191	6.57903	–5.28872	–5.12096	–2.00331
5	2.95018	7.4637	7.60287	–5.50601	–5.37765	–2.04169
10	3.44136	8.82339	9.02368	–5.89122	–5.76215	–2.10164
15	3.94905	10.3187	10.6569	–6.47061	–6.25529	–2.17918
GR_c						
1	2.57139	6.52653	6.67111	–5.50591	–5.35993	–2.03506
5	2.95018	7.4637	7.60287	–5.50601	–5.37765	–2.04169
10	3.42371	8.65246	8.79515	–5.50617	–5.40786	–2.053
15	3.89731	9.86041	10.0181	–5.50639	–5.44702	–2.06763
α						
0.1	2.95172	7.54163	7.78628	–5.50601	–5.3802	–2.04478
0.5	2.95135	7.52136	7.73191	–5.50601	–5.37953	–2.04383
1	2.95018	7.4637	7.60287	–5.50601	–5.37765	–2.04169
1.5	2.94824	7.38318	7.46623	–5.50601	–5.37512	–2.03964
Br						
0	2.92909	7.33706	7.58184	–5.39674	–5.19449	–2.01023
0.1	2.95135	7.52136	7.73191	–5.50601	–5.37953	–2.04383
0.5	3.04072	8.26187	8.30113	–5.94309	–6.11029	–2.16754
1	3.15195	9.13437	8.86018	–6.48945	–7.00079	–2.30426
M						
4	2.95018	7.4637	7.60287	–5.50601	–5.37765	–2.04169
6	4.57901	8.47224	8.41997	–6.95583	–6.98652	–3.71814
8	6.58056	9.89097	9.82926	–8.69979	–8.74509	–5.77982
10	8.78968	11.5243	11.4728	–10.552	–10.5923	–8.0517
E						
–2	4.52116	11.7102	11.7408	–9.66328	–9.63963	–3.61414
–1	2.95018	7.4637	7.60287	–5.50601	–5.37765	–2.04169
1	–0.08577	–0.18818	–0.32186	2.45737	2.30527	0.99718
2	–1.55073	–3.59361	–4.10864	6.26348	5.72623	2.4636

Table 15.14 Skin friction

	Nu_1			Nu_2		
GR_T	$y^* = -0.8$	$y^* = 0$	$y^* = 0.8$	$y^* = -0.8$	$y^* = 0$	$y^* = 0.8$
1	0.2535	0.54673	0.35596	−1.1001	−1.42956	−1.46421
5	0.3172	0.71448	−1.60179	−1.27357	−1.55062	−1.60179
10	0.4691	0.99858	−1.82031	−1.57949	−1.7604	−1.82031
15	0.70134	1.36538	−2.09053	−1.98439	−2.03513	−2.09053
GR_c						
1	0.31115	0.5101	0.31827	−1.27325	−1.48418	−1.38767
5	0.3172	0.71448	0.60472	−1.27357	−1.55062	−1.60179
10	0.32741	1.06058	1.08118	−1.2741	−1.66388	−1.96631
15	0.34057	1.50739	1.68918	−1.2748	−1.81071	−2.43845
α						
0.1	0.31723	0.74361	0.73338	−1.27357	−1.56017	−1.70141
0.5	0.31722	0.73593	0.69384	−1.27357	−1.55765	−1.67072
1	0.3172	0.71448	0.60472	−1.27357	−1.55062	−1.60179
1.5	0.31715	0.68549	0.5184	−1.27357	−1.54113	−1.5354
Br						
0	−0.5	−0.5	−0.5	−0.5	−0.5	−0.5
0.1	0.31722	0.73593	0.69384	−1.27357	−1.55765	−1.67072
0.5	3.58597	5.57239	5.02362	−4.36784	−5.75312	−6.00897
1	7.67152	1.3549	9.68395	−8.23565	−10.9113	−10.854
M						
4	0.3172	0.71448	0.60472	−1.27357	−1.55062	−1.60179
6	0.72938	0.94208	0.52327	−1.36814	−1.86045	−1.87879
8	1.26914	1.25106	0.63165	−1.54831	−2.20863	−2.36426
10	1.5	1.60032	0.80544	−1.75148	−2.57817	−2.5
E						
−2	2.20731	3.00672	1.97429	−2.69175	−3.88572	−3.54052
−1	0.3172	0.71448	0.60472	−1.27357	−1.55062	−1.60179
1	0.42631	0.75733	0.59284	−1.16445	−1.50777	−1.61368
2	2.42554	3.09243	1.95052	−2.47352	−3.80001	−3.5643

Table 15.15 Nomenclature

h	Channel width	C_0	Reference concentration
h^*	Width of passage	$\overline{U_1}$	Reference velocity
y^*	Baffle position	D	Diffusion coefficients
C_p	Dimensionless specific heat at constant pressure	T_1, T_2	Dimensional Temperature distributions
g	Acceleration due to gravity	T_{w_1}, T_{w_2}	Temperatures of the boundaries
Gr	Grashof number $\left(\frac{h^3 g\beta_T \Delta T}{\nu^2}\right)$	U_1, U_2	Dimensional velocity distributions
β_T	Coefficients of thermal expansion	u_1, u_2	Non-dimensional velocities in Stream-I, Stream-II
β_c	Coefficients of concentration expansion	G_c	Modified Grashoff Number $\left(\frac{g\beta_c \Delta C h^3}{\upsilon^2}\right)$
C_1	Concentration in Stream-I	GR_T & GR_C	Dimensionless parameters $(GR_T = \frac{GR}{Re})\&(GR_C = \frac{Gc}{Re})$
K_1	Thermal conductivity of fluid	Re	Reynolds number $\frac{\overline{U_1}h}{\upsilon}$
Br	Brinkman number	p	Non-dimensional pressure gradient $\left(\frac{h^2}{\overline{U_1}\mu}\frac{dp}{dX}\right)$
α	Chemical reaction parameters	C_p	Specific heat at constant pressure
M	Hartmann number $\left(B_0 h_1\sqrt{\frac{\sigma_0}{\mu_1}}\right)$	E	Electric field load parameter
B_0	Magnetic field	E_0	Applied electric field
σ_e	Electrical conductivity		
Greek symbols			
$\Delta T, \Delta C$	Difference in temperatures and concentration	ρ	Density
μ	Viscosity	υ	Kinematics viscosity
θ_i	Non-dimensional temperature $\left(\frac{T_i - T_{w_2}}{T_{w_1} - T_{w_2}}\right)$	ϕ_1	Non-dimensional concentrations
Subscripts			
i	Refers to quantities for the fluids in Stream-I and Stream-II, respectively		

15.5 Conclusion

The characteristics of heat and mass transfer of electrically conducting fluid in a vertical double passage channel with a perfectly conducting baffle was studied. The solutions of governing equations and the associated boundary conditions have been obtained by using regular the PM method valid for small values of the Brinkman number and by the DTM valid for all values of Br. The following conclusions are made

1. Increase in thermal Grashof number, mass Grashof number and Brinkman number enhances the flow in both passages at different baffle positions.
2. The maximum velocity profiles are obtained in Stream-II when the baffle is near the left wall and in Stream-I when the baffle position is in the middle of the channel and at the right wall.
3. Increase in chemical reaction parameter decreases the velocity, temperature and concentration in Stream-I and remains unaltered in Stream-II.
4. The effect of the Hartman number is to reduce the flow at all baffle positions. The flow profiles are flat in the wider passage when compared to the narrow one. The negative electric field load parameter is to increase the velocity field and opposite effect is observed for the positive one. The temperature field is enhanced for both positive and negative values of the electric field load parameter at all baffle positions.
5. An exact agreement was obtained with the results of the DTM and the PM in the absence of the Brinkman number error increases between DTM and PM as the Brinkman number increases.
6. The results of the present model agree with the results obtained by Fasogbon [7], Srinivas Mutturajan [31] and Umavathi [33] in the absence of the baffle and the first order chemical reaction.

Acknowledgements One of the authors, J.C. Umavathi, is thankful for the financial support under the UGC-MRP F.43-66/2014 (SR) Project, and also to Prof. Maurizio Sasso, supervisor and Prof. Matteo Savino co-coordinator of the ERUSMUS MUNDUS "Featured Europe and South/south-east Asia mobility Network FUSION" for their support to do Post-Doctoral Research.

References

1. Blum, E.L., Zake, M.V., Ivanov, U.I.: Mikhailov, YuA: Heat and Mass Transfer in the Presence of an Electromagnetic Field. Zinatne, Riga (1967). (in Russian)
2. Boricic, Z., Nikodijevic, D., Obrovic, B., Stamenkovic, Z.: Universal equations of unsteady two-dimensional MHD boundary layer whose temperature varies with time. Theor. Appl. Mech. **36**(2), 119–135 (2009)
3. Cheng, C.Y.: Natural convection heat and mass transfer near a vertical wavy surface with constant wall temperature and concentration in a porous medium. Int. Commun. Heat Mass Transf. **127**, 1143–1154 (2000)

4. Cheng, C.-H., Kou, H.-S., Huang, W.-H.: Laminar fully developed forced convective flow within an asymmetrically heated horizontal double passage channel. Appl. Energy **33**, 265–286 (1989)
5. Davidson, P.A.: Pressure forces in the MHD propulsion of submersibles. Magnetohydrodynamics. **29(3)**, 49–58 (1993) (in Russian)
6. Fan, J.R., Shi, J.M., Xu, X.A.: Similarity solution of mixed convection with diffusion and chemical reaction over a horizontal moving plate. Acta Mechanica **126**, 59–69 (1998)
7. Fasogbon, P.F.: Analytical study of heat and mass transfer by free convection in a two-dimensional irregular channel. Int. J. Appl. Math. Mech. **6**(4), 17–37 (2010)
8. Hossain, M.A., Rees, D.A.S.: Combined heat mass transfer in natural convection flow from a vertical wavy surface. Acta Mechanica **136**, 133–149 (1999)
9. Jang, M.-J., Yeh, Y.-L., Chen, C.-L., Yeh, W.-C.: Differential transformation approach to thermal conductive problems with discontinuous boundary condition. Appl. Math. Comput. **216**, 2339–2350 (2010)
10. Kandasamy, R., Anjalidevi, S.P.: Effects of chemical reaction, heat and mass transfer on non-linear laminar boundary later flow over a wedge with suction or injection. J. Comput. Appl. Mech. **5**, 21–31 (2004)
11. Kessel, C.E., Meade, D., Jardin, S.C.: Physics basis and simulation of burning plasma physics for the fusion ignition research experiment (FIRE). Fus. Eng. Des. **2002**(64), 559–567 (2002)
12. Kumar, J.P., Umavathi, J.C.: Dispersion of a solute in magnetohydrodynamic two fluid flow with homogeneous and heterogeneous chemical reactions. Int. J. Math. Archive **3**(5), 1920–1939 (2012)
13. Kumar, J.P., Umavathi, J.C., Basavaraj, A.: Use of Taylor dispersion of a solute for immiscible viscous fluids between two plates. Int. J. Appl. Mech. Eng. **16**(2), 399–410 (2011)
14. Kumar, J.P., Umavathi, J.C., Madhavarao, S.: Dispersion in composite porous medium with homogeneous and heterogenous chemical reactions. Heat Tans. Asian Res. **40**(7), 608–640 (2011)
15. Kumar, J.P., Umavathi, J.C., Madhavarao, S.: Effect of homogeneous and heterogeneous reactions on the solute dispersion in composite porous medium. Int. J. Eng. Sci. Technol. **4**(2), 58–76 (2012)
16. Kumar, J.P., Umavathi, J.C., Chamkha, A.J., Basawaraj, A.: Solute dispersion between two parallel plates containing porous and fluid layers. J. Porous Media **15**(11), 1031–1047 (2012)
17. Makinde, O.D., Sibanda, P.: MHD mixed–convective flow and heat and mass transfer past a vertical plate in a porous medium with constant wall suction. J. Heat Transf. **130**, 112602/1–8 (2008)
18. Malashetty, M.S., Umavathi, J.C.: Two-phase magnetohydrodynamic flow and heat transfer in an inclined channel. Int. J. Multiph. Flow **23**(3), 545–560 (1997)
19. Malashetty, M.S., Umavathi, J.C., Kumar, J.P.: Two-fluid magnetoconvection flow in an inclined channel. Int. J. Trans. Phenom. **3**, 73–84 (2001)
20. Malashetty, M.S., Umavathi, J.C., Kumar, J.P.: Convective magnetohydrodynamic two fluid flow and heat transfer in an inclined channel. Heat Mass Transf. **37**, 259–264 (2001)
21. Muthucumaraswamy, R., Ganesan, P.: First order chemical reaction on flow past an impulsively started vertical plate with uniform heat and mass flux. Acta Mechanica **147**, 45–57 (2001)
22. Ni, Q., Zhang, Z.L., Wang, L.: Application of the differential transformation method to vibration analysis of pipes conveying fluid. Appl. Math. Comput. **217**, 7028–7038 (2011)
23. Nikodijevic, D., Boricic, Z., Blagojevic, B., Stamenkovic, Z.: Universal solutions of unsteady two-dimensional MHD boundary layer on the body with temperature gradient along surface WSEAS. Trans. Fluid Mech. **4**(3), 97–106 (2009)
24. Obrovic, B., Nikodijevic, D., Savic, S.: Boundary layer of dissociated gas on bodies of revolution of a porous contour. Strojniski Vestnik - J. Mech. Eng. **55**(4), 244–253 (2009)
25. Rashidi, M.M.: The modified differential transform method for solving MHD boundary-layer equations. Comput. Phys. Commun. **180**, 2210–2217 (2009)
26. Ravi Kanth, A.S.V., Aruna, K.: Solution of singular two-point boundary value problems using differential transformation method. Phy. Let. A. **372**, 4671–4673 (2008)

27. Rees, D.A.S., Pop, I.: A note on a free convective along a vertical wavy surface in a porous medium. ASME. J. Heat Transf. **115**, 505–508 (1994)
28. El-Din, Salah: M.M.: Fully developed laminar convection in a vertical double-passage channel. Appl. Energy **47**, 69–75 (1994)
29. Sattar, M.A.: Free and forced convection boundary layer flow through a porous medium with large suction. Int. J. Energy Res. **17**, 1–7 (1993)
30. Singh, A.K.: MHD free convection and mass transfer flow with heat source and thermal diffusion. J. Energy Heat Mass Transf. **23**, 167–178 (2001)
31. Srinivas, S., Muthuraj, R.: Effect of chemical reaction and space porosity on MHD mixed convective flow in a vertical asymmetric channel with peristalsis. Math. Comput. Model. **54**, 1213–1227 (2011)
32. Tsinober, A., Bushnell, D.M., Hefner, J.N.: MHD-drag reduction, in viscous drag reduction in boundary layers. AIAA Prog. Aeron. Astron. **123**, 327–349 (1990)
33. Umavathi, J.C.: A note on magnetoconvection in a vertical enclosure. Int. J. Nonlinear Mech. **31**(3), 371–376 (1996)
34. Umavathi, J.C.: Free convection flow of couple stress fluid for radiating medium in a vertical channel. AMSE Model. Meas. Control B **69**(8), 1–20 (2000)
35. Umavathi, J.C., Malashetty, M.S.: Magneto hydrodynamic mixed convection in a vertical channel. Int. J. Nonlinear Mech. **40**(1), 91–101 (2005)
36. Umavathi, J.C., Shekar, M.: Mixed convective flow of two immiscible viscous fluids in a vertical wavy channel with traveling thermal waves. Heat Transf. Asian Res. **40**(8), 721–743 (2011)
37. Umavathi, J.C., Chamkha, A.J., Mateen, A., Kumar, J.P.: Unsteady magneto hydrodynanmic two fluid flow and heat transfer in a horizontal channel. Int. J. Heat Tech. **26**(2), 121–133 (2008)
38. Umavathi, J.C., Liu, I.C., Kumar, J.P., Pop, I.: Fully developed magneto convection flow in a vertical rectangular duct. Heat Mass Transf. **47**, 1–11 (2011)
39. Xu, H., Liao, S.J., Pop, I.: Series solutions of unsteady three-dimensional MHD flow and heat transfer in the boundary layer over an impulsively stretching plate. Eur. J. Mech. B/Fluids **26**, 15–27 (2007)
40. Yaghoobi, H., Torabi, M.: The application of differential transformation method to nonlinear equations arising in heat transfer. Int. Commun. Heat Mass Transf. **38**, 815–820 (2011)
41. Yao, L.S.: Natural convection along a wavy surface. ASME J. Heat Transf. **105**, 465–468 (1983)
42. Zhou, J.K.: Differential transformation and its applications for electrical circuits. Huarjung University Press (1986) (in Chinese)

Chapter 16
Spectral Expansion of Three-Dimensional Elasticity Tensor Random Fields

Anatoliy Malyarenko and Martin Ostoja-Starzewski

Abstract We consider a random field model of the 21-dimensional elasticity tensor. Representation theory is used to obtain the spectral expansion of the model in terms of stochastic integrals with respect to random measures.

Keywords Random field · Elasticity · Spectral expansion

16.1 Introduction

16.1.1 Motivation

In many problems of continuum physics there is a need to account for spatially random phenomena using the random field (RF) concept. This implies working with tensor-valued RFs or, simply, tensor random fields (TRFs). The key to using TRFs is the ability to write their main characteristics explicitly, to determine restrictions implied by the underlying physics, and also to generate the fields involved. Working in the setting of second-order RFs, rather than the strict-sense stationary RFs, the focus is on the first and second moments. Thus, to proceed further, explicit representations of TRFs have to be determined.

In this chapter we consider the 4th rank TRF of elasticity (or stiffness) tensor in classical elasticity. The motivation for treating this tensor as a random field is that almost all the materials encountered in nature as well those produced by man, except for the purest crystals, possess some degree of disorder or inhomogeneity. At the same time, elasticity is the starting point for any solid mechanics model.

A. Malyarenko (✉)
Division of Applied Mathematics School of Education Culture
and Communication, Mälardalen University, 883, 721 23 Västerås, Sweden
e-mail: anatoliy.malyarenko@mdh.se

M. Ostoja-Starzewski
University of Illinois at Urbana-Champaign, Champaign, USA
e-mail: martinos@illinois.edu

S. Silvestrov and M. Rančić (eds.), *Engineering Mathematics I*,
Springer Proceedings in Mathematics & Statistics 178,
DOI 10.1007/978-3-319-42082-0_16

To be more specific, we consider a deformable body occupying the region D in the space domain $\mathbb{R}^3$. The *stress tensor* $\sigma(\mathbf{x})\colon D \to \mathsf{S}^2(\mathbb{R}^3)$ describes the state of stress at a point $\mathbf{x}$ inside the body. As a consequence of the linear momentum balance, the stress field satisfies the equation

$$\sigma_{ji,j} = \rho \ddot{u}_i, \tag{16.1}$$

where ρ is the mass density and $\ddot{u}_i$ is the material time derivative of u_i. As a consequence of the angular momentum balance,

$$\sigma_{ji} = \sigma_{ij}. \tag{16.2}$$

Here $\varepsilon(\mathbf{x})$ is the *strain tensor*: a function defined on D and taking values in the space $\mathsf{S}^2(\mathbb{R}^3)$ of symmetric rank 2 tensors over $\mathbb{R}^3$.

Next, the deformation of the body around a point $\mathbf{x}$ is described by the symmetrised gradient of the displacement field u:

$$\varepsilon_{ij} = u_{(i,j)}. \tag{16.3}$$

Assuming the body to be *linear hyperelastic*, at each point $\mathbf{x}$ there exists a strain energy function $w(\varepsilon) = \frac{1}{2}\varepsilon_{ij}C_{ijkl}\varepsilon_{kl}$ such that

$$\sigma_{ij} = \frac{\partial w}{\partial \varepsilon_{ij}}. \tag{16.4}$$

In view of (16.2)–(16.4), the *elasticity tensor* C has these well-known symmetries

$$C_{ijkl} = C_{jikl} = C_{jilk} = C_{klji}, \tag{16.5}$$

which implies that of all the $3^4 = 81$ components only 21 are independent. In the case (16.4) does not apply, the last equality in (16.5) does not hold, and there are 36 independent components – this is called *Cauchy elasticity*. Whether we have a hyperelastic or a Cauchy elastic body, the stress-strain relation is written in the form of Hooke's law

$$\sigma(\mathbf{x}) = C(\mathbf{x})\varepsilon(\mathbf{x}), \qquad \mathbf{x} \in D, \tag{16.6}$$

where $\mathbf{x}$ indicates the dependence of all the fields on the position in $D \subset \mathbb{R}^3$. Thus, $C(\mathbf{x})$ is a function on D with values in the linear space $\mathsf{S}^2(\mathsf{S}^2(\mathbb{R}^3))$ of symmetric linear operators over $\mathsf{S}^2(\mathbb{R}^3)$.

16.1.2 Elasticity Tensor as a Rank 4 Tensor Random Field

The tensor field $C(\mathbf{x})$ may depend on time, temperature, pressure, microstructure, and other physical variables. We suppose that $C(\mathbf{x})$ is the restriction to D of a *second-*

order mean-square continuous random field. We denote the above field by the same symbol $C(\mathbf{x})$, $\mathbf{x} \in \mathbb{R}^3$. This means the following:

1. Random field: C is such a function of two variables $C(\mathbf{x}, \omega)\colon \mathbb{R}^3 \times \Omega \to \mathsf{S}^2(\mathsf{S}^2(\mathbb{R}^3))$ that for any $\mathbf{x}_0 \in \mathbb{R}^3$ the function of one variable $C(\mathbf{x}_0, \omega)$ is a random tensor on a probability space $(\Omega, \mathfrak{F}, \mathsf{P})$.
2. Second-order: $\mathsf{E}[\|C(\mathbf{x})\|^2] < \infty$, where $\|\cdot\|$ is the norm on the space $\mathsf{S}^2(\mathsf{S}^2(\mathbb{R}^3))$ generated by the standard norm on $\mathbb{R}^3$.
3. Mean-square continuous:

$$\lim_{\mathbf{x}\to\mathbf{x}_0} \mathsf{E}[\|C(\mathbf{x}) - C(\mathbf{x}_0)\|^2] = 0, \qquad \mathbf{x}_0 \in \mathbb{R}^3.$$

If one shifts the origin of a coordinate system in the space domain, the random field $C(\mathbf{x})$ does not change its value. It follows that $C(\mathbf{x})$ is *strictly homogeneous*: for any positive integer n, for any distinct points $\mathbf{x}_1, \dots, \mathbf{x}_n$ in the space domain, and for any shift $\mathbf{x} \in \mathbb{R}^3$, the random vectors $(C(\mathbf{x}_1), \dots, C(\mathbf{x}_n))^\top$ and $(C(\mathbf{x}_1 + \mathbf{x}), \dots, C(\mathbf{x}_n + \mathbf{x}))^\top$ are identically distributed. In particular, the mean value

$$E(\mathbf{x}) = \mathsf{E}[C(\mathbf{x})]$$

is a constant rank 4 tensor, while the correlation function

$$B(\mathbf{x}, \mathbf{y}) = \mathsf{E}[(C(\mathbf{x}) - E(\mathbf{x})) \otimes (C(\mathbf{y}) - E(\mathbf{y}))]$$

is a rank 8 tensor that depends only on the difference $\mathbf{x} - \mathbf{y}$. Such a field is called *wide-sense homogeneous*. In what follows, we consider only wide-sense homogeneous random fields and omit words "wide-sense".

Apply an arbitrary rotation k to the tensor field $C(\mathbf{x})$. After the rotation k the point $\mathbf{x}$ becomes the point $k\mathbf{x}$. Evidently, the tensor $C(\mathbf{x})$ is transformed by the rotation into the tensor $\mathsf{S}^2(\mathsf{S}^2(k))C(\mathbf{x})$, where $\mathsf{S}^2(k)$ is the shortcut for $\mathsf{S}^2(\theta_1(k))$. It follows that for any positive integer n, for any distinct points $\mathbf{x}_1, \dots, \mathbf{x}_n$ in the space domain, and for any rotation k, the random vectors $(C(k\mathbf{x}_1), \dots, C(k\mathbf{x}_n))^\top$ and $(\mathsf{S}^2(\mathsf{S}^2(k))C(\mathbf{x}_1), \dots, \mathsf{S}^2(\mathsf{S}^2(k))C(\mathbf{x}_n))^\top$ are identically distributed. Such a rank 4 tensor field is called *strictly isotropic*.

In particular, the expected value of the rotated field is not changed:

$$E(k\mathbf{x}) = \mathsf{E}[C(k\mathbf{x})] = \mathsf{E}[\mathsf{S}^2(\mathsf{S}^2(k))C(\mathbf{x})] = \mathsf{S}^2(\mathsf{S}^2(k))\mathsf{E}[C(\mathbf{x})] = \mathsf{S}^2(\mathsf{S}^2(k))E(\mathbf{x}).$$

The correlation function is not changed as well:

$$\begin{aligned} B(k\mathbf{x}, k\mathbf{y}) &= \mathsf{E}[(C(k\mathbf{x}) - E(k\mathbf{x})) \otimes (C(k\mathbf{y}) - E(k\mathbf{y}))] = \\ &= \mathsf{E}[(\mathsf{S}^2(\mathsf{S}^2(k))(C(\mathbf{x}) - E(\mathbf{x}))) \otimes (\mathsf{S}^2(\mathsf{S}^2(k))(C(\mathbf{y}) - E(\mathbf{y})))] = \\ &= [\mathsf{S}^2(\mathsf{S}^2(k)) \otimes \mathsf{S}^2(\mathsf{S}^2(k))]B(\mathbf{x}, \mathbf{y}). \end{aligned}$$

We call the random field $C(\mathbf{x})$ *wide-sense isotropic* if for any rotation k we have

$$\begin{aligned} E(k\mathbf{x}) &= \mathsf{S}^2(\mathsf{S}^2(k))E(\mathbf{x}), \\ B(k\mathbf{x}, k\mathbf{y}) &= [\mathsf{S}^2(\mathsf{S}^2(k)) \otimes \mathsf{S}^2(\mathsf{S}^2(k))]B(\mathbf{x}, \mathbf{y}). \end{aligned} \tag{16.7}$$

Again, we consider only wide-sense isotropic random fields and omit words "wide-sense".

We would like to find spectral expansion of a homogeneous and isotropic elasticity tensor field $C(\mathbf{x})$.

16.1.3 Local Versus Wide-Sense Isotropy

Before we proceed, it is important to observe the difference between wide-sense isotropy introduced above and the *local isotropy* of C well known in the theory of elasticity. The latter is expressed by this very specific form

$$C_{ijkl} = \lambda\delta_{ij}\delta_{kl} + \mu\left(\delta_{ik}\delta_{jl} + \delta_{il}\delta_{jk}\right), \tag{16.8}$$

where λ and μ are the so-called *Lamé constants*. Of course, in a random elastic medium, λ and μ are scalar random fields related to C_{ijkl} by

$$C_{1122} = \lambda, \quad C_{1212} = \mu, \tag{16.9}$$

with the corresponding strain energy being $w(\varepsilon) = \frac{1}{2}\lambda\left(\varepsilon_{ii}\right)^2 + \mu\varepsilon_{ij}\varepsilon_{ij}$.

In a symbolic (bold letter) notation, the local isotropy is expressed as

$$\mathbf{C} = \lambda\left(\mathbf{1} \otimes \mathbf{1}\right) + 2\mu\mathbf{1}_4^s, \tag{16.10}$$

where $\mathbf{1} = \delta_{ij}\delta_{kl}$ and $\mathbf{1}_4^s$ is the symmetric part of $\mathbf{1}_4 = \mathbf{1}\overline{\otimes}\mathbf{1}$. For reference, we give the key relations pertaining to isotropic rank 4 tensors

$$\begin{aligned} &\mathbf{1} \otimes \mathbf{1} \Leftrightarrow \delta_{ij}\delta_{kl}, \quad \mathbf{1}_4 = \mathbf{1}\overline{\otimes}\mathbf{1} \Leftrightarrow 1_{ijkl} = \delta_{ik}\delta_{jl}, \quad \mathbf{1}\underline{\otimes}\mathbf{1} \Leftrightarrow \delta_{il}\delta_{jk}, \\ &\mathbf{1}_4 = \mathbf{1}_4^s + \mathbf{1}_4^a, \quad \mathbf{1}_4^s = \left(\mathbf{1}\overline{\otimes}\mathbf{1} + \mathbf{1}\underline{\otimes}\mathbf{1}\right)/2, \quad \mathbf{1}_4^a = \left(\mathbf{1}\overline{\otimes}\mathbf{1} - \mathbf{1}\underline{\otimes}\mathbf{1}\right)/2. \end{aligned} \tag{16.11}$$

Note that the tensor random field C may either be wide-sense isotropic or anisotropic and, simultaneously, locally isotropic or anisotropic. There are four distinct possibilities, but which one is actually the case depends on the microstructure and random morphology of the body.

16.2 The Results

Lomakin [4] found the correlation tensor of an elasticity random field. To formulate his result, we introduce the *Ogden tensors* after Ogden [6]. Let ν be a nonnegative integer. The Ogden tensor I^{ν} of rank $2\nu+2$ is determined inductively as

$$
\begin{aligned}
\mathsf{I}^0_{ij} &= \delta_{ij}, \qquad \mathsf{I}^1_{ijk\ell} = \frac{1}{2}(\delta_{ik}\delta_{j\ell}+\delta_{i\ell}\delta_{jk}),\\
\mathsf{I}^{\nu}_{i_1\dots i_{2\nu+2}} &= \nu^{-1}(\mathsf{I}^1_{i_1pi_3i_4}\mathsf{I}^{\nu-1}_{pi_2i_5\dots i_{2\nu+2}}+\dots+\mathsf{I}^1_{i_1pi_{2\nu+1}i_{2\nu+2}}\mathsf{I}^{\nu-1}_{pi_2\dots i_{2\nu-1}i_{2\nu}}),
\end{aligned}
\tag{16.12}
$$

where there is a summation over p.

Lomakin [4] stated without proof that the correlation tensor of an elasticity random field has the form

$$
B_{ijk\ell i'j'k'\ell'}(\mathbf{x},\mathbf{y})=\sum L^m_{ijk\ell i'j'k'\ell'}(\mathbf{x}-\mathbf{y})B_m(\|\mathbf{x}-\mathbf{y}\|),
$$

where the 15 functions $L^m_{ijk\ell i'j'k'\ell'}$ are shown in Table 16.1 in Roman.

In fact, Lomakin missed 14 functions. Our first result is

Theorem 16.1 *The correlation tensor of an elasticity random field has the form*

$$
B_{ijk\ell i'j'k'\ell'}(\mathbf{x},\mathbf{y})=\sum_{m=1}^{29} L^m_{ijk\ell i'j'k'\ell'}(\mathbf{x}-\mathbf{y})B_m(\|\mathbf{x}-\mathbf{y}\|),
$$

where all 29 functions $L^m_{ijk\ell i'j'k'\ell'}$ *are shown in Table 16.1.*

The complete description of the 29 functions $B_m(\|\mathbf{x}-\mathbf{y}\|)$ occupies about 20 pages of formulae and will be published elsewhere. Here we consider a subclass of the class of elasticity random fields for which the formulae become simpler. The next result is

Theorem 16.2 *The expected value of the elasticity random field has the form*

$$
E_{ijk\ell}(\mathbf{x})=C_1\delta_{ij}\delta_{k\ell}+C_2(\delta_{ik}\delta_{j\ell}+\delta_{i\ell}\delta_{jk}),\qquad C_1,C_2\in\mathbb{R}.
$$

The tensor-valued function

$$
B_{ijk\ell i'j'k'\ell'}(\mathbf{x},\mathbf{y})=\sum_{n=1}^{13}\int_0^\infty\sum_{q=1}^{29}N_{nq}(\lambda,\|\mathbf{y}-\mathbf{x}\|)L^q_{ijk\ell i'j'k'\ell'}(\mathbf{y}-\mathbf{x})\,\mathrm{d}\Phi_n(\lambda),
$$

where the functions $N_{nq}(\lambda,\rho)$ *are given in Table 16.2, and where* Φ_n *are finite Radon measures on the interval* $[0,\infty)$ *satisfying the conditions*

Table 16.1 Lomakin's functions

m	$\mathbf{L}^{\mathbf{m}}_{\mathbf{ijk}\ell\mathbf{i'j'k'}\ell'}(\mathbf{x})$
1	2
1	$\delta_{ij}\delta_{kl}\delta_{i'j'}\delta_{k'\ell'}$
2	$2(\delta_{ij}\delta_{k\ell}I_{i'j'k'\ell'}+\delta_{i'j'}\delta_{k'\ell'}I_{ijk\ell})$
3	$\boldsymbol{2(\delta_{ij}(\delta_{i'j'}I_{k\ell k'\ell'}+\delta_{k'\ell'}I_{k\ell i'j'})+\delta_{k\ell}(\delta_{i'j'}I_{ijk'\ell'}+\delta_{k'\ell'}I_{iji'j'}))}$
4	$4I_{ijk\ell}I_{i'j'k'\ell'}$
5	$\boldsymbol{8(\delta_{ij}I_{k\ell i'j'k'\ell'}+\delta_{k\ell}I_{iji'j'k'\ell'}+\delta_{i'j'}I_{ijk\ell k'\ell'}+\delta_{k'\ell'}I_{ijk\ell i'j'})}$
6	$\boldsymbol{4(I_{iji'j'}I_{k\ell k'\ell'}+I_{ijk'\ell'}I_{k\ell i'j'})}$
7	$\boldsymbol{4(I_{iji'k'}I_{k\ell j'\ell'}+I_{iji'\ell'}I_{k\ell j'k'}+I_{ijj'k'}I_{k\ell i'\ell'}+I_{ijj'\ell'}I_{k\ell i'k'})}$
8	$\delta_{ij}\delta_{k\ell}(\delta_{i'j'}x_{k'}x_{\ell'}+\delta_{k'\ell'}x_{i'}x_{j'})+\delta_{i'j'}\delta_{k'\ell'}(\delta_{ij}x_kx_\ell+\delta_{k\ell}x_ix_j)$
9	$2(I_{ijk\ell}(\delta_{i'j'}x_{k'}x_{\ell'}+\delta_{k'\ell'}x_{i'}x_{j'})+I_{i'j'k'\ell'}(\delta_{ij}x_kx_\ell+\delta_{k\ell}x_ix_j))$
10	$\delta_{ij}\delta_{k\ell}(\delta_{i'k'}x_{j'}x_{\ell'}+\delta_{i'\ell'}x_{j'}x_{k'}+\delta_{j'k'}x_{i'}x_{\ell'}+\delta_{j'\ell'}x_{i'}x_{k'})$ $+\delta_{i'j'}\delta_{k'\ell'}(\delta_{ik}x_jx_\ell+\delta_{i\ell}x_jx_k+\delta_{jk}x_ix_\ell+\delta_{j\ell}x_ix_k)$
11	$\boldsymbol{\delta_{ij}\delta_{i'j'}(\delta_{kk'}x_\ell x_{\ell'}+\delta_{k\ell'}x_\ell x_{k'}+\delta_{\ell k'}x_kx_{\ell'}+\delta_{\ell\ell'}x_kx_{k'})}$ $\boldsymbol{+\delta_{ij}\delta_{k'\ell'}(\delta_{ki'}x_\ell x_{j'}+\delta_{kj'}x_\ell x_{j'}+\delta_{\ell i'}x_kx_{j'}+\delta_{\ell j'}x_kx_{i'})}$ $\boldsymbol{+\delta_{k\ell}\delta_{i'j'}(\delta_{ik'}x_jx_{\ell'}+\delta_{jk'}x_jx_{\ell'}+\delta_{i\ell'}x_jx_{k'}+\delta_{j\ell'}x_ix_{k'})}$ $\boldsymbol{+\delta_{k\ell}\delta_{k'\ell'}(\delta_{ii'}x_jx_{j'}+\delta_{ij'}x_jx_{i'}+\delta_{ji'}x_ix_{j'}+\delta_{jj'}x_ix_{i'})}$
12	$2(I_{ijk\ell}(\delta_{i'k'}x_{j'}x_{\ell'}+\delta_{i'\ell'}x_{j'}x_{k'}+\delta_{j'k'}x_{i'}x_{\ell'}+\delta_{j'\ell'}x_{i'}x_{k'})$ $+I_{i'j'k'\ell'}(\delta_{ik}x_jx_\ell+\delta_{i\ell}x_jx_k+\delta_{jk}x_ix_\ell+\delta_{j\ell}x_ix_k))$
13	$\boldsymbol{2((\delta_{ij}I_{k\ell i'j'}+\delta_{k\ell}I_{iji'j'})x_{k'}x_{\ell'}+(\delta_{ij}I_{k\ell k'\ell'}+\delta_{k\ell}I_{ijk'\ell'})x_{i'}x_{j'}}$ $\boldsymbol{+(\delta_{i'j'}I_{ijk'\ell'}+\delta_{k'\ell'}I_{iji'j'})x_kx_\ell+(\delta_{i'j'}I_{k\ell k'\ell'}+\delta_{k'\ell'}I_{k\ell i'j'})x_ix_j)}$
14	$\boldsymbol{2((\delta_{ij}I_{k\ell i'k'}+\delta_{k\ell}I_{iji'k'})x_{j'}x_{\ell'}+(\delta_{ij}I_{k\ell i'\ell'}+\delta_{k\ell}I_{iji'\ell'})x_{j'}x_{k'}}$ $\boldsymbol{+(\delta_{ij}I_{k\ell j'k'}+\delta_{k\ell}I_{ijj'k'})x_{i'}x_{\ell'}+(\delta_{ij}I_{k\ell j'\ell'}+\delta_{k\ell}I_{ijj'\ell'})x_{i'}x_{k'}}$ $\boldsymbol{+(\delta_{i'j'}I_{ikk'\ell'}+\delta_{k'\ell'}I_{iki'j'})x_jx_\ell+(\delta_{i'j'}I_{i\ell k'\ell'}+\delta_{k'\ell'}I_{i\ell i'j'})x_jx_k}$ $\boldsymbol{+(\delta_{i'j'}I_{jkk'\ell'}+\delta_{k'\ell'}I_{jki'j'})x_ix_\ell+(\delta_{i'j'}I_{j\ell k'\ell'}+\delta_{k'\ell'}I_{j\ell i'j'})x_ix_k)}$
15	$\boldsymbol{8(I_{ijk\ell i'j'}x_{k'}x_{\ell'}+I_{ijk\ell k'\ell'}x_{i'}x_{j'}+I_{iji'j'k'\ell'}x_kx_\ell+I_{k\ell i'j'k'\ell'}x_ix_j)}$
16	$\boldsymbol{8(I_{ijk\ell i'k'}x_{j'}x_{\ell'}+I_{ijk\ell i'\ell'}x_{j'}x_{k'}+I_{ijk\ell j'k'}x_{i'}x_{\ell'}+I_{ijk\ell j'\ell'}x_{i'}x_{k'}}$ $\boldsymbol{+I_{iki'j'k'\ell'}x_jx_\ell+I_{i\ell i'j'k'\ell'}x_jx_k+I_{jki'j'k'\ell'}x_ix_\ell+I_{j\ell i'j'k'\ell'}x_ix_k)}$
17	$\boldsymbol{2(I_{iji'j'}(\delta_{kk'}x_\ell x_{\ell'}+\delta_{k\ell'}x_\ell x_{k'}+\delta_{\ell k'}x_kx_{\ell'}+\delta_{\ell\ell'}x_kx_{k'})}$ $\boldsymbol{+I_{ijk'\ell'}(\delta_{ki'}x_\ell x_{j'}+\delta_{kj'}x_\ell x_{i'}+\delta_{\ell i'}x_kx_{j'}+\delta_{\ell j'}x_kx_{i'}}$ $\boldsymbol{+I_{k\ell i'j'}(\delta_{ik'}x_jx_{\ell'}+\delta_{i\ell'}x_jx_{k'}+\delta_{jk'}x_ix_{\ell'}+\delta_{j\ell'}x_ix_{k'})}$ $\boldsymbol{+I_{k\ell k'\ell'}(\delta_{ii'}x_jx_{j'}+\delta_{ij'}x_jx_{i'}+\delta_{ji'}x_ix_{j'}+\delta_{jj'}x_ix_{i'}))}$
18	$\delta_{ij}\delta_{k\ell}x_{i'}x_{j'}x_{k'}x_{\ell'}+\delta_{i'j'}\delta_{k'\ell'}x_ix_jx_kx_\ell$
19	$(\delta_{ij}x_kx_\ell+\delta_{k\ell}x_ix_j)(\delta_{i'j'}x_{k'}x_{\ell'}+\delta_{k'\ell'}x_{i'}x_{j'})$
20	$2(I_{ijk\ell}x_{i'}x_{j'}x_{k'}x_{\ell'}+I_{i'j'k'\ell'}x_ix_jx_kx_\ell)$
21	$(\delta_{ij}x_kx_\ell+\delta_{k\ell}x_ix_j)(\delta_{i'k'}x_{j'}x_{\ell'}+\delta_{i'\ell'}x_{j'}x_{k'}+\delta_{j'k'}x_{i'}x_{\ell'}+\delta_{j'\ell'}x_{i'}x_{k'})$ $+(\delta_{i'j'}x_{k'}x_{\ell'}+\delta_{k'\ell'}x_{i'}x_{j'})(\delta_{ik}x_jx_\ell+\delta_{i\ell}x_jx_k+\delta_{jk}x_ix_\ell+\delta_{j\ell}x_ix_k)$

(continued)

Table 16.1 (continued)

m	$\mathbf{L}^{\mathbf{m}}_{\mathbf{ijk\ell i'j'k'\ell'}}(\mathbf{x})$
1	2
22	$\delta_{\mathbf{ij}}(\delta_{\mathbf{ki'}}\mathbf{x}_\ell\mathbf{x}_{\mathbf{j'}}\mathbf{x}_{\mathbf{k'}}\mathbf{x}_{\ell'}+\delta_{\mathbf{kj'}}\mathbf{x}_\ell\mathbf{x}_{\mathbf{i'}}\mathbf{x}_{\mathbf{k'}}\mathbf{x}_{\ell'}+\delta_{\mathbf{kk'}}\mathbf{x}_\ell\mathbf{x}_{\mathbf{i'}}\mathbf{x}_{\mathbf{j'}}\mathbf{x}_{\ell'}+\delta_{\mathbf{k}\ell'}\mathbf{x}_\ell\mathbf{x}_{\mathbf{i'}}\mathbf{x}_{\mathbf{j'}}\mathbf{x}_{\mathbf{k'}}$ $+\delta_{\ell\mathbf{i'}}\mathbf{x}_{\mathbf{k}}\mathbf{x}_{\mathbf{j'}}\mathbf{x}_{\mathbf{k'}}\mathbf{x}_{\ell'}+\delta_{\ell\mathbf{j'}}\mathbf{x}_{\mathbf{k}}\mathbf{x}_{\mathbf{i'}}\mathbf{x}_{\mathbf{k'}}\mathbf{x}_{\ell'}+\delta_{\ell\mathbf{k'}}\mathbf{x}_{\mathbf{k}}\mathbf{x}_{\mathbf{i'}}\mathbf{x}_{\mathbf{j'}}\mathbf{x}_{\ell'}+\delta_{\ell\ell'}\mathbf{x}_{\mathbf{k}}\mathbf{x}_{\mathbf{i'}}\mathbf{x}_{\mathbf{j'}}\mathbf{x}_{\mathbf{k'}})$ $+\delta_{\mathbf{k}\ell}(\delta_{\mathbf{ii'}}\mathbf{x}_{\mathbf{j}}\mathbf{x}_{\mathbf{j'}}\mathbf{x}_{\mathbf{k'}}\mathbf{x}_{\ell'}+\delta_{\mathbf{ij'}}\mathbf{x}_{\mathbf{j}}\mathbf{x}_{\mathbf{i'}}\mathbf{x}_{\mathbf{k'}}\mathbf{x}_{\ell'}+\delta_{\mathbf{ik'}}\mathbf{x}_{\mathbf{j}}\mathbf{x}_{\mathbf{i'}}\mathbf{x}_{\mathbf{j'}}\mathbf{x}_{\ell'}+\delta_{\mathbf{i}\ell'}\mathbf{x}_{\mathbf{j}}\mathbf{x}_{\mathbf{i'}}\mathbf{x}_{\mathbf{j'}}\mathbf{x}_{\mathbf{k'}}$ $+\delta_{\mathbf{ji'}}\mathbf{x}_{\mathbf{i}}\mathbf{x}_{\mathbf{j'}}\mathbf{x}_{\mathbf{k'}}\mathbf{x}_{\ell'}+\delta_{\mathbf{jj'}}\mathbf{x}_{\mathbf{i}}\mathbf{x}_{\mathbf{i'}}\mathbf{x}_{\mathbf{k'}}\mathbf{x}_{\ell'}+\delta_{\mathbf{jk'}}\mathbf{x}_{\mathbf{i}}\mathbf{x}_{\mathbf{i'}}\mathbf{x}_{\mathbf{j'}}\mathbf{x}_{\ell'}+\delta_{\mathbf{j}\ell'}\mathbf{x}_{\mathbf{i}}\mathbf{x}_{\mathbf{i'}}\mathbf{x}_{\mathbf{j'}}\mathbf{x}_{\mathbf{k'}})$ $+\delta_{\mathbf{i'j'}}(\delta_{\mathbf{ik'}}\mathbf{x}_{\mathbf{j}}\mathbf{x}_{\mathbf{k}}\mathbf{x}_\ell\mathbf{x}_{\ell'}+\delta_{\mathbf{jk'}}\mathbf{x}_{\mathbf{i}}\mathbf{x}_{\mathbf{k}}\mathbf{x}_\ell\mathbf{x}_{\ell'}+\delta_{\mathbf{kk'}}\mathbf{x}_{\mathbf{i}}\mathbf{x}_{\mathbf{j}}\mathbf{x}_\ell\mathbf{x}_{\ell'}+\delta_{\ell\mathbf{k'}}\mathbf{x}_{\mathbf{i}}\mathbf{x}_{\mathbf{j}}\mathbf{x}_{\mathbf{k}}\mathbf{x}_{\ell'}$ $+\delta_{\mathbf{i}\ell'}\mathbf{x}_{\mathbf{k}}\mathbf{x}_{\ell'}\mathbf{x}_{\mathbf{j'}}\mathbf{x}_{\mathbf{k'}}+\delta_{\mathbf{j}\ell'}\mathbf{x}_{\mathbf{k}}\mathbf{x}_{\ell'}\mathbf{x}_{\mathbf{i'}}\mathbf{x}_{\mathbf{k'}}+\delta_{\mathbf{k}\ell'}\mathbf{x}_{\mathbf{i}}\mathbf{x}_{\mathbf{j}}\mathbf{x}_{\mathbf{k'}}\mathbf{x}_{\ell'}+\delta_{\ell\ell'}\mathbf{x}_{\mathbf{i}}\mathbf{x}_{\mathbf{j}}\mathbf{x}_{\mathbf{k}}\mathbf{x}_{\mathbf{k'}})$ $+\delta_{\mathbf{k'}\ell'}(\delta_{\mathbf{ii'}}\mathbf{x}_{\mathbf{j}}\mathbf{x}_{\mathbf{k}}\mathbf{x}_\ell\mathbf{x}_{\mathbf{j'}}+\delta_{\mathbf{ji'}}\mathbf{x}_{\mathbf{i}}\mathbf{x}_{\mathbf{k}}\mathbf{x}_{\mathbf{j'}}\mathbf{x}_{\ell'}+\delta_{\mathbf{ki'}}\mathbf{x}_{\mathbf{j}}\mathbf{x}_{\mathbf{j}}\mathbf{x}_\ell\mathbf{x}_{\mathbf{j'}}+\delta_{\ell\mathbf{i'}}\mathbf{x}_{\mathbf{i}}\mathbf{x}_{\mathbf{j}}\mathbf{x}_{\mathbf{k}}\mathbf{x}_{\mathbf{j'}}$ $+\delta_{\mathbf{ij'}}\mathbf{x}_{\mathbf{j}}\mathbf{x}_{\mathbf{k}}\mathbf{x}_\ell\mathbf{x}_{\mathbf{i'}}+\delta_{\mathbf{jj'}}\mathbf{x}_{\mathbf{i}}\mathbf{x}_{\mathbf{k}}\mathbf{x}_\ell\mathbf{x}_{\mathbf{i'}}+\delta_{\mathbf{kj'}}\mathbf{x}_{\mathbf{i}}\mathbf{x}_{\mathbf{j}}\mathbf{x}_\ell\mathbf{x}_{\mathbf{i'}}+\delta_{\ell\mathbf{j'}}\mathbf{x}_{\mathbf{i}}\mathbf{x}_{\mathbf{j}}\mathbf{x}_{\mathbf{k}}\mathbf{x}_{\mathbf{i'}})$
23	$(\delta_{ik}x_jx_\ell+\delta_{i\ell}x_jx_k+\delta_{jk}x_ix_\ell+\delta_{j\ell}x_ix_k)$ $\times(\delta_{i'k'}x_{j'}x_{\ell'}+\delta_{i'\ell'}x_{j'}x_{k'}+\delta_{j'k'}x_{i'}x_{\ell'}+\delta_{j'\ell'}x_{i'}x_{k'})$
24	$\mathbf{2}(\mathbf{I}_{\mathbf{iji'j'}}\mathbf{x}_{\mathbf{k}}\mathbf{x}_\ell\mathbf{x}_{\mathbf{k'}}\mathbf{x}_{\ell'}+\mathbf{I}_{\mathbf{ijk'}\ell'}\mathbf{x}_{\mathbf{k}}\mathbf{x}_\ell\mathbf{x}_{\mathbf{i'}}\mathbf{x}_{\mathbf{j'}}+\mathbf{I}_{\mathbf{k}\ell\mathbf{i'j'}}\mathbf{x}_{\mathbf{i}}\mathbf{x}_{\mathbf{j}}\mathbf{x}_{\mathbf{k'}}\mathbf{x}_{\ell'}+\mathbf{I}_{\mathbf{k}\ell\mathbf{k'}\ell'}\mathbf{x}_{\mathbf{i}}\mathbf{x}_{\mathbf{j}}\mathbf{x}_{\mathbf{i'}}\mathbf{x}_{\mathbf{j'}})$
25	$\mathbf{2}[(\mathbf{I}_{\mathbf{iji'k'}}\mathbf{x}_{\mathbf{j'}}\mathbf{x}_{\ell'}+\mathbf{I}_{\mathbf{iji'}\ell'}\mathbf{x}_{\mathbf{j'}}\mathbf{x}_{\mathbf{k'}}+\mathbf{I}_{\mathbf{ijj'k'}}\mathbf{x}_{\mathbf{i'}}\mathbf{x}_{\ell'}+\mathbf{I}_{\mathbf{ijj'}\ell'}\mathbf{x}_{\mathbf{i'}}\mathbf{x}_{\mathbf{k'}})\mathbf{x}_{\mathbf{k}}\mathbf{x}_\ell$ $+(\mathbf{I}_{\mathbf{iki'j'}}\mathbf{x}_{\mathbf{k'}}\mathbf{x}_{\ell'}+\mathbf{I}_{\mathbf{ikk'}\ell'}\mathbf{x}_{\mathbf{i'}}\mathbf{x}_{\mathbf{j'}})\mathbf{x}_{\mathbf{j}}\mathbf{x}_\ell+(\mathbf{I}_{\mathbf{i}\ell\mathbf{i'j'}}\mathbf{x}_{\mathbf{k'}}\mathbf{x}_{\ell'}+\mathbf{I}_{\mathbf{i}\ell\mathbf{k'}\ell'}\mathbf{x}_{\mathbf{i'}}\mathbf{x}_{\mathbf{j'}})\mathbf{x}_{\mathbf{j}}\mathbf{x}_{\mathbf{k}}$ $+(\mathbf{I}_{\mathbf{jki'j'}}\mathbf{x}_{\mathbf{k'}}\mathbf{x}_{\ell'}+\mathbf{I}_{\mathbf{jkk'}\ell'}\mathbf{x}_{\mathbf{i'}}\mathbf{x}_{\mathbf{j'}})\mathbf{x}_{\mathbf{i}}\mathbf{x}_\ell+(\mathbf{I}_{\mathbf{j}\ell\mathbf{i'j'}}\mathbf{x}_{\mathbf{k'}}\mathbf{x}_{\ell'}+\mathbf{I}_{\mathbf{j}\ell\mathbf{k'}\ell'}\mathbf{x}_{\mathbf{i'}}\mathbf{x}_{\mathbf{j'}})\mathbf{x}_{\mathbf{j}}\mathbf{x}_{\mathbf{k}}$ $+(\mathbf{I}_{\mathbf{k}\ell\mathbf{i'k'}}\mathbf{x}_{\mathbf{j'}}\mathbf{x}_{\ell'}+\mathbf{I}_{\mathbf{k}\ell\mathbf{i'}\ell'}\mathbf{x}_{\mathbf{j'}}\mathbf{x}_{\mathbf{k'}}+\mathbf{I}_{\mathbf{k}\ell\mathbf{j'k'}}\mathbf{x}_{\mathbf{i'}}\mathbf{x}_{\ell'}+\mathbf{I}_{\mathbf{k}\ell\mathbf{j'}\ell'}\mathbf{x}_{\mathbf{i'}}\mathbf{x}_{\mathbf{k'}})\mathbf{x}_{\mathbf{i}}\mathbf{x}_{\mathbf{j}}]$
26	$(\delta_{ij}x_kx_\ell+\delta_{k\ell}x_ix_j)x_{i'}x_{j'}x_{k'}x_{\ell'}+(\delta_{i'j'}x_{k'}x_{\ell'}+\delta_{k'\ell'}x_{i'}x_{j'})x_ix_jx_kx_\ell$
27	$(\delta_{ik}x_jx_\ell+\delta_{i\ell}x_jx_k+\delta_{jk}x_ix_\ell+\delta_{j\ell}x_ix_k)x_{i'}x_{j'}x_{k'}x_{\ell'}$ $+(\delta_{i'k'}x_{j'}x_{\ell'}+\delta_{i'\ell'}x_{j'}x_{k'}+\delta_{j'k'}x_{i'}x_{\ell'}+\delta_{j'\ell'}x_{i'}x_{k'})x_ix_jx_kx_\ell$
28	$(\delta_{\mathbf{ii'}}\mathbf{x}_{\mathbf{j'}}\mathbf{x}_{\mathbf{k'}}\mathbf{x}_{\ell'}+\delta_{\mathbf{ij'}}\mathbf{x}_{\mathbf{i'}}\mathbf{x}_{\mathbf{k'}}\mathbf{x}_{\ell'}+\delta_{\mathbf{ik'}}\mathbf{x}_{\mathbf{i'}}\mathbf{x}_{\mathbf{j'}}\mathbf{x}_{\ell'}+\delta_{\mathbf{i}\ell'}\mathbf{x}_{\mathbf{i'}}\mathbf{x}_{\mathbf{j'}}\mathbf{x}_{\mathbf{k'}})\mathbf{x}_{\mathbf{j}}\mathbf{x}_{\mathbf{k}}\mathbf{x}_\ell$ $+(\delta_{\mathbf{ji'}}\mathbf{x}_{\mathbf{j'}}\mathbf{x}_{\mathbf{k'}}\mathbf{x}_{\ell'}+\delta_{\mathbf{jj'}}\mathbf{x}_{\mathbf{i'}}\mathbf{x}_{\mathbf{k'}}\mathbf{x}_{\ell'}+\delta_{\mathbf{jk'}}\mathbf{x}_{\mathbf{i'}}\mathbf{x}_{\mathbf{j'}}\mathbf{x}_{\ell'}+\delta_{\mathbf{j}\ell'}\mathbf{x}_{\mathbf{i'}}\mathbf{x}_{\mathbf{j'}}\mathbf{x}_{\mathbf{k'}})\mathbf{x}_{\mathbf{i}}\mathbf{x}_{\mathbf{k}}\mathbf{x}_\ell$ $+(\delta_{\mathbf{ki'}}\mathbf{x}_{\mathbf{j'}}\mathbf{x}_{\mathbf{k'}}\mathbf{x}_{\ell'}+\delta_{\mathbf{kj'}}\mathbf{x}_{\mathbf{i'}}\mathbf{x}_{\mathbf{k'}}\mathbf{x}_{\ell'}+\delta_{\mathbf{kk'}}\mathbf{x}_{\mathbf{i'}}\mathbf{x}_{\mathbf{j'}}\mathbf{x}_{\ell'}+\delta_{\mathbf{k}\ell'}\mathbf{x}_{\mathbf{i'}}\mathbf{x}_{\mathbf{j'}}\mathbf{x}_{\mathbf{k'}})\mathbf{x}_{\mathbf{i}}\mathbf{x}_{\mathbf{j}}\mathbf{x}_\ell$ $+(\delta_{\ell\mathbf{i'}}\mathbf{x}_{\mathbf{j'}}\mathbf{x}_{\mathbf{k'}}\mathbf{x}_{\ell'}+\delta_{\ell\mathbf{j'}}\mathbf{x}_{\mathbf{i'}}\mathbf{x}_{\mathbf{k'}}\mathbf{x}_{\ell'}+\delta_{\ell\mathbf{k'}}\mathbf{x}_{\mathbf{i'}}\mathbf{x}_{\mathbf{j'}}\mathbf{x}_{\ell'}+\delta_{\ell\ell'}\mathbf{x}_{\mathbf{i'}}\mathbf{x}_{\mathbf{j'}}\mathbf{x}_{\mathbf{k'}})\mathbf{x}_{\mathbf{i}}\mathbf{x}_{\mathbf{j}}\mathbf{x}_{\mathbf{k}}$
29	$x_ix_jx_kx_\ell x_{i'}x_{j'}x_{k'}x_{\ell'}$

$$\Phi_1(0)=2\Phi_2(0),\quad \Phi_3(0)=2\Phi_4(0),\quad \Phi_5(0)=2\Phi_6(0),$$
$$\Phi_7(0)=2\Phi_{11}(0),\quad \Phi_8(0)=2\Phi_{13}(0),\quad \Phi_9(0)=\Phi_{10}(0)=\Phi_{12}(0),$$

is a correlation tensor of an elasticity random field.

We recognise C_1 and C_2 as the Lamé constants.

Here $j_m(\lambda\rho)$ denote the spherical Bessel functions.

To formulate the result concerning the spectral expansion of the elasticity random field, we need to introduce more notation. Put $m_0=7$, $m_2=10$, $m_4=8$, $m_6=3$, and $m_8=1$. Let a_{nqm}, $0\le n\le 4$, $1\le q\le m_{2n}$, $1\le m\le 13$ be the numbers shown in Table 16.3.

Let $g^{m[m_1,m_2]}_{\ell[\ell_1,\ell_2]}$ be the Godunov–Gordienko coefficients, introduced and calculated by Godunov and Gordienko in [2]. They correspond to real representations of the

Table 16.2 The nonzero functions $N_{nq}(\lambda, \rho)$

n	q	$N_{nq}(\lambda, \rho)$	n	q	$N_{nq}(\lambda, \rho)$	n	q	$N_{nq}(\lambda, \rho)$
1	1	$-\frac{15523}{66150} j_0(\lambda\rho)$	1	2	$-\frac{19452949}{57726900} j_0(\lambda\rho)$	1	3	$\frac{6623}{26460} j_0(\lambda\rho)$
1	4	$\frac{16677923}{70515900} j_0(\lambda\rho)$	1	5	$-\frac{221}{185220} j_0(\lambda\rho)$	1	6	$-\frac{6607}{52920} j_0(\lambda\rho)$
1	7	$\frac{239}{92610} j_0(\lambda\rho)$	1	8	$-\frac{4631}{5145} j_2(\lambda\rho)$	1	9	$\frac{2}{147} j_2(\lambda\rho)$
1	11	$\frac{1}{4} j_2(\lambda\rho)$	1	12	$\frac{748751}{5885880} j_2(\lambda\rho)$	1	13	$\frac{3505}{7546} j_2(\lambda\rho)$
1	14	$-\frac{1037}{8232} j_2(\lambda\rho)$	1	15	$-\frac{283}{2744} j_2(\lambda\rho)$	1	16	$-\frac{949}{5488} j_2(\lambda\rho)$
1	17	$-\frac{2987}{24024} j_2(\lambda\rho)$	1	18	$-\frac{2603}{2541} j_4(\lambda\rho)$	1	19	$-j_4(\lambda\rho)$
1	20	$\frac{1}{4} j_4(\lambda\rho)$	1	21	$\frac{239}{924} j_4(\lambda\rho)$	1	22	$\frac{31}{132} j_4(\lambda\rho)$
1	23	$\frac{36}{7007} j_4(\lambda\rho)$	1	24	$\frac{1}{2} j_4(\lambda\rho)$	1	25	$-\frac{15679}{60060} j_4(\lambda\rho)$
1	26	$-\frac{2}{21} j_6(\lambda\rho)$	1	27	$\frac{1}{4} j_6(\lambda\rho)$	1	29	$\frac{1}{2} j_8(\lambda\rho)$
2	1	$-\frac{8944}{99225} j_0(\lambda\rho)$	2	2	$-\frac{20456}{2061675} j_0(\lambda\rho)$	2	3	$\frac{4}{6615} j_0(\lambda\rho)$
2	4	$\frac{196054}{17628975} j_0(\lambda\rho)$	2	5	$-\frac{4}{6615} j_0(\lambda\rho)$	2	6	$\frac{2}{6615} j_0(\lambda\rho)$
2	7	$-\frac{1}{46305} j_0(\lambda\rho)$	2	8	$\frac{748}{2205} j_2(\lambda\rho)$	2	9	$-\frac{8}{735} j_2(\lambda\rho)$
2	12	$\frac{2099}{147147} j_2(\lambda\rho)$	2	13	$-\frac{268}{11319} j_2(\lambda\rho)$	2	14	$-\frac{2}{1029} j_2(\lambda\rho)$
2	16	$\frac{4}{21} j_2(\lambda\rho)$	2	17	$\frac{4}{3003} j_2(\lambda\rho)$	2	18	$\frac{604}{2541} j_4(\lambda\rho)$
2	21	$-\frac{4}{231} j_4(\lambda\rho)$	2	22	$\frac{1}{33} j_4(\lambda\rho)$	2	23	$\frac{72}{7007} j_4(\lambda\rho)$
2	25	$-\frac{332}{15015} j_4(\lambda\rho)$	2	26	$\frac{10}{21} j_6(\lambda\rho)$	2	29	$j_8(\lambda\rho)$
3	1	$-\frac{128}{11025} j_0(\lambda\rho)$	3	2	$\frac{150512}{2061675} j_0(\lambda\rho)$	3	3	$\frac{32}{6615} j_0(\lambda\rho)$
3	4	$-\frac{1117888}{17628975} j_0(\lambda\rho)$	3	5	$-\frac{32}{6615} j_0(\lambda\rho)$	3	6	$\frac{16}{6615} j_0(\lambda\rho)$
3	7	$\frac{89}{92610} j_0(\lambda\rho)$	3	9	$\frac{16}{245} j_2(\lambda\rho)$	3	12	$-\frac{96584}{735735} j_2(\lambda\rho)$
3	13	$\frac{536}{3773} j_2(\lambda\rho)$	3	14	$-\frac{16}{1029} j_2(\lambda\rho)$	3	16	$-\frac{8}{7} j_2(\lambda\rho)$
3	17	$\frac{32}{3003} j_2(\lambda\rho)$	3	18	$\frac{640}{847} j_4(\lambda\rho)$	3	23	$\frac{8159}{14014} j_4(\lambda\rho)$
3	25	$-\frac{2656}{15015} j_4(\lambda\rho)$	3	26	$\frac{8}{7} j_6(\lambda\rho)$	3	27	$2 j_6(\lambda\rho)$
3	29	$8 j_8(\lambda\rho)$	4	1	$\frac{3908}{3969} j_0(\lambda\rho)$	4	2	$\frac{2890}{1029} j_0(\lambda\rho)$
4	3	$-2 j_0(\lambda\rho)$	4	4	$-j_0(\lambda\rho)$	4	5	$\frac{22}{3077} j_0(\lambda\rho)$
4	6	$j_0(\lambda\rho)$	4	7	$-\frac{85}{3087} j_0(\lambda\rho)$	4	8	$\frac{18040}{3087} j_2(\lambda\rho)$
4	11	$-2 j_2(\lambda\rho)$	4	12	$-\frac{212}{1029} j_2(\lambda\rho)$	4	13	$-\frac{3580}{1029} j_2(\lambda\rho)$

(continued)

Table 16.2 (continued)

n	q	$N_{nq}(\lambda, \rho)$	n	q	$N_{nq}(\lambda, \rho)$	n	q	$N_{nq}(\lambda, \rho)$
4	14	$j_2(\lambda\rho)$	4	15	$\frac{283}{343} j_2(\lambda\rho)$	4	16	$-\frac{1073}{2058} j_2(\lambda\rho)$
4	17	$j_2(\lambda\rho)$	4	18	$8 j_4(\lambda\rho)$	4	19	$8 j_4(\lambda\rho)$
4	21	$-2 j_4(\lambda\rho)$	4	22	$-2 j_4(\lambda\rho)$	4	23	$j_4(\lambda\rho)$
4	24	$-4 j_4(\lambda\rho)$	4	25	$2 j_4(\lambda\rho)$	5	1	$\frac{1216}{14175} j_0(\lambda\rho)$
5	2	$-\frac{40912}{2061675} j_0(\lambda\rho)$	5	3	$\frac{8}{6615} j_0(\lambda\rho)$	5	4	$\frac{392108}{17628975} j_0(\lambda\rho)$
5	5	$-\frac{8}{6615} j_0(\lambda\rho)$	5	6	$\frac{4}{6615} j_0(\lambda\rho)$	5	7	$-\frac{2}{46305} j_0(\lambda\rho)$
5	8	$-\frac{748}{2205} j_2(\lambda\rho)$	5	9	$-\frac{16}{735} j_2(\lambda\rho)$	5	12	$\frac{4198}{147147} j_2(\lambda\rho)$
5	13	$-\frac{536}{11319} j_2(\lambda\rho)$	5	14	$-\frac{4}{1029} j_0(\lambda\rho)$	5	16	$\frac{8}{21} j_2(\lambda\rho)$
5	17	$\frac{8}{3003} j_2(\lambda\rho)$	5	18	$-\frac{1564}{2541} j_4(\lambda\rho)$	5	19	$\frac{1}{2} j_4(\lambda\rho)$
5	21	$\frac{4}{231} j_4(\lambda\rho)$	5	22	$-\frac{1}{33} j_4(\lambda\rho)$	5	23	$\frac{144}{7007} j_4(\lambda\rho)$
5	25	$-\frac{664}{15015} j_4(\lambda\rho)$	5	26	$-\frac{1}{21} j_6(\lambda\rho)$	5	29	$2 j_8(\lambda\rho)$
6	1	$\frac{1408}{1323} j_0(\lambda\rho)$	6	2	$-\frac{1237}{1029} j_0(\lambda\rho)$	6	3	$\frac{1}{2} j_0(\lambda\rho)$
6	4	$\frac{1}{2} j_0(\lambda\rho)$	6	5	$-\frac{11}{6174} j_0(\lambda\rho)$	6	6	$-\frac{1}{4} j_0(\lambda\rho)$
6	7	$\frac{1}{343} j_0(\lambda\rho)$	6	8	$-\frac{1423}{3087} j_2(\lambda\rho)$	6	9	$-\frac{1}{2} j_2(\lambda\rho)$
6	10	$-\frac{1}{2} j_2(\lambda\rho)$	6	11	$\frac{1}{2} j_2(\lambda\rho)$	6	12	$\frac{1241}{4116} j_2(\lambda\rho)$
6	13	$\frac{895}{1029} j_2(\lambda\rho)$	6	14	$-\frac{1}{4} j_2(\lambda\rho)$	6	15	$-\frac{283}{1372} j_2(\lambda\rho)$
6	16	$\frac{1073}{8232} j_2(\lambda\rho)$	6	17	$-\frac{1}{4} j_2(\lambda\rho)$	6	18	$-2 j_4(\lambda\rho)$
6	19	$-j_4(\lambda\rho)$	6	22	$\frac{1}{2} j_4(\lambda\rho)$	6	24	$j_4(\lambda\rho)$
6	25	$-\frac{1}{2} j_4(\lambda\rho)$	7	1	$-\frac{352}{1323} j_0(\lambda\rho)$	7	2	$-\frac{25856}{5145} j_0(\lambda\rho)$
7	4	$\frac{8464}{47355} j_0(\lambda\rho)$	7	5	$\frac{22}{3087} j_0(\lambda\rho)$	7	7	$\frac{271}{2058} j_0(\lambda\rho)$
7	8	$-\frac{15212}{1715} j_2(\lambda\rho)$	7	9	$\frac{1342}{735} j_2(\lambda\rho)$	7	10	$2 j_2(\lambda\rho)$
7	11	$j_2(\lambda\rho)$	7	12	$-\frac{17461}{18865} j_2(\lambda\rho)$	7	13	$\frac{8082}{3773} j_2(\lambda\rho)$
7	14	$-j_2(\lambda\rho)$	7	15	$-\frac{403}{343} j_2(\lambda\rho)$	7	16	$\frac{2419}{686} j_2(\lambda\rho)$
7	17	$-\frac{1}{2} j_2(\lambda\rho)$	7	18	$-\frac{12512}{2541} j_4(\lambda\rho)$	7	19	$-\frac{94}{77} j_4(\lambda\rho)$
7	21	$\frac{788}{231} j_4(\lambda\rho)$	7	22	$-\frac{101}{231} j_4(\lambda\rho)$	7	23	$\frac{18}{77} j_4(\lambda\rho)$
7	24	$-\frac{146}{77} j_4(\lambda\rho)$	7	25	$-\frac{92}{77} j_4(\lambda\rho)$	7	26	$-\frac{8}{21} j_6(\lambda\rho)$

(continued)

Table 16.2 (continued)

n	q	$N_{nq}(\lambda, \rho)$	n	q	$N_{nq}(\lambda, \rho)$	n	q	$N_{nq}(\lambda, \rho)$
7	28	$-2j_6(\lambda\rho)$	8	1	$-\frac{3265}{1323}j_0(\lambda\rho)$	8	2	$\frac{7493}{2058}j_0(\lambda\rho)$
8	3	$j_0(\lambda\rho)$	8	4	$\frac{1797}{12628}j_0(\lambda\rho)$	8	5	$-\frac{1571}{6174}j_0(\lambda\rho)$
8	6	$-\frac{1}{4}j_0(\lambda\rho)$	8	7	$\frac{313}{8232}j_0(\lambda\rho)$	8	8	$-\frac{15167}{10290}j_2(\lambda\rho)$
8	9	$\frac{505}{588}j_2(\lambda\rho)$	8	10	$\frac{3}{4}j_2(\lambda\rho)$	8	11	$\frac{1}{4}j_2(\lambda\rho)$
8	12	$\frac{2369}{30184}j_2(\lambda\rho)$	8	13	$\frac{17525}{15092}j_2(\lambda\rho)$	8	14	$-\frac{3}{8}j_2(\lambda\rho)$
8	15	$-\frac{729}{2744}j_2(\lambda\rho)$	8	16	$-\frac{10037}{5488}j_2(\lambda\rho)$	8	17	$-\frac{1}{8}j_2(\lambda\rho)$
8	18	$-\frac{34}{2541}j_4(\lambda\rho)$	8	19	$-\frac{663}{154}j_4(\lambda\rho)$	8	20	$j_4(\lambda\rho)$
8	21	$\frac{593}{462}j_4(\lambda\rho)$	8	22	$\frac{1}{84}j_4(\lambda\rho)$	8	23	$\frac{43}{77}j_4(\lambda\rho)$
8	24	$\frac{81}{154}j_4(\lambda\rho)$	8	25	$-\frac{169}{308}j_4(\lambda\rho)$	8	26	$\frac{26}{21}j_6(\lambda\rho)$
8	27	$2j_6(\lambda\rho)$	8	28	$-\frac{1}{2}j_6(\lambda\rho)$	9	1	$-\frac{90809}{33075}j_0(\lambda\rho)$
9	2	$-\frac{69048493}{28863450}j_0(\lambda\rho)$	9	3	$-\frac{6647}{6615}j_0(\lambda\rho)$	9	4	$-\frac{13157423}{70515900}j_0(\lambda\rho)$
9	5	$\frac{11924}{46305}j_0(\lambda\rho)$	9	6	$\frac{6551}{26460}j_0(\lambda\rho)$	9	7	$-\frac{28121}{370440}j_0(\lambda\rho)$
9	8	$\frac{16213}{3430}j_2(\lambda\rho)$	9	9	$-\frac{799}{980}j_2(\lambda\rho)$			
9	10	$-\frac{4}{4}j_2(\lambda\rho)$	9	11	$-\frac{3}{4}j_2(\lambda\rho)$	9	12	$-\frac{417803}{5885880}j_2(\lambda\rho)$
9	13	$-\frac{18061}{15092}j_2(\lambda\rho)$	9	14	$\frac{3215}{8232}j_2(\lambda\rho)$	9	15	$\frac{849}{2744}j_2(\lambda\rho)$
9	16	$\frac{5885}{5488}j_2(\lambda\rho)$	9	17	$\frac{2747}{24024}j_2(\lambda\rho)$	9	18	$\frac{2747}{24024}j_4(\lambda\rho)$
9	19	$3j_4(\lambda\rho)$	9	20	$-j_4(\lambda\rho)$	9	21	$\frac{247}{154}j_4(\lambda\rho)$
9	22	$-\frac{25}{44}j_4(\lambda\rho)$	9	23	$-\frac{576}{7007}j_4(\lambda\rho)$	9	24	$-\frac{3}{2}j_4(\lambda\rho)$
9	25	$\frac{55669}{60060}j_4(\lambda\rho)$	9	26	$\frac{6}{7}j_6(\lambda\rho)$	9	27	$-2j_6(\lambda\rho)$
9	29	$-8j_8(\lambda\rho)$	10	1	$-\frac{4016}{33075}j_0(\lambda\rho)$	10	2	$\frac{18020416}{14431725}j_0(\lambda\rho)$
10	3	$-\frac{32}{6615}j_0(\lambda\rho)$	10	4	$-\frac{8587732}{193918725}j_0(\lambda\rho)$	10	5	$\frac{283}{92610}j_0(\lambda\rho)$
10	6	$-\frac{16}{6615}j_0(\lambda\rho)$	10	7	$-\frac{3509}{92610}j_0(\lambda\rho)$	10	8	$-\frac{4208}{5145}j_2(\lambda\rho)$
10	9	$\frac{32}{735}j_2(\lambda\rho)$	10	12	$\frac{5519}{735735}j_2(\lambda\rho)$	10	13	$-\frac{134}{3773}j_2(\lambda\rho)$
10	14	$\frac{16}{1029}j_2(\lambda\rho)$	10	15	$\frac{15}{343}j_2(\lambda\rho)$	10	16	$-\frac{519}{686}j_2(\lambda\rho)$
10	17	$-\frac{32}{3003}j_2(\lambda\rho)$	10	18	$-\frac{2416/2541}{j}{}_4(\lambda\rho)$	10	19	$\frac{107}{154}j_4(\lambda\rho)$
10	21	$-\frac{26}{231}j_4(\lambda\rho)$	10	22	$-\frac{97}{231}j_4(\lambda\rho)$	10	23	$\frac{3337}{7007}j_4(\lambda\rho)$

(continued)

Table 16.2 (continued)

n	q	$N_{nq}(\lambda,\rho)$	n	q	$N_{nq}(\lambda,\rho)$	n	q	$N_{nq}(\lambda,\rho)$
10	24	$-\frac{75}{77}j_4(\lambda\rho)$	10	25	$\frac{11357}{30030}j_4(\lambda\rho)$	10	26	$-\frac{40}{21}j_6(\lambda\rho)$
10	28	$-\frac{1}{2}j_6(\lambda\rho)$	10	29	$-8j_8(\lambda\rho)$	11	2	$\frac{3712}{1715}j_0(\lambda\rho)$
11	4	$\frac{11936}{47355}j_0(\lambda\rho)$	11	5	$\frac{22}{1029}j_0(\lambda\rho)$	11	7	$-\frac{17}{343}j_0(\lambda\rho)$
11	8	$-\frac{2540}{1029}j_2(\lambda\rho)$	11	9	$-\frac{554}{245}j_2(\lambda\rho)$	11	10	$-2j_2(\lambda\rho)$
11	12	$\frac{309}{385}j_2(\lambda\rho)$	11	13	$-\frac{542}{539}j_2(\lambda\rho)$	11	14	$j_2(\lambda\rho)$
11	15	$\frac{163}{343}j_2(\lambda\rho)$	11	16	$\frac{2749}{686}j_4(\lambda\rho)$	11	18	$-\frac{2560}{847}j_4(\lambda\rho)$
11	19	$\frac{188}{77}j_4(\lambda\rho)$	11	21	$-\frac{28}{11}j_4(\lambda\rho)$	11	22	$\frac{184}{77}j_4(\lambda\rho)$
11	23	$-\frac{344}{77}j_4(\lambda\rho)$	11	24	$\frac{292}{77}j_4(\lambda\rho)$	11	25	$\frac{30}{77}j_4(\lambda\rho)$
11	26	$-\frac{32}{7}j_6(\lambda\rho)$	11	27	$-8j_6(\lambda\rho)$	11	28	$4j_6(\lambda\rho)$
12	2	$\frac{237}{245}j_0(\lambda\rho)$	12	3	$-\frac{1}{2}j_0(\lambda\rho)$	12	4	$-\frac{45863}{47355}j_0(\lambda\rho)$
12	6	$\frac{1}{2}j_0(\lambda\rho)$	12	7	$\frac{2}{343}j_0(\lambda\rho)$	12	8	$2j_2(\lambda\rho)$
12	9	$-\frac{8}{245}j_2(\lambda\rho)$	12	11	$-\frac{1}{2}j_2(\lambda\rho)$	12	12	$-\frac{16881}{37730}j_2(\lambda\rho)$
12	13	$-\frac{11855}{7546}j_2(\lambda\rho)$	12	14	$\frac{1}{2}j_2(\lambda\rho)$	12	15	$\frac{1}{2}j_2(\lambda\rho)$
12	16	$\frac{9}{28}j_2(\lambda\rho)$	12	17	$\frac{1}{2}j_2(\lambda\rho)$	12	18	$\frac{1374}{847}j_4(\lambda\rho)$
12	19	$\frac{355}{154}j_4(\lambda\rho)$	12	20	$-j_4(\lambda\rho)$	12	21	$-\frac{9}{11}j_4(\lambda\rho)$
12	22	$-\frac{31}{154}j_4(\lambda\rho)$	12	23	$-\frac{43}{77}j_4(\lambda\rho)$	12	24	$-\frac{81}{154}j_4(\lambda\rho)$
12	25	$\frac{123}{154}j_4(\lambda\rho)$	12	26	$-\frac{4}{7}j_6(\lambda\rho)$	12	27	$-j_6(\lambda\rho)$
12	28	$\frac{1}{2}j_6(\lambda\rho)$	13	1	$-\frac{704}{567}j_0(\lambda\rho)$	13	2	$-\frac{3712}{5145}j_0(\lambda\rho)$
13	4	$-\frac{4232}{47355}j_0(\lambda\rho)$	13	5	$-\frac{22}{3087}j_0(\lambda\rho)$	13	7	$\frac{100}{3087}j_0(\lambda\rho)$
13	8	$\frac{54224}{15435}j_2(\lambda\rho)$	13	9	$\frac{64}{735}j_2(\lambda\rho)$	13	12	$\frac{76}{1155}j_2(\lambda\rho)$
13	13	$-\frac{536}{1617}j_2(\lambda\rho)$	13	15	$\frac{60}{343}j_2(\lambda\rho)$	13	16	$-\frac{1546}{1029}j_2(\lambda\rho)$
13	18	$\frac{6256}{2541}j_4(\lambda\rho)$	13	19	$-\frac{261}{77}j_4(\lambda\rho)$	13	21	$\frac{68}{231}j_4(\lambda\rho)$
13	22	$\frac{166}{231}j_4(\lambda\rho)$	13	23	$-\frac{86}{77}j_4(\lambda\rho)$	13	24	$\frac{227}{77}j_4(\lambda\rho)$
13	25	$-\frac{31}{77}j_4(\lambda\rho)$	13	26	$\frac{4}{21}j_6(\lambda\rho)$	13	28	$j_6(\lambda\rho)$

Table 16.3 The nonzero numbers a_{nqm}

n	q	m	a_{nqm}	n	q	m	a_{nqm}	n	q	m	a_{nqm}	n	q	m	a_{nqm}
0	1	2	$\frac{1}{9}$	0	1	4	$\frac{16}{9}$	0	1	5	$\frac{8}{9}$	0	1	13	$-\frac{16}{9}$
0	2	2	$\frac{2\sqrt{10}}{45}$	0	2	4	$\frac{56\sqrt{10}}{45}$	0	2	5	$-\frac{8\sqrt{10}}{45}$	0	2	6	$-\frac{8\sqrt{10}}{15}$
0	2	13	$\frac{16\sqrt{10}}{45}$	0	3	1	$\frac{4\sqrt{5}}{15}$	0	3	2	$\frac{4\sqrt{5}}{45}$	0	3	4	$-\frac{16\sqrt{5}}{9}$
0	3	5	$\frac{2\sqrt{5}}{45}$	0	3	6	$\frac{8\sqrt{5}}{15}$	0	3	7	$-\frac{16\sqrt{5}}{15}$	0	3	8	$\frac{8\sqrt{5}}{15}$
0	3	9	$\frac{8\sqrt{5}}{15}$	0	3	10	$\frac{4\sqrt{5}}{15}$	0	3	13	$\frac{4\sqrt{5}}{9}$	0	4	1	$\frac{2}{5}$
0	4	2	$\frac{4}{45}$	0	4	3	$\frac{8}{5}$	0	4	4	$\frac{52}{45}$	0	4	5	$\frac{8}{45}$
0	4	6	$\frac{4}{15}$	0	4	11	$-\frac{16}{5}$	0	4	12	$-\frac{8}{5}$	0	4	13	$-\frac{16}{45}$
0	5	1	$\frac{8\sqrt{35}}{105}$	0	5	2	$\frac{8\sqrt{35}}{315}$	0	5	4	$-\frac{16\sqrt{35}}{45}$	0	5	5	$-\frac{8\sqrt{35}}{315}$
0	5	7	$\frac{16\sqrt{35}}{105}$	0	5	8	$-\frac{32\sqrt{35}}{105}$	0	5	9	$\frac{16\sqrt{35}}{105}$	0	5	10	$\frac{8\sqrt{35}}{105}$
0	5	13	$-\frac{16\sqrt{35}}{63}$	0	6	1	$\frac{4\sqrt{5}}{21}$	0	6	2	$\frac{8\sqrt{5}}{315}$	0	6	3	$\frac{4\sqrt{5}}{35}$
0	6	4	$-\frac{8\sqrt{5}}{9}$	0	6	5	$\frac{16\sqrt{5}}{315}$	0	6	6	$\frac{8\sqrt{5}}{15}$	0	6	7	$\frac{136\sqrt{5}}{105}$
0	6	8	$-\frac{8\sqrt{5}}{105}$	0	6	9	$\frac{16\sqrt{5}}{105}$	0	6	10	$\frac{8\sqrt{5}}{105}$	0	6	11	$\frac{8\sqrt{5}}{7}$
0	6	12	$-\frac{16\sqrt{5}}{35}$	0	6	13	$\frac{32\sqrt{5}}{63}$	0	7	1	$-\frac{59}{105}$	0	7	2	$\frac{8}{105}$
0	7	3	$\frac{64}{105}$	0	7	4	$\frac{24}{5}$	0	7	5	$\frac{16}{105}$	0	7	6	$-\frac{6}{5}$
0	7	7	$\frac{16}{7}$	0	7	8	$\frac{4}{7}$	0	7	9	$\frac{4}{21}$	0	7	10	$\frac{16}{21}$

(continued)

Table 16.3 (continued)

n	q	m	a_{nqm}	n	q	m	a_{nqm}	n	q	m	a_{nqm}	n	q	m	a_{nqm}
0	7	11	$\frac{64}{35}$	0	7	12	$\frac{92}{35}$	0	7	13	$\frac{16}{35}$	1	1	2	$\frac{2\sqrt{2}}{9}$
1	1	4	$-\frac{16\sqrt{2}}{9}$	1	1	5	$\frac{4\sqrt{2}}{9}$	1	1	13	$-\frac{8\sqrt{2}}{9}$	1	2	2	$\frac{4\sqrt{10}}{45}$
1	2	4	$-\frac{56\sqrt{10}}{45}$	1	2	5	$-\frac{4\sqrt{10}}{45}$	1	2	6	$\frac{8\sqrt{10}}{15}$	1	2	13	$\frac{8\sqrt{10}}{45}$
1	3	2	$\frac{4\sqrt{7}}{63}$	1	3	4	$-\frac{80\sqrt{7}}{63}$	1	3	5	$-\frac{16\sqrt{7}}{63}$	1	3	6	$\frac{16\sqrt{7}}{21}$
1	3	13	$\frac{32\sqrt{7}}{63}$	1	4	1	$-\frac{4\sqrt{14}}{21}$	1	4	2	$\frac{4\sqrt{14}}{63}$	1	4	4	$\frac{16\sqrt{14}}{9}$
1	4	5	$\frac{2\sqrt{14}}{63}$	1	4	6	$-\frac{8\sqrt{14}}{21}$	1	4	7	$-\frac{8\sqrt{14}}{21}$	1	4	8	$\frac{4\sqrt{14}}{21}$
1	4	9	$\frac{4\sqrt{14}}{21}$	1	4	10	$\frac{2\sqrt{14}}{21}$	1	4	13	$-\frac{4\sqrt{14}}{9}$	1	5	1	$-\frac{4\sqrt{35}}{35}$
1	5	2	$\frac{8\sqrt{35}}{315}$	1	5	3	$\frac{8\sqrt{35}}{35}$	1	5	4	$\frac{8\sqrt{35}}{315}$	1	5	5	$\frac{16\sqrt{35}}{315}$
1	5	6	$-\frac{8\sqrt{35}}{35}$	1	5	11	$-\frac{16\sqrt{35}}{35}$	1	5	12	$\frac{16\sqrt{35}}{35}$	1	5	13	$-\frac{32\sqrt{35}}{315}$
1	6	1	$-\frac{2\sqrt{10}}{21}$	1	6	2	$\frac{8\sqrt{10}}{105}$	1	6	4	$\frac{16\sqrt{10}}{35}$	1	6	5	$-\frac{8\sqrt{10}}{105}$
1	6	6	$-\frac{4\sqrt{10}}{35}$	1	6	7	$\frac{16\sqrt{10}}{21}$	1	6	8	$-\frac{4\sqrt{10}}{21}$	1	6	9	$-\frac{4\sqrt{10}}{7}$
1	6	10	$\frac{8\sqrt{10}}{21}$	1		6 13	$\frac{8\sqrt{10}}{15}$	1	7	1	$-\frac{8\sqrt{2}}{21}$	1	7	2	$\frac{8\sqrt{2}}{63}$
1	7	4	$\frac{272\sqrt{2}}{63}$	1	7	5	$-\frac{8\sqrt{2}}{63}$	1	7	6	$-\frac{32\sqrt{2}}{21}$	1	7	7	$\frac{8\sqrt{2}}{21}$
1	7	8	$-\frac{16\sqrt{2}}{21}$	1	7	9	$\frac{8\sqrt{2}}{21}$	1	7	10	$\frac{4\sqrt{2}}{21}$	1	7	13	$\frac{16\sqrt{2}}{9}$
1	8	1	$\frac{4\sqrt{14}}{147}$	1	8	2	$\frac{8\sqrt{14}}{441}$	1	8	3	$\frac{4\sqrt{14}}{49}$	1	8	4	$\frac{104\sqrt{14}}{441}$

(continued)

Table 16.3 (continued)

n	q	m	a_{nqm}	n	q	m	a_{nqm}	n	q	m	a_{nqm}	n	q	m	a_{nqm}
1	8	5	$\frac{16\sqrt{14}}{441}$	1	8	6	$\frac{8\sqrt{14}}{49}$	1	8	7	$\frac{68\sqrt{14}}{147}$	1	8	8	$-\frac{4\sqrt{14}}{147}$
1	8	9	$\frac{8\sqrt{14}}{147}$	1	8	10	$\frac{4\sqrt{14}}{147}$	1	8	11	$-\frac{8\sqrt{14}}{7}$	1	8	12	$-\frac{16\sqrt{14}}{49}$
1	8	13	$-\frac{32\sqrt{14}}{63}$	1	9	1	$-\frac{32\sqrt{35}}{735}$	1	9	2	$\frac{16\sqrt{35}}{735}$	1	9	3	$-\frac{32\sqrt{35}}{245}$
1	9	4	$\frac{32\sqrt{35}}{245}$	1	9	5	$\frac{32\sqrt{35}}{735}$	1	9	6	$-\frac{8\sqrt{35}}{245}$	1	9	7	$-\frac{64\sqrt{35}}{147}$
1	9	8	$\frac{40\sqrt{35}}{147}$	1	9	9	$-\frac{8\sqrt{35}}{49}$	1	9	10	$\frac{16\sqrt{35}}{147}$	1	9	11	$\frac{32\sqrt{35}}{35}$
1	9	12	$\frac{16\sqrt{35}}{245}$	1	9	13	$-\frac{32\sqrt{35}}{105}$	1	10	1	$\frac{206\sqrt{77}}{1617}$	1	10	2	$\frac{16\sqrt{77}}{1617}$
1	10	3	$\frac{128\sqrt{77}}{1617}$	1	10	4	$-\frac{48\sqrt{77}}{49}$	1	10	5	$\frac{32\sqrt{77}}{1617}$	1	10	6	$\frac{12\sqrt{77}}{49}$
1	10	7	$-\frac{8\sqrt{77}}{49}$	1	10	8	$-\frac{2\sqrt{77}}{49}$	1	10	9	$\frac{202\sqrt{77}}{1617}$	1	10	10	$\frac{136\sqrt{77}}{1617}$
1	10	12	$-\frac{24\sqrt{77}}{49}$	2	1	2	$\frac{4\sqrt{35}}{105}$	2	1	4	$\frac{32\sqrt{35}}{105}$	2	1	5	$-\frac{16\sqrt{35}}{105}$
2	1	6	$-\frac{8\sqrt{35}}{105}$	2	1	13	$\frac{32\sqrt{35}}{105}$	2	2	1	$\frac{2\sqrt{70}}{105}$	2	2	2	$\frac{4\sqrt{70}}{105}$
2	2	5	$\frac{2\sqrt{70}}{105}$	2	2	6	$\frac{4\sqrt{70}}{105}$	2	2	7	$\frac{32\sqrt{70}}{105}$	2	2	8	$-\frac{16\sqrt{70}}{105}$
2	2	9	$-\frac{16\sqrt{70}}{105}$	2	2	10	$-\frac{8\sqrt{70}}{105}$	2	3	1	$\frac{2\sqrt{7}}{35}$	2	3	2	$\frac{8\sqrt{7}}{105}$
2	3	3	$-\frac{32\sqrt{7}}{35}$	2	3	4	$\frac{64\sqrt{7}}{105}$	2	3	5	$\frac{16\sqrt{7}}{105}$	2	3	6	$-\frac{16\sqrt{7}}{105}$
2	3	11	$\frac{64\sqrt{7}}{35}$	2	3	12	$-\frac{8\sqrt{7}}{35}$	2	3	13	$-\frac{32\sqrt{7}}{105}$	2	4	1	$\frac{4\sqrt{10}}{105}$
2	4	2	$\frac{8\sqrt{10}}{105}$	2	4	4	$\frac{16\sqrt{10}}{35}$	2	4	5	$-\frac{8\sqrt{10}}{105}$	2	4	6	$-\frac{8\sqrt{10}}{21}$

(continued)

Table 16.3 (continued)

n	q	m	a_{nqm}	n	q	m	a_{nqm}	n	q	m	a_{nqm}	n	q	m	a_{nqm}
2	4	7	$-\frac{32\sqrt{10}}{105}$	2	4	8	$\frac{64\sqrt{10}}{105}$	2	4	9	$-\frac{32\sqrt{10}}{105}$	2	4	10	$-\frac{16\sqrt{10}}{105}$
2	5	1	$\frac{12\sqrt{11}}{77}$	2	5	2	$\frac{16\sqrt{11}}{231}$	2	5	4	$-\frac{32\sqrt{11}}{21}$	2	5	5	$-\frac{16\sqrt{11}}{231}$
2	5	6	$\frac{8\sqrt{11}}{21}$	2	5	7	$\frac{16\sqrt{11}}{77}$	2	5	8	$-\frac{4\sqrt{11}}{77}$	2	5	9	$-\frac{12\sqrt{11}}{77}$
2	5	10	$\frac{8\sqrt{11}}{77}$	2	5	13	$-\frac{16\sqrt{11}}{33}$	2	6	1	$\frac{8\sqrt{70}}{147}$	2	6	2	$\frac{8\sqrt{70}}{735}$
2	6	3	$\frac{12\sqrt{70}}{245}$	2	6	4	$-\frac{8\sqrt{70}}{49}$	2	6	5	$\frac{16\sqrt{70}}{735}$	2	6	6	$\frac{128\sqrt{70}}{735}$
2	6	7	$-\frac{272\sqrt{70}}{735}$	2	6	8	$\frac{16\sqrt{70}}{735}$	2	6	9	$-\frac{32\sqrt{70}}{735}$	2	6	10	$-\frac{16\sqrt{70}}{735}$
2	6	12	$-\frac{48\sqrt{70}}{245}$	2	7	1	$\frac{8\sqrt{154}}{539}$	2	7	2	$\frac{16\sqrt{154}}{1617}$	2	7	3	$-\frac{32\sqrt{154}}{539}$
2	7	4	$-\frac{32\sqrt{154}}{147}$	2	7	5	$\frac{32\sqrt{154}}{1617}$	2	7	6	$\frac{8\sqrt{154}}{147}$	2	7	7	$-\frac{32\sqrt{154}}{539}$
2	7	8	$\frac{20\sqrt{154}}{539}$	2	7	9	$-\frac{12\sqrt{154}}{539}$	2	7	10	$\frac{8\sqrt{154}}{539}$	2	7	11	$-\frac{32\sqrt{154}}{77}$
2	7	12	$\frac{16\sqrt{154}}{539}$	2	7	13	$\frac{32\sqrt{154}}{231}$	2	8	1	$\frac{536\sqrt{2002}}{35035}$	2	8	2	$\frac{72\sqrt{2002}}{35035}$
2	8	3	$\frac{576\sqrt{2002}}{35035}$	2	8	4	$\frac{32\sqrt{2002}}{245}$	2	8	5	$\frac{144\sqrt{2002}}{35035}$	2	8	6	$-\frac{8\sqrt{2002}}{245}$
2	8	7	$-\frac{48\sqrt{2002}}{539}$	2	8	8	$-\frac{12\sqrt{2002}}{539}$	2	8	9	$\frac{228\sqrt{2002}}{7007}$	2	8	10	$\frac{72\sqrt{2002}}{7007}$
2	8	11	$-\frac{32\sqrt{2002}}{385}$	2	8	12	$\frac{148\sqrt{2002}}{2695}$	2	8	13	$-\frac{8\sqrt{2002}}{385}$	3	1	1	$-\frac{4\sqrt{77}}{231}$
3	1	2	$\frac{8\sqrt{77}}{231}$	3	1	5	$-\frac{8\sqrt{77}}{231}$	3	1	7	$-\frac{64\sqrt{77}}{231}$	3	1	8	$\frac{16\sqrt{77}}{231}$
3	1	9	$\frac{16\sqrt{77}}{77}$	3	1	10	$-\frac{32\sqrt{77}}{231}$	3	1	13	$\frac{32\sqrt{77}}{231}$	3	2	1	$-\frac{10\sqrt{22}}{231}$

(continued)

Table 16.3 (continued)

n	q	m	a_{nqm}	n	q	m	a_{nqm}	n	q	m	a_{nqm}	n	q	m	a_{nqm}
3	2	2	$\frac{8\sqrt{22}}{231}$	3	2	3	$-\frac{16\sqrt{22}}{77}$	3	2	5	$\frac{16\sqrt{22}}{231}$	3	2	7	$\frac{128\sqrt{22}}{231}$
3	2	8	$-\frac{80\sqrt{22}}{231}$	3	2	9	$\frac{16\sqrt{22}}{77}$	3	2	10	$-\frac{32\sqrt{22}}{231}$	3	2	11	$\frac{64\sqrt{22}}{77}$
3	2	12	$\frac{8\sqrt{22}}{77}$	3	2	13	$-\frac{64\sqrt{22}}{231}$	3	3	1	$\frac{8\sqrt{55}}{1155}$	3	3	2	$\frac{16\sqrt{55}}{1155}$
3	3	3	$\frac{128\sqrt{55}}{1155}$	3	3	5	$\frac{32\sqrt{55}}{1155}$	3	3	7	$\frac{32\sqrt{55}}{77}$	3	3	8	$\frac{8\sqrt{55}}{77}$
3	3	9	$-\frac{128\sqrt{55}}{1155}$	3	3	10	$-\frac{8\sqrt{55}}{1155}$	3	3	11	$-\frac{64\sqrt{55}}{77}$	3	3	12	$-\frac{8\sqrt{55}}{77}$
3	3	13	$-\frac{16\sqrt{55}}{77}$	4	1	1	$\frac{4\sqrt{1430}}{2145}$	4	1	2	$\frac{8\sqrt{1430}}{2145}$	4	1	3	$\frac{64\sqrt{1430}}{2145}$
4	1	5	$\frac{16\sqrt{1430}}{2145}$	4	1	9	$-\frac{64\sqrt{1430}}{2145}$	4	1	10	$-\frac{64\sqrt{1430}}{2145}$				

group $O(3)$, while the celebrated Clebsch–Gordan coefficients, see [7], correspond to complex representations of the above group. Introduce the following notation.

$$
\begin{aligned}
\mathsf{T}^{0,1}_{ijk\ell} &= \frac{1}{3}\delta_{ij}\delta_{k\ell},\\
\mathsf{T}^{0,2}_{ijk\ell} &= \frac{1}{\sqrt{5}}\sum_{m=-2}^{2} g^{m[i,j]}_{2[1,1]} g^{m[k,\ell]}_{2[1,1]},\\
\mathsf{T}^{2,1,q}_{ijk\ell} &= \frac{1}{\sqrt{6}}(\delta_{ij} g^{q[k,\ell]}_{2[1,1]} + \delta_{k\ell} g^{q[i,j]}_{2[1,1]}),\\
\mathsf{T}^{2,2,q}_{ijk\ell} &= \sum_{m,n=-2}^{2} g^{q[m,n]}_{2[2,2]} g^{m[i,j]}_{2[1,1]} g^{n[k,\ell]}_{2[1,1]},\\
\mathsf{T}^{4,1,q}_{ijk\ell} &= \sum_{m,n=-2}^{2} g^{q[m,n]}_{4[2,2]} g^{m[i,j]}_{2[1,1]} g^{n[k,\ell]}_{2[1,1]}.
\end{aligned}
$$

Introduce the shortcut $i\dots\ell' = ijk\ell i'j'k'\ell'$. Let $\mathsf{T}^{2n,q,v}_{i\dots\ell'}$ be the rank 8 tensors shown in Table 16.4.

Let $<$ be the lexicographic order on the sequences $tuijk\ell$, where $ijk\ell$ are indices that numerate the 21 component of the elasticity tensor, $t \geq 0$, and $-t \leq u \leq t$. Consider the infinite symmetric positive definite matrices given by

$$
b^{t'u'i'j'k'\ell'}_{tuijk\ell}(m) = \mathrm{i}^{t'-t}\sqrt{(2t+1)(2t'+1)}\sum_{n=0}^{4}\frac{1}{4n+1} g^{0[0,0]}_{2n[t,t']}\sum_{q=1}^{m_n} a_{nqm}\sum_{v=-2n}^{2n}\mathsf{T}^{2n,q,v}_{i\dots\ell'} g^{v[u,u']}_{2n[t,t']}
$$

with $1 \leq m \leq 13$. Let $L(m)$ be the infinite lower triangular matrices of the Cholesky factorisation of the matrices $b^{t'u'i'j'k'\ell'}_{tuijk\ell}(m)$ constructed by Hansen in [3]. Let $Z'_{mtuijk\ell}$ be the sequence of centred scattered random measures on the interval $[0,\infty)$ satisfying the following condition: for any Borel sets A_1 and A_2 we have

$$
\mathsf{E}[Z'_{mtuijk\ell}(A_1)Z'_{m't'u'i'j'k'\ell'}(A_2)] = \delta_{mm'}\delta_{tt'}\delta_{uu'}\delta_{ii'}\delta_{jj'}\delta_{kk'}\delta_{\ell\ell'}\Phi_m(A_1\cap A_2).
$$

Define

$$
Z_{mtuijk\ell} = \sum_{(t'u'i'j'k'\ell')\leq(tuijk\ell)} L^{t'u'i'j'k'\ell'}_{tuijk\ell}(m) Z'_{mtuijk\ell}.
$$

Finally, let (ρ,θ,φ) be the spherical coordinates in $\mathbb{R}^3$, and let $S^u_t(\theta,\varphi)$ be the real-valued spherical harmonics, see [1].

Theorem 16.3 *The spectral expansion of the class of elasticity random fields under consideration has the form*

Table 16.4 The tensors $\mathsf{T}^{2n,q,v}_{i\dots\ell'}$

Tensor	Value
$\mathsf{T}^{0,1,0}_{i\dots\ell'}$	$\mathsf{T}^{0,1}_{ijk\ell}\mathsf{T}^{0,1}_{i'j'k'\ell'}$
$\mathsf{T}^{0,2}_{i\dots\ell'}$	$\frac{1}{\sqrt{2}}(\mathsf{T}^{0,1}_{ijk\ell}\mathsf{T}^{0,2}_{i'j'k'\ell'}+\mathsf{T}^{0,1}_{i'j'k'\ell'}\mathsf{T}^{0,2}_{ijk\ell})$
$\mathsf{T}^{0,3,0}_{i\dots\ell'}$	$\sum_{q,q'=-2}^{2} g^{0[q,q']}_{0[2,2]}\mathsf{T}^{2,1,q}_{ijk\ell}\mathsf{T}^{2,1,q'}_{i'j'k'\ell'}$
$\mathsf{T}^{0,4,0}_{i\dots\ell'}$	$\mathsf{T}^{0,2}_{ijk\ell}\mathsf{T}^{0,2}_{i'j'k'\ell'}$
$\mathsf{T}^{0,5,0}_{i\dots\ell'}$	$\frac{1}{\sqrt{2}}\left(\sum_{q,q'=-2}^{2} g^{0[q,q']}_{0[2,2]}\mathsf{T}^{2,1,q}_{ijk\ell}\mathsf{T}^{2,2,q'}_{i'j'k'\ell'}+\sum_{q,q'=-2}^{2} g^{0[q',q]}_{0[2,2]}\mathsf{T}^{2,1,q'}_{i'j'k'\ell'}\mathsf{T}^{2,2,q}_{ijk\ell}\right)$
$\mathsf{T}^{0,6,0}_{i\dots\ell'}$	$\sum_{q,q'=-2}^{2} g^{0[q,q']}_{0[2,2]}\mathsf{T}^{2,2,q}_{ijk\ell}\mathsf{T}^{2,2,q'}_{i'j'k'\ell'}$
$\mathsf{T}^{0,7,0}_{i\dots\ell'}$	$\sum_{q,q'=-4}^{4} g^{0[q,q']}_{0[4,4]}\mathsf{T}^{4,1,q}_{ijk\ell}\mathsf{T}^{4,1,q'}_{i'j'k'\ell'}$
$\mathsf{T}^{2,1,v}_{i\dots\ell'}$	$\frac{1}{\sqrt{2}}(\mathsf{T}^{0,1}_{ijk\ell}\mathsf{T}^{2,1,v}_{i'j'k'\ell'}+\mathsf{T}^{0,1}_{i'j'k'\ell'}\mathsf{T}^{2,1,v}_{ijk\ell})$
$\mathsf{T}^{2,2,v}_{i\dots\ell'}$	$\frac{1}{\sqrt{2}}(\mathsf{T}^{0,2}_{ijk\ell}\mathsf{T}^{2,1,v}_{i'j'k'\ell'}+\mathsf{T}^{0,2}_{i'j'k'\ell'}\mathsf{T}^{2,1,v}_{ijk\ell})$
$\mathsf{T}^{2,3,v}_{i\dots\ell'}$	$\frac{1}{\sqrt{2}}(\mathsf{T}^{0,1}_{ijk\ell}\mathsf{T}^{2,2,v}_{i'j'k'\ell'}+\mathsf{T}^{0,1}_{i'j'k'\ell'}\mathsf{T}^{2,2,v}_{ijk\ell})$
$\mathsf{T}^{2,4,v}_{i\dots\ell'}$	$\sum_{q,q'=-2}^{2} g^{v[q,q']}_{2[2,2]}\mathsf{T}^{2,1,q}_{ijk\ell}\mathsf{T}^{2,1,q'}_{i'j'k'\ell'}$
$\mathsf{T}^{2,5,v}_{i\dots\ell'}$	$\frac{1}{\sqrt{2}}(\mathsf{T}^{0,2}_{ijk\ell}\mathsf{T}^{2,2,v}_{i'j'k'\ell'}+\mathsf{T}^{0,2}_{i'j'k'\ell'}\mathsf{T}^{2,2,v}_{ijk\ell})$
$\mathsf{T}^{2,6,v}_{i\dots\ell'}$	$\frac{1}{\sqrt{2}}\left(\sum_{q=-2}^{2}\sum_{q'=-4}^{4} g^{v[q,q']}_{2[2,4]}\mathsf{T}^{2,1,q}_{ijk\ell}\mathsf{T}^{4,1,q'}_{i'j'k'\ell'}+\sum_{q'=-2}^{2}\sum_{q=-4}^{4} g^{v[q',q]}_{2[2,4]}\mathsf{T}^{2,1,q'}_{i'j'k'\ell'}\mathsf{T}^{4,1,q}_{ijk\ell}\right)$
$\mathsf{T}^{2,7,v}_{i\dots\ell'}$	$\frac{1}{\sqrt{2}}\left(\sum_{q,q'=-2}^{2} g^{v[q,q']}_{2[2,2]}\mathsf{T}^{2,2,q}_{ijk\ell}\mathsf{T}^{2,1,q'}_{i'j'k'\ell'}+\sum_{q',q=-2}^{2} g^{v[q',q]}_{2[2,2]}\mathsf{T}^{2,2,q'}_{i'j'k'\ell'}\mathsf{T}^{2,1,q}_{ijk\ell}\right)$
$\mathsf{T}^{2,8,v}_{i\dots\ell'}$	$\sum_{q,q'=-2}^{2} g^{v[q,q']}_{2[2,2]}\mathsf{T}^{2,2,q}_{ijk\ell}\mathsf{T}^{2,2,q'}_{i'j'k'\ell'}$
$\mathsf{T}^{2,9,v}_{i\dots\ell'}$	$\frac{1}{\sqrt{2}}\left(\sum_{q=-2}^{2}\sum_{q'=-4}^{4} g^{v[q,q']}_{2[2,4]}\mathsf{T}^{2,2,q}_{ijk\ell}\mathsf{T}^{4,1,q'}_{i'j'k'\ell'}+\sum_{q'=-2}^{2}\sum_{q=-4}^{4} g^{v[q',q]}_{2[2,4]}\mathsf{T}^{2,2,q'}_{i'j'k'\ell'}\mathsf{T}^{4,1,q}_{ijk\ell}\right)$
$\mathsf{T}^{2,10,v}_{i\dots\ell'}$	$\sum_{q,q'=-4}^{4} g^{v[q,q']}_{2[4,4]}\mathsf{T}^{4,1,q}_{ijk\ell}\mathsf{T}^{4,1,q'}_{i'j'k'\ell'}$
$\mathsf{T}^{4,1,v}_{i\dots\ell'}$	$\frac{1}{\sqrt{2}}(\mathsf{T}^{0,1}_{ijk\ell}\mathsf{T}^{4,1,v}_{i'j'k'\ell'}+\mathsf{T}^{0,1}_{i'j'k'\ell'}\mathsf{T}^{4,1,v}_{ijk\ell})$

(continued)

Table 16.4 (continued)

Tensor	Value
$\mathsf{T}^{4,2,v}_{i\dots\ell'}$	$\sum_{q,q'=-4}^{4} g^{v[q,q']}_{4[2,2]}\mathsf{T}^{2,1,q}_{ijk\ell}\mathsf{T}^{2,1,q'}_{i'j'k'\ell'}$
$\mathsf{T}^{4,3,v}_{i\dots\ell'}$	$\frac{1}{\sqrt{2}}(\mathsf{T}^{0,2}_{ijk\ell}\mathsf{T}^{4,1,v}_{i'j'k'\ell'}+\mathsf{T}^{0,2}_{i'j'k'\ell'}\mathsf{T}^{4,1,v}_{ijk\ell})$
$\mathsf{T}^{4,4,v}_{i\dots\ell'}$	$\frac{1}{\sqrt{2}}\left(\sum_{q,q'=-4}^{4} g^{v[q,q']}_{4[2,2]}\mathsf{T}^{2,2,q}_{ijk\ell}\mathsf{T}^{2,1,q'}_{i'j'k'\ell'}+\sum_{q',q=-4}^{4} g^{v[q',q]}_{4[2,2]}\mathsf{T}^{2,2,q'}_{i'j'k'\ell'}\mathsf{T}^{2,1,q}_{ijk\ell}\right)$
$\mathsf{T}^{4,5,v}_{i\dots\ell'}$	$\frac{1}{\sqrt{2}}\left(\sum_{q=-2}^{2}\sum_{q'=-4}^{4} g^{v[q,q']}_{4[2,4]}\mathsf{T}^{2,1,q}_{ijk\ell}\mathsf{T}^{4,1,q'}_{i'j'k'\ell'}+\sum_{q'=-2}^{2}\sum_{q=-4}^{4} g^{v[q',q]}_{4[2,4]}\mathsf{T}^{2,1,q'}_{i'j'k'\ell'}\mathsf{T}^{4,1,q}_{ijk\ell}\right)$
$\mathsf{T}^{4,6,v}_{i\dots\ell'}$	$\sum_{q,q'=-4}^{4} g^{v[q,q']}_{4[2,2]}\mathsf{T}^{2,2,q}_{ijk\ell}\mathsf{T}^{2,2,q'}_{i'j'k'\ell'}$
$\mathsf{T}^{4,7,v}_{i\dots\ell'}$	$\frac{1}{\sqrt{2}}\left(\sum_{q=-2}^{2}\sum_{q'=-4}^{4} g^{v[q,q']}_{4[2,4]}\mathsf{T}^{2,2,q}_{ijk\ell}\mathsf{T}^{4,1,q'}_{i'j'k'\ell'}+\sum_{q'=-2}^{2}\sum_{q=-4}^{4} g^{v[q',q]}_{4[2,4]}\mathsf{T}^{2,2,q'}_{i'j'k'\ell'}\mathsf{T}^{4,1,q}_{ijk\ell}\right)$
$\mathsf{T}^{4,8,v}_{i\dots\ell'}$	$\sum_{q,q'=-4}^{4} g^{v[q,q']}_{4[4,4]}\mathsf{T}^{4,1,q}_{ijk\ell}\mathsf{T}^{4,1,q'}_{i'j'k'\ell'}$
$\mathsf{T}^{6,1,v}_{i\dots\ell'}$	$\frac{1}{\sqrt{2}}\left(\sum_{q=-2}^{2}\sum_{q'=-4}^{4} g^{v[q,q']}_{6[2,4]}\mathsf{T}^{2,1,q}_{ijk\ell}\mathsf{T}^{4,1,q'}_{i'j'k'\ell'}+\sum_{q'=-2}^{2}\sum_{q=-4}^{4} g^{v[q',q]}_{6[2,4]}\mathsf{T}^{2,1,q'}_{i'j'k'\ell'}\mathsf{T}^{4,1,q}_{ijk\ell}\right)$
$\mathsf{T}^{6,2,v}_{i\dots\ell'}$	$\frac{1}{\sqrt{2}}\left(\sum_{q=-2}^{2}\sum_{q'=-4}^{4} g^{v[q,q']}_{6[2,4]}\mathsf{T}^{2,2,q}_{ijk\ell}\mathsf{T}^{4,1,q'}_{i'j'k'\ell'}+\sum_{q'=-2}^{2}\sum_{q=-4}^{4} g^{v[q',q]}_{6[2,4]}\mathsf{T}^{2,2,q'}_{i'j'k'\ell'}\mathsf{T}^{4,1,q}_{ijk\ell}\right)$
$\mathsf{T}^{6,3,v}_{i\dots\ell'}$	$\sum_{q,q'=-4}^{4} g^{v[q,q']}_{6[4,4]}\mathsf{T}^{4,1,q}_{ijk\ell}\mathsf{T}^{4,1,q'}_{i'j'k'\ell'}$
$\mathsf{T}^{8,1,v}_{i\dots\ell'}$	$\sum_{q,q'=-4}^{4} g^{v[q,q']}_{8[4,4]}\mathsf{T}^{4,1,q}_{ijk\ell}\mathsf{T}^{4,1,q'}_{i'j'k'\ell'}$

$$
\begin{aligned}
\mathsf{C}_{ijk\ell}(\rho,\theta,\varphi) &= C_1\delta_{ij}\delta_{k\ell}+C_2(\delta_{ik}\delta_{j\ell}+\delta_{i\ell}\delta_{jk})+\\
&+4\pi\sum_{m=1}^{13}\sum_{t=0}^{\infty}\sum_{u=-t}^{t}\int_0^\infty j_t(\lambda\rho)\,\mathrm{d}Z_{mtuijk\ell}(\lambda)S^u_t(\theta,\varphi).
\end{aligned}
$$

Proofs of the above theorems are long but straightforward. They are similar to those of the two-dimensional case considered in [5].

References

1. Erdélyi, A., Magnus, W., Oberhettinger, F., Tricomi, F.G.: Higher Transcendental Functions, vol. II. Robert E. Krieger Publishing Co., Inc., Melbourne, Fla (1981). Based on notes left by Harry Bateman, Reprint of the 1953 edition
2. Godunov, S.K., Gordienko, V.M.: Clebsch-Gordan coefficients in the case of various choices of bases of unitary and orthogonal representations of the groups SU(2) and SO(3). Sibirsk. Mat. Zh. **45**(3), 540–557 (2004)
3. Hansen, A.C.: Infinite-dimensional numerical linear algebra: theory and applications. Proc. R. Soc. Lond. Ser. A Math. Phys. Eng. Sci. **466**(2124), 3539–3559 (2010)
4. Lomakin, V.: Deformation of microscopically nonhomogeneous elastic bodies. J. Appl. Math. Mech. **29**(5), 1048–1054 (1965)
5. Malyarenko, A., Ostoja-Starzewski, M.: The spectral expansion of the elasticity random field. In: S. Sivasundaram (ed.) 10th International Conference on Mathematical problems in Engineering, Aerospace, and Sciences ICNPAA 2014, *AIP Conference Proceedings*, vol. 1637, pp. 647–655 (2014)
6. Ogden, R.W.: On isotropic tensors and elastic moduli. Proc. Cambridge Philos. Soc. **75**, 427–436 (1974)
7. Varshalovich, D.A., Moskalev, A.N., Khersonskiĭ, V.K.: Quantum Theory of Angular Momentum. Irreducible tensors, spherical harmonics, vector coupling coefficients, $3nj$ symbols. World Scientific Publishing Co., Inc., Teaneck, NJ (1988). Translated from the Russian

Chapter 17
Sensitivity Analysis of Catastrophe Bond Price Under the Hull–White Interest Rate Model

Anatoliy Malyarenko, Jan Röman and Oskar Schyberg

Abstract We consider a model, where the natural risk index is described by the Merton jump-diffusion while the risk-free interest rate is governed by the Hull–White stochastic differential equation. We price a catastrophe bond with payoff depending on finitely many values of the underlying index. The sensitivities of the bond price with respect to the initial condition, volatility of the diffusion component, and jump amplitude, are calculated using the Malliavin calculus approach.

Keywords Catastrophe bond · Hedging · Malliavin calculus

17.1 Introduction

Insurance companies require reinsurance to cover damages due to catastrophes like earthquakes, hurricanes, tornadoes, fall of a big meteorite, etc. Capital in reinsurance industry is limited relative to the magnitude of damages due to catastrophes. As a response to these challenges, reinsurance companies issue risk-linked securities in different forms.

In this paper, we consider *catastrophe bonds*. This financial instrument is working as follows. A hedger seeks to transfer the risk of catastrophe to investors who accept the risk for higher expected returns. The hedger pays the insurance premium in exchange for a pre-specified coverage if a catastrophic event of a certain magnitude takes place. Investors buy an insurance-linked security for cash. The total

A. Malyarenko (✉) · O. Schyberg
Division of Applied Mathematics, School of Education, Culture and Communication, Mälardalen University, Box 883, 721 23 Västerås, Sweden
e-mail: anatoliy.malyarenko@mdh.se

O. Schyberg
e-mail: oskar.schyberg@mdh.se

J. Röman
Swedbank, Stockholm 105 34, Sweden
e-mail: jan.roman@swedbankrobur.se

S. Silvestrov and M. Rančić (eds.), *Engineering Mathematics I*,
Springer Proceedings in Mathematics & Statistics 178,
DOI 10.1007/978-3-319-42082-0_17

amount, i.e., premium and cash, is directed to a special fund, the so called *special purpose vehicle*. The fund issues the catastrophe bonds to investors and buys safe securities, like Treasury bonds. If a catastrophic event of a certain magnitude takes place before the bond matures, the fund compensates the hedger and there is a full or partial forgiveness of the repayment of principal and/or interest. Otherwise, the investors receive their principal plus risk-free interest rate plus a risk-premium. Different payoff functions of the catastrophe bond will be considered below. See also more detailed description in Ma and Ma [20] and Vaugirard [31, 32].

Pricing models for catastrophe bonds can be divided into two classes according to the financial methodology used. *Econometric models* follow the theory of equilibrium pricing and will not be considered there. For their description, see Aase [1], Burnecki and Kukla [3], Cox et al. [5], Cox and Pedersen [6], Loubergé et al. [19]. *No-arbitrage models* follow the no-arbitrage valuation framework, see Cox and Schwebach [7], Cummins and Geman [8], Dassios and Jang [9], Geman and Yor [13], Ma and Ma [20], Nowak and Romaniuk [22], Romaniuk [26], Sondermann [28], Vaugirard [31] and especially the review paper by Burnecki et al. [4].

In no-arbitrage models, the price of a bond driven by a stochastic process $X(t)$ and defined by the payoff function φ involving the times $t_1, \dots, t_n$ is usually presented as the expectation of a random variable

$$\mathsf{E}[\varphi(X(t_1), \dots, X(t_n)) \mid X(0) = x]$$

under the risk-neutral probability measure P^*_0 and can be computed by Monte Carlo methods. Applications require not only to compute the price of a bond, but also to compute its derivatives with respect to the initial condition, drift coefficient, volatility coefficient, etc. A natural approach is to compute the finite-difference approximation of the derivative by Monte Carlo simulation. The convergence rate of both price and sensitivity estimators is $n^{-1/2}$, where n is the number of simulations. However, by the result of Glynn [15], the best possible convergence rate is $n^{-1/3}$, if the central finite difference is used.

Fournié et al. [11, 12] initiated calculation of Greeks using *Malliavin calculus*. They showed that the derivatives of interest can be expressed as

$$\mathsf{E}[\pi\varphi(X(t_1), \dots, X(t_n)) \mid X(0) = x],$$

where π is another random variable. This approach was applied to sensitivity analysis in insurance by Privault and Wei [25] and Roumelioti et al. [27].

More precisely, changes in market come as changes of the probability measure P^*_0. The variation in price becomes

$$\begin{aligned}\mathsf{E}^{\mathsf{P}^*}[\varphi(X(t_1), \dots, X(t_n)) \mid X(0) = x] - \mathsf{E}^{\mathsf{P}^*_0}[\varphi(X(t_1), \dots, X(t_n)) \mid X(0) = x]\\ = \mathsf{E}[\pi\varphi(X(t_1), \dots, X(t_n)) \mid X(0) = x],\end{aligned}$$

where

$$\pi = \frac{\mathrm{d}\mathsf{P}^* - \mathrm{d}\mathsf{P}_0^*}{\mathrm{d}\mathsf{P}_0^*}.$$

Let $\mathsf{P}^* = \mathsf{P}_\lambda^*$, where λ runs over a neighbourhood of 0. Then the sensitivity is

$$\frac{\mathrm{d}}{\mathrm{d}\lambda}\mathsf{E}[\varphi(X(t_1), \ldots, X(t_n)) \mid X(0) = x] = \mathsf{E}[\pi\varphi(X(t_1), \ldots, X(t_n)) \mid X(0) = x],$$

where

$$\pi = \frac{\mathrm{d}f}{\mathrm{d}\lambda}, \qquad f = \frac{\mathrm{d}\mathsf{P}_\lambda^*}{\mathrm{d}\mathsf{P}_0^*},$$

i.e., π is the logarithmic derivative of P^* at P_0^*. It is the Malliavin calculus that can give exact sense to above considerations. A primer on Malliavin calculus may be found in Bichteler et al. [2], Di Nunno et al. [10], and Petrou [24]. Our model combines the approach of Nowak and Romaniuk [22] with that of Petrou [24].

Swedbank uses the Hull–White model to value instruments with some embedding optionality. It is a difficult task to value and calculate the risk for such instruments in continuous time, especially if one has the right to exercise the embedded options during the lifetime of the contract.

Typical instrument is Cancellable Swaps and Bermudan Swaptions. Such instruments can have a high nautili amount and a maturity of 30 years from the day that contract was bought or sold.

Like in the equity market, American (or Bermudan) option are normally prices with a Binomial tree and not by use of the Black–Scholes–Merton model. The reason is the problem due to the boundary condition, since one can exercise such options at any time during the lifetime.

Similar situation exists in the Fixed Income market. A simple Cancellable Swap can be seen (replicated) as a plain vanilla Interest Rate Swap (an IRS) and a Bermudan Swaption. Many Cancellable Swaps are also based on a Cross Currency Swap (CCS or XCCS) where the notional amount is exchanged when entering the contract and also at the redemption of the contract. If the holder of the optionality decide to call (or put) back the Swap, i.e., to use the right to exercise the option, the nationals will be exchanged as well.

The Hull–White model can be represented as a trinomial tree, which made it easy to handle the optionality described above. The model can also be represented as a Partial Differential Equation (a PDE) which can be solved by a Crank–Nicholson method. The PDE can also be transformed to speed up the solution. This is a similar technique as is done when solving the Black–Scholes PDE. By instead expressing the PDE in terms of the logarithm of the underlying equity price, the coefficients will be independent of the underlying price.

The rest of the paper is organised as follows. In Sect. 17.2 we give a complete description of our model as a system of stochastic differential equations driven by Wiener processes and an independent compound Poisson process under risk-neutral probability measure. In Sect. 17.3 we calculate both the price of a catastrophe bond and its sensitivities with respect to the initial condition, volatility of the diffusion component, and jump amplitude. Section 17.4 concludes.

17.2 The Model

Let P be a *historical* probability measure on a σ-field $\mathfrak{F}$ of events on a set Ω of possible states of economy. Assume that the probability space $(\Omega, \mathfrak{F}, \mathsf{P})$ carries four independent sources of randomness: a set of independent and identically distributed random variables $\{U_j : j \geq 1\}$ with probability distribution μ and finite second moment taking values in $(-1, \infty)$, two Wiener processes $W_1(t)$ and $W_2(t)$, and a Poisson process $N(t)$ with intensity $\lambda > 0$. The stochastic process $W_1(t)$ accounts for non-catastrophic economical risk, the process $W_2(t)$ accounts for the uncertainty of interest rates, the process $N(t)$ for the occurrences of catastrophes, and the set $\{U_j : j \geq 1\}$ for the size of catastrophes. The processes are defined on a *trading horizon* $[0, T]$, where T is the maturity of a catastrophe bond. For all $t \in [0, T]$, let $\mathfrak{F}_t$ be the σ-field generated by $W_1(s)$, $W_2(s)$, $N(s)$ for $s \in [0, t]$, $U_j \mathbf{1}_{\{j \leq N(t)\}}$, and all P-null events. The filtration $\{\mathfrak{F}_t : 0 \leq t \leq T\}$ satisfies standard assumptions and represents the information flow available to market players. We also define $U_0 = 0$.

Let

$$\tau_n = \inf\{t \in [0, T] : N(t) = n\}, \qquad n \geq 0,$$

be the time moment when the nth catastrophe occurs. The aggregated catastrophe losses up to and including the time moment t are then given by the compound Poisson process

$$J(t) = \sum_{n=1}^{\infty} U_n \mathbf{1}_{(\tau_{n-1}, \tau_n]}(t). \tag{17.1}$$

We assume that the *natural risk index* $I(t)$ is a right-continuous stochastic process that satisfies the *Merton jump-diffusion model* [21], i.e., the stochastic differential equation

$$\frac{\mathrm{d}I(t)}{I(t-)} = \mu(t)\,\mathrm{d}t + \sigma_I(t)\,\mathrm{d}W_1(t) + J(t)\,\mathrm{d}N(t),$$

where $\mu(t)$ is the drift parameter, and where $\sigma_I(t)$ is the volatility of the Wiener component of the natural risk index. In particular, the jump of $I(t)$ at time τ_n is

$$I(\tau_n) - I(\tau_n-) = I(\tau_n-)U_n,$$

or $I(\tau_n) = I(\tau_n-)(1 + U_n)$. A popular choice for the distribution μ is that $1 + U_n$ are log-normally distributed.

Denote by $F(0, t)$ the instantaneous forward rate at time 0 for the maturity t. Assume that the instantaneous risk-free interest rate, $r(t)$, is given under the historical probability P by the *Hull–White model*, see Hull and White [17]:

$$\mathrm{d}r(t) = (\theta(t) - a(t)r(t) + \lambda(t, r)\sigma_R(t))\,\mathrm{d}t + \sigma_R(t)\,\mathrm{d}W_2(t), \tag{17.2}$$

where $\lambda(t, r)$ is the market price of *interest rate* risk. In what follows we assume that $a(t)$ is a constant. Hull and White [18] proved that the model fits the yield curve if and only if

$$\theta(t) = \frac{\partial F(0, t)}{\partial t} + aF(0, t) + \frac{1}{2a}\sigma_R^2(t)(1 - \mathrm{e}^{-2at}).$$

The next assumption follows Merton [21]:

…the jumps represent "pure" non-systematic risk.

In other words, investors are neutral toward jump risk. The overall economy is only marginally influenced by localised catastrophes, and the catastrophe losses pertain to idiosyncratic shocks to the capital markets. The catastrophic risk associated with jumps can be diversified away, i.e., the catastrophic shocks will represent "non-systematic" and have a zero risk premium. On the other hand, we suppose that the non-catastrophic changes both in the risk index and in interest rates can be replicated by existing quoted securities.

In mathematical terms, the above assumptions are formulated as follows. Let

$$D(t, T) = \exp\left(-\int_t^T r(u)\,\mathrm{d}u\right)$$

be the stochastic discount factor. Let C_I be a quoted contingent claim written on I (in particular, a catastrophe bond). There exist equivalent martingale measures under which the discounted time t value of the claim, $D(t, T)C_I(T)$, is a martingale. All such measures have the unique restriction, say P^*, to the σ-field generated by $W_1(t)$ and $W_2(t)$, $0 \le t \le T$. The no-arbitrage price for C is

$$C_I(t) = \mathsf{E}^{\mathsf{P}^*}[D(t, T)C_I(T) \mid \mathfrak{F}_t]. \tag{17.3}$$

The dynamics of I and r under P^* is described by the following system:

$$\begin{aligned} \frac{\mathrm{d}I(t)}{I(t-)} &= (\mu(t) - \lambda(t)\sigma_I)\,\mathrm{d}t + \sigma_I(t)\,\mathrm{d}W_1^{\mathsf{P}^*}(t) + J(t)\,\mathrm{d}N(t), \quad & I(0) = I_0, \\ \mathrm{d}r(t) &= (\theta(t) - ar(t))\,\mathrm{d}t + \sigma_R(t)\,\mathrm{d}W_2^{\mathsf{P}^*}(t), \quad & r(0) = r_0, \end{aligned} \tag{17.4}$$

where $\lambda(t)$ is the market price of *nature* risk, and where $W_1^{\mathsf{P}^*}(t)$ and $W_2^{\mathsf{P}^*}(t)$ are the P^*-Wiener processes that correspond to P-Wiener processes $W_1(t)$ and $W_2(t)$. Specifically,

$$\begin{aligned} \mathrm{d}W_1^{\mathsf{P}^*}(t) &= \lambda(t)\,\mathrm{d}t + \mathrm{d}W_1(t), \\ \mathrm{d}W_2(t) &= \lambda(t,r)\,\mathrm{d}t + \mathrm{d}W_2^{\mathsf{P}^*}(t), \end{aligned}$$

by the Girsanov theorem [14]. Note that in the case of the natural risk index we replace the historical probability by the risk-neutral one. The Hull–White model is usually formulated in the risk-neutral setting, and we used the Girsanov theorem in opposite direction to obtain (17.2). See discussion in the recent paper by Hull et al. [16].

To investigate existence and uniqueness of solutions of the system (17.4), write it in the form

$$\mathrm{d}\mathbf{X}(t) = \mathbf{a}(t,\mathbf{X}(t))\,\mathrm{d}t + \sigma(t,\mathbf{X}(t))\,\mathrm{d}\mathbf{W}^{\mathsf{P}^*}(t) + \int_0^\infty (sI(t-),0)^\top \tilde{N}(\mathrm{d}t,\mathrm{d}s),$$

where

$$\mathbf{X}(t) = \begin{pmatrix} I(t-) \\ r(t) \end{pmatrix}, \qquad \mathbf{a}(t,\mathbf{X}(t)) = \begin{pmatrix} I(t-)(\mu(t) - \lambda(t)\sigma_I(t)) \\ \theta(t) - ar(t) \end{pmatrix},$$
$$\sigma(t,\mathbf{X}(t)) = \begin{pmatrix} I(t-)\sigma_I(t) & 0 \\ 0 & \sigma_R(t) \end{pmatrix}, \qquad \mathbf{W}^{\mathsf{P}^*}(t) = \begin{pmatrix} W_1^{\mathsf{P}^*}(t) \\ W_2^{\mathsf{P}^*}(t) \end{pmatrix},$$

and where

$$\tilde{N}(\mathrm{d}t,\mathrm{d}s) = N(\mathrm{d}t,\mathrm{d}s) - \lambda\mu(\mathrm{d}s)\,\mathrm{d}t$$

is the compensated Poisson random measure of the Lévy process (17.1).

By Øksendal and Salem [23, Theorem 1.19], the above system has the unique solution, if its coefficients satisfy the conditions of *at most linear growth*: there exists a positive finite constant C_1 such that

$$\begin{aligned} x_1^2(\mu(t) - \lambda(t)\sigma_I^2)^2 + (\theta(t) - ax_2)^2 + x_1^2\sigma_I^2(t) + \sigma_R^2(t) \\ + \lambda\int_0^\infty t^2\nu(\mathrm{d}t) \le C_1(1 + \|\mathbf{x}\|^2), \end{aligned}$$

for all $\mathbf{x} = (x_1, x_2)^\top \in \mathbb{R}^2$, and *Lipschitz continuity*: there exists a positive finite constant C_2 such that

$$\begin{aligned} (x_1 - y_1)^2(\mu(t) - \lambda(t)\sigma_I)^2(t) + (\theta(t) - a(x_2 - y_2))^2 + (x_1 - y_1)^2\sigma_I^2(t) \\ + \sigma_R^2(t) \le C_2\|\mathbf{x} - \mathbf{y}\|^2, \end{aligned}$$

for all $\mathbf{x}, \mathbf{y} \in \mathbb{R}^2$.

17.3 Calculating Sensitivities of Bond Price

Currently, there exist essentially two general approaches to calculate the no-arbitrage price (17.3): the partial integro-differential equations method and the probabilistic simulation method.

Let $\mathscr{L}$ be the infinitesimal generator of the process $\mathbf{X}(t)$. Then the function

$$v(\mathbf{x}, t) = \mathsf{E}\left[f(\mathbf{X}(T)) \exp\left(\lambda \int_t^T g(\mathbf{X}(s), s)\,\mathrm{d}s \right) \mid \mathbf{X}(t) = \mathbf{x} \right]$$

satisfies the equation

$$\frac{\partial v}{\partial t} + \mathscr{L}v + \lambda g(\mathbf{x}, t)v = 0$$

subject to the final condition

$$\lim_{t \uparrow T} v(\mathbf{x}, t) = f(\mathbf{x}(T)).$$

The payoff function of the catastrophe bond, however, depends on the whole trajectory of the natural risk index process and therefore is not of a form $f(\mathbf{x}(T))$. Moreover, the operator $\mathscr{L}$ is integro-differential, therefore we obtain a complicated computational problem. A strategy for pricing components of a Bermudan-style callable catastrophe bond by sophisticated numerical methods has been realised by Unger [29, 30]. We will not use this approach here. Instead, we will combine some closed-type formulae with probabilistic simulation method, following Nowak and Romaniuk [22] and Vaugirard [31].

Let F be the face value of a zero-coupon catastrophe bond with maturity T, and let $v = v(T, F, \{\, J(t)\colon 0 \le t \le T \,\})$ be its payoff function. By Nowak and Romaniuk [22, Theorem 1], the no-arbitrage bond price $B(0)$ is

$$B(0) = \mathsf{E}^{\mathsf{P}^*}[D(0, T)]\mathsf{E}^{\mathsf{P}^*}[v(T, F, \{\, J(t)\colon 0 \le t \le T \,\})].$$

Note that the first term is the time 0 zero-coupon bond price, $P(0, T)$.

In the case of constant volatility, $\sigma_R(t) = \sigma$, the the zero-coupon bond price has been calculated by Hull and White [18]. If the volatility is not constant, we have the following result.

Lemma 17.1 *The time t zero-coupon price in the Hull–White model with variable volatility is given by*

$$P(t, T) = \mathrm{e}^{A(t,T) - B(t,T)r(t)},$$

where

$$A(t,T) = \ln\frac{P(0,T)}{P(0,t)} + F(0,t)B(t,T) + \frac{1}{2}\int_t^T B^2(s,T)\sigma_R^2(s)\,\mathrm{d}s,$$

$$B(t,T) = \frac{1}{a}(1 - \mathrm{e}^{(T-t)}).$$

Proof of Lemma 17.1 is similar to Hull and White [18]. Note that in the case of constant volatility, the integral may be easily calculated and leads to the classical formula

$$P(t,T) = \ln\frac{P(0,T)}{P(0,t)} + F(0,t)B(t,T) - \frac{\sigma_R^2}{4a^3}(\mathrm{e}^{-aT} - \mathrm{e}^{-at})^2(\mathrm{e}^{2at} - 1).$$

In other words, in place of a σ_R, there appears a time integral that includes $\sigma_R(s)$.

Vaugirard [31] assumed that the triggering point of a catastrophe bond is the first passage of the natural risk index $I(t)$ through a level K. If the triggering point does not occur, then is paid the face value, F, otherwise she receives $F(1-w)$ with $w \in (0,1)$.

Nowak and Romaniuk [22] considered two payoff functions $v_i(T, F, \{\, J(t)\colon 0 \le t \le T\,\})$, $i = 1, 2$. They calculated the expectations $\mathsf{E}^{\mathsf{P}^*}[v_i(T, F, \{\, J(t)\colon 0 \le t \le T\,\})]$ in a closed form, see [22, Lemma 1, Lemma 2]. As they note, some terms in their formulae are often analytically intractable and require probabilistic simulation.

Specifically, the payoff function $v_1(T, F, \{\, J(t)\colon 0 \le t \le T\,\})$ is a stepwise one and may be described as follows. Let $n \ge 2$ be a positive integer. Let $0 \le K_0 < K_1 < \cdots < K_n$ be a set of reals. Define the set of stopping times $\{\, \tau_i(\omega)\colon 1 \le i \le n\,\}$ by

$$\tau_n(\omega) = \min\{\inf\{\, t \in [0,T]\colon J(t,\omega) > K_i\,\}, T\}.$$

Finally, let $0 < w_1 < w_2 < \cdots < w_n$ be a set of reals with $w_1 + \cdots + w_n \le 1$.

If $\tau_1 > T$, then the bondholder is paid the face value, F. If $\tau_n \le T$, then she receives $F(1 - w_1 - \cdots - w_n)$. If $\tau_{k-1} \le T < \tau_k$, $1 < k \le n$, then she receives $F(1 - w_1 - \cdots - w_{k-1})$. Cash payments are made at maturity T. The price of this bond is

$$B(0) = P(0,T)F\left(1 - \sum_{i=1}^{n} w_i\Phi_i(T)\right),$$

where

$$\Phi_i(T) = 1 - \mathrm{e}^{-\lambda T}\sum_{j=0}^{\infty}\frac{(\lambda T)^j}{j!}\tilde{\Phi}_j(K_i),$$

and where $\tilde{\Phi}_j$ is the cumulative distribution function of the sum $U_0 + \cdots + U_j$. Note that Malliavin calculus cannot be applied here to calculate the bond's sensitivities, because the payoff function depends on *stochastic* time moments τ_1, …, τ_n.

The second payoff function considered by Nowak and Romaniuk in [22] is piecewise linear:

$$v_2(T) = F\left(1 - \sum_{j=0}^{n-1} \frac{\min\{J(T), K_{j+1}\} - \min\{J(T), K_j\}}{K_{j+1} - K_j} w_{j+1}\right). \tag{17.5}$$

Denote

$$\begin{aligned} \varphi_m &= \mathsf{P}\{J(T) \leq K_m\}, \qquad 0 \leq m \leq n, \\ e_m &= \mathsf{E}[J(T)\mathbf{1}_{\{K_m < J(T) \leq K_{m+1}\}}], \qquad 0 \leq m \leq n-1. \end{aligned}$$

The price of this bond is

$$\begin{aligned} B(0) = P(0,T)F \Bigg\{ 1 - (1-\varphi_n)\sum_{j=1}^{n} w_j - \sum_{m=0}^{n-1} \Bigg[(\varphi_{m+1} - \varphi_m) \sum_{j=1}^{m} w_j \\ + \frac{e_n - K_m(\varphi_{m+1} - \varphi_m)}{K_{m+1} - K_m} \Bigg] \Bigg\}. \end{aligned}$$

Note that this payoff function depends only on $J(T)$. We consider a payoff function similar to (17.5):

$$v(T) = F\left(1 - \sum_{j=0}^{n-1} \frac{\min\{I(T), K_{j+1}\} - \min\{I(T), K_j\}}{K_{j+1} - K_j} w_{j+1}\right).$$

We calculate Δ, the derivative of the bond price with respect to the initial condition I_0, Γ, the second derivative, ν or *vega*, the derivative with respect to the volatility σ_I, and α, the derivative with respect to the jump amplitude.

Theorem 17.1 *Assume that* $0 < \inf\{\sigma_I(t) \colon 0 \leq t \leq T\} \leq \sup\{\sigma_I(t) \colon 0 \leq t \leq T\} < \infty$. *Then we have*

$$\begin{aligned} \Delta &= \frac{P(0,T)}{TI(0)} \mathsf{E}\left[v(I(T)) \int_0^T \sigma_I^{-1}(t) I(t-) \, \mathrm{d}W^{\mathsf{P}^*}_1(t) \right], \qquad (17.6) \\ \Gamma &= \frac{P(0,T)}{T^2 I^2(0)} \mathsf{E}\left[v(I(T)) \left(\int_0^T \sigma_I^{-1}(t) I(t-) \, \mathrm{d}W^{\mathsf{P}^*}_1(t) \right)^2 \right], \\ \nu &= \frac{P(0,T)}{T} \mathsf{E}\Bigg\{ v(I(T)) \Bigg[\left(W_1^{\mathsf{P}^*}(T) - \int_0^T \sigma_I(s)\,\mathrm{d}s \right) \int_0^T \sigma_I^{-1}(t) I(t-) \, \mathrm{d}W^{\mathsf{P}^*}_1(t) \\ &\qquad - \int_0^T \sigma_I^{-1}(t) I(t-) \, \mathrm{d}t \Bigg] \Bigg\}, \\ \alpha &= \frac{P(0,T)}{T} \mathsf{E}\Bigg[v(I(T)) \left(\int_0^t \int_0^\infty s \tilde{N}(\mathrm{d}u, \mathrm{d}s) - \lambda \int_0^t \int_0^\infty s^2 I(u-) \mu(\mathrm{d}s) \, \mathrm{d}u \right) \\ &\qquad \times \int_0^T \sigma_I^{-1}(t) I(t-) \, \mathrm{d}W^{\mathsf{P}^*}_1(t) \Bigg]. \end{aligned}$$

Proof Denote by

$$\begin{aligned} b(t, I(t-)) &= (\mu(t) - \lambda(t)\sigma_I(t))I(t-), \\ \sigma(t, I(t-)) &= \sigma_I(t)I(t-), \\ \gamma(s, t, I(t-)) &= sI(t-), \end{aligned}$$

the coefficients of the first equation in (17.4). The first variation of the process $I(t)$ [24], $Y(t) = \nabla I(t)$, satisfies the following equation

$$\begin{aligned} \mathrm{d}Y(t) = \frac{\partial b(t, I(t-))}{\partial I(t-)} Y(t-)\,\mathrm{d}t &+ \frac{\partial \sigma(t, I(t-))}{\partial I(t-)} Y(t-)\,\mathrm{d}W_1^{\mathsf{P}^*}(t) \\ &+ Y(t-)\int_0^\infty \frac{\partial \gamma(s, t, I(t-))}{\partial I(t-)} \tilde{N}(\mathrm{d}t, \mathrm{d}s), \\ &Y(0) = 1, \end{aligned}$$

which becomes

$$\begin{aligned} \mathrm{d}Y(t) = (\mu(t) - \lambda(t)\sigma_I(t))Y(t-)\,\mathrm{d}t + \sigma_I(t)Y(t-)\,\mathrm{d}W_1^{\mathsf{P}^*}(t) + Y(t-)\int_0^\infty s\tilde{N}(\mathrm{d}t, \mathrm{d}s), \\ Y(0) = 1. \end{aligned}$$

It follows that

$$Y(t) = \frac{I(t)}{I(0)}.$$

Note that Condition 1 of Theorem 17.1 means that the diffusion coefficient $\sigma_I(t)$ is *uniformly elliptic* in the sense of Petrou [24]. The first formula in (17.6) now follows from [24, Proposition 9].

Note that the second variation of $I(t)$ is equal to $Y(t)$. The second equation in (17.6) follows.

Consider the perturbed process $I^\varepsilon(t)$ satisfying the following equation:

$$\begin{aligned} \mathrm{d}I^\varepsilon(t) = (\mu(t) - \lambda(t)\sigma_I(t))I^\varepsilon(t-)\,\mathrm{d}t + (\sigma_I(t)I^\varepsilon(t-) + \varepsilon\xi(t, I(t-)))\,\mathrm{d}W_1^{\mathsf{P}^*}(t) \\ + I^\varepsilon(t-)\int_0^\infty s\tilde{N}(\mathrm{d}t, \mathrm{d}s), \\ I^\varepsilon(0) = I(0). \end{aligned}$$

where ε is a real number and $\xi(t, x)$ is a continuously differentiable function with bounded gradient. Its first variation,

$$Z^\varepsilon(t) = \frac{\partial I^\varepsilon(t)}{\partial \varepsilon}$$

satisfies the equation

$$\begin{aligned}\mathrm{d}Z^{\varepsilon}(t)=&\frac{\partial b(t,I^{\varepsilon}(t-))}{\partial I^{\varepsilon}(t-)}Z^{\varepsilon}(t-)\,\mathrm{d}t+\frac{\partial[\sigma_I(t)I^{\varepsilon}(t-)+\varepsilon\xi(t,I^{\varepsilon}(t-))]}{\partial I^{\varepsilon}(t-)}Z^{\varepsilon}(t-)\,\mathrm{d}W_1^{\mathsf{P}^*}(t)\\&+\xi(t,I^{\varepsilon}(t-))\,\mathrm{d}W_1^{\mathsf{P}^*}(t)+Z^{\varepsilon}(t-)\int_0^{\infty}\frac{\partial\gamma(s,t,I^{\varepsilon}(t-))}{\partial I^{\varepsilon}(t-)}\tilde{N}(\mathrm{d}t,\mathrm{d}s),\\&Z^{\varepsilon}(0)=0,\end{aligned}$$

which becomes

$$\begin{aligned}\mathrm{d}Z^{\varepsilon}(t)=&(\mu(t)-\lambda(t)\sigma_I(t))Z^{\varepsilon}(t-)\,\mathrm{d}t+\left[\sigma_I(t)+\varepsilon\frac{\partial\xi(t,I^{\varepsilon}(t-))}{\partial I^{\varepsilon}(t-)}\right]Z^{\varepsilon}(t-)\,\mathrm{d}W_1^{\mathsf{P}^*}(t)\\&+\xi(t,I^{\varepsilon}(t-))\,\mathrm{d}W_1^{\mathsf{P}^*}(t)+Z^{\varepsilon}(t-)\int_0^{\infty}s\tilde{N}(\mathrm{d}t,\mathrm{d}s),\\&Z^{\varepsilon}(0)=0.\end{aligned}$$

Put $\xi(t,I^{\varepsilon}(t-))=I^{\varepsilon}(t-)$. The stochastic differential equation for $Z^{\varepsilon}(t)$ takes the form

$$\begin{aligned}\mathrm{d}Z^{\varepsilon}(t)=&(\mu(t)-\lambda(t)\sigma_I(t))Z^{\varepsilon}(t-)\,\mathrm{d}t+[\sigma_I(t)+\varepsilon]Z^{\varepsilon}(t-)\,\mathrm{d}W_1^{\mathsf{P}^*}(t)\\&+I^{\varepsilon}(t-)\,\mathrm{d}W_1^{\mathsf{P}^*}(t)+Z^{\varepsilon}(t-)\int_0^{\infty}s\tilde{N}(\mathrm{d}t,\mathrm{d}s),\\&Z^{\varepsilon}(0)=0.\end{aligned}$$

Denote $Z(t)=Z^0(t)$. The solution to this equation when $\varepsilon=0$ has the form

$$Z(t)=I(t)\left[W_1^{\mathsf{P}^*}(t)-\int_0^t\sigma_I(s)\,\mathrm{d}s\right],$$

which is easy to check using the multidimensional Itô formula, see Øksendal and Sulem [23, Theorem 1.16].

Define $\beta(t)=Y^{-1}(t)Z(t)$. We have

$$\beta(t)=I(0)\left[W_1^{\mathsf{P}^*}(t)-\int_0^t\sigma_I(s)\,\mathrm{d}s\right].$$

To use [24, Proposition 10], we have to calculate the *Malliavin derivative* of $\beta(T)$. By Di Nunno et al. [10, Eq. (3.8)], we have

$$D_t\beta(T)=I(0)\mathbf{1}_{[0,T]}(t).$$

All conditions of [24, Proposition 10] are satisfied. The above proposition gives the third equation in (17.6), if we put $\zeta(t)=T^{-1}$.

Perturb the process $I(t)$ in the following way:

$$\begin{aligned} \mathrm{d}I^{\varepsilon}(t) &= (\mu(t) - \lambda(t)\sigma_I(t))I^{\varepsilon}(t-)\,\mathrm{d}t + \sigma_I(t)I^{\varepsilon}(t-)\,\mathrm{d}W_1^{\mathsf{P}^*}(t) \\ &\quad + I^{\varepsilon}(t-)\int_0^{\infty}(s + \varepsilon\xi(t, I^{\varepsilon}(t-)))\tilde{N}(\mathrm{d}t, \mathrm{d}s), \\ I^{\varepsilon}(0) &= I(0). \end{aligned}$$

where ε and ξ have the same meaning as before. The first variation of the perturbed process $I^{\varepsilon}(t)$ with respect to ε, $Z^{\varepsilon}(t)$, satisfies the equation

$$\begin{aligned} \mathrm{d}Z^{\varepsilon}(t) &= \frac{\partial b(t, I^{\varepsilon}(t-))}{\partial I^{\varepsilon}(t-)} Z^{\varepsilon}(t-)\,\mathrm{d}t + \frac{\partial \sigma_I(t) I^{\varepsilon}(t-)}{\partial I^{\varepsilon}(t-)} Z^{\varepsilon}(t-)\,\mathrm{d}W_1^{\mathsf{P}^*}(t) \\ &\quad + Z^{\varepsilon}(t-)\int_0^{\infty}\frac{\partial[\gamma(s, t, I^{\varepsilon}(t-)) + \varepsilon\xi(t, I^{\varepsilon}(t-))]}{\partial I^{\varepsilon}(t-)}\tilde{N}(\mathrm{d}t, \mathrm{d}s) \\ &\quad + \int_0^{\infty}\xi(t, I^{\varepsilon}(t-))\tilde{N}(\mathrm{d}t, \mathrm{d}s), \\ Z^{\varepsilon}(0) &= 0. \end{aligned}$$

Put $\xi(t, I^{\varepsilon}(t-)) = I^{\varepsilon}(t-)$. The above equation becomes

$$\begin{aligned} \mathrm{d}Z^{\varepsilon}(t) &= (\mu(t) - \lambda(t)\sigma_I(t))Z^{\varepsilon}(t-)\,\mathrm{d}t + \sigma_I(t)Z^{\varepsilon}(t-)\,\mathrm{d}W_1^{\mathsf{P}^*}(t) \\ &\quad + Z^{\varepsilon}(t-)\int_0^{\infty}(s + \varepsilon)\tilde{N}(\mathrm{d}t, \mathrm{d}s) + \int_0^{\infty} I^{\varepsilon}(t-)\tilde{N}(\mathrm{d}t, \mathrm{d}s), \\ Z^{\varepsilon}(0) &= 0. \end{aligned}$$

The solution to this equation when $\varepsilon = 0$ is as follows:

$$Z(t) = I(t)\left(\int_0^t\int_0^{\infty} s\tilde{N}(\mathrm{d}u, \mathrm{d}s) - \lambda\int_0^t\int_0^{\infty} s^2 I(u-)\mu(\mathrm{d}s)\,\mathrm{d}u\right),$$

To check this, we use the the multidimensional Itô formula cited above. The stochastic process $\beta(t)$ takes the form

$$\beta(t) = I(0)\left(\int_0^t\int_0^{\infty} s\tilde{N}(\mathrm{d}u, \mathrm{d}s) - \lambda\int_0^t\int_0^{\infty} s^2 I(u-)\mu(\mathrm{d}s)\,\mathrm{d}u\right).$$

To use [24, Proposition 11], we have to calculate the *Wiener directional derivative* $D_t^{(0)}\beta(T)$, see its definition in [24, Definition 1]. By [24, Proposition 2], the Wiener directional derivative of a random variable that depends only on the Poisson random measure, is 0. Using [24, Proposition 11] with $\zeta(t) = T^{-1}$, we arrive at the last equation in (17.6).

To calculate the expectations in (17.6), one can use probabilistic simulation methods.

17.4 Concluding Remarks

The above approach may also price catastrophe bonds and calculate sensitivities in the case when the payoff depends on the values $I(t_1)$, $I(t_2)$, …, $I(t_n)$ of the natural risk index at finitely many *deterministic* time moments.

A catastrophe bond with coupon payments may be represented as a portfolio of zero-coupon bonds with different maturities. Our approach may be used for each portfolio component independently.

In a forthcoming paper, we will report the results of numerical experiments based on historical data.

References

1. Aase, K.: An equilibrium model of catastrophe insurance futures and spreads. The GENEVA Pap. Risk and Insur. Theory **24**(1), 69–96 (1999)
2. Bichteler, K., Gravereaux, J.B., Jacod, J.: Malliavin Calculus for Processes with Jumps, Stochastics Monographs, vol. 2. Gordon and Breach Science Publishers, New York (1987)
3. Burnecki, K., Kukla, G.: Pricing of zero-coupon and coupon CAT bonds. Appl. Math. (Warsaw) **30**(3), 315–324 (2003)
4. Burnecki, K., Kukla, G., Taylor, D.: Pricing of catastrophe bonds. In: Čížek, P., Härdle, W.K., Weron, R. (eds.) Statistical Tools for Finance and Insurance, 2nd edn, pp. 371–391. Springer, Heidelberg (2011)
5. Cox, S.H., Fairchild, J.R., Pedersen, H.W.: Economic aspects of securitization of risk. Astin Bull. **30**(1), 157–193 (2000)
6. Cox, S.H., Pedersen, H.W.: Catastrophe risk bonds. N. Am. Actuar. J. **4**(4), 56–82 (2000)
7. Cox, S.H., Schwebach, R.G.: Insurance futures and hedging insurance price risk. The J. of Risk and Insur. **59**(4), 628–644 (1992)
8. Cummins, J.D., Geman, H.: Pricing catastrophe insurance futures and call spreads: An arbitrage approach. The J. of Fixed Income **4**(4), 46–57 (1995)
9. Dassios, A., Jang, J.W.: Pricing of catastrophe reinsurance and derivatives using the Cox process with shot noise intensity. Financ. Stoch. **7**(1), 73–95 (2003)
10. Di Nunno, G., Øksendal, B., Proske, F.: Malliavin Calculus for Lévy Processes with Applications to Finance. Springer, Berlin (2009)
11. Fournié, E., Lasry, J.M., Lebuchoux, J., Lions, P.L.: Applications of Malliavin calculus to Monte-Carlo methods in finance. II. Financ. Stoch. **5**(2), 201–236 (2001)
12. Fournié, E., Lasry, J.M., Lebuchoux, J., Lions, P.L., Touzi, N.: Applications of Malliavin calculus to Monte Carlo methods in finance. Financ. Stoch. **3**(4), 391–412 (1999)
13. Geman, H., Yor, M.: Stochastic time changes in catastrophe option pricing. Insur. Math. Econ. **21**(3), 185–193 (1997)
14. Girsanov, I.: On transforming a certain class of stochastic processes by absolutely continuous substitution of measures. Theory Probab. Appl. **5**(3), 285–301 (1960)
15. Glynn, P.: Optimization of stochastic systems via simulation. In: Proceedings of the 1989 Winter simulation Conference, pp. 90–105. (1989)
16. Hull, J., Sokol, A., White, A.: Short-rate joint-measure models. Risk **10**, 59–63 (2014)
17. Hull, J., White, A.: Pricing interest-rate-derivative securities. Rev. of Financ. Stud. **3**(4), 573–592 (1990)
18. Hull, J., White, A.: Bond option pricing based on a model for the evolution of bond prices. Adv. in Futures and Options Res. **3**, 1–13 (1993)

19. Loubergé, H., Kellezi, E., Gilli, M.: Using catastrophe-linked securities to diversity insurance risk: A financial analysis of cat bonds. J. of Insur. Issues **22**(2), 125–146 (1999)
20. Ma, Z.G., Ma, C.Q.: Pricing catastrophe risk bonds: A mixed approximation method. Insur. Math. Econ. **52**(2), 243–254 (2013)
21. Merton, R.C.: Option pricing when underlying stock returns are discontinuous. J. of Financ. Econ. **3**(1), 125–144 (1976)
22. Nowak, P., Romaniuk, M.: Pricing and simulations of catastrophe bonds. Insur. Math. Econ. **52**(1), 18–28 (2013)
23. Øksendal, B., Sulem, A.: Applied Stochastic Control of Jump Diffusions, 2nd edn. Springer, Berlin (2007)
24. Petrou, E.: Malliavin calculus in Lévy spaces and applications to finance. Electron. J. Probab. **13**(27), 852–879 (2008)
25. Privault, N., Wei, X.: A Malliavin calculus approach to sensitivity analysis in insurance. Insur. Math. Econ. **35**(3), 679–690 (2004)
26. Romaniuk, M.: Pricing the risk-transfer financial instruments via Monte Carlo methods. Syst. Anal. Model. Simul. **43**(8), 1043–1064 (2003)
27. Roumelioti, E.E., Zazanis, M.A., Frangos, N.E.: Sensitivity of the joint survival probability for reinsurance schemes. Math. Methods Appl. Sci. **37**(2), 289–295 (2014)
28. Sondermann, D.: Reinsurance in arbitrage-free markets. Insur. Math. Econ. **10**(3), 191–202 (1991)
29. Unger, A.J.: Pricing index-based catastrophe bonds: Part 1: Formulation and discretization issues using a numerical PDE approach. Comput. Geosci. **36**(2), 139–149 (2010)
30. Unger, A.J.: Pricing index-based catastrophe bonds: Part 2: Object-oriented design issues and sensitivity analysis. Comput. Geosci. **36**(2), 150–160 (2010)
31. Vaugirard, V.E.: Pricing catastrophe bonds by an arbitrage approach. The Q. Rev. of Econ. Financ. **43**(1), 119–132 (2003)
32. Vaugirard, V.E.: Valuing catastrophe bonds by Monte Carlo simulations. Appl. Math. Financ. **10**(1), 75–90 (2003)

Chapter 18
Pricing European Options Under Stochastic Volatilities Models

Betuel Canhanga, Anatoliy Malyarenko, Jean-Paul Murara and Sergei Silvestrov

Abstract Interested by the volatility behavior, different models have been developed for option pricing. Starting from constant volatility model which did not succeed on capturing the effects of volatility smiles and skews; stochastic volatility models appear as a response to the weakness of the constant volatility models. Constant elasticity of volatility, Heston, Hull and White, Schöbel–Zhu, Schöbel–Zhu–Hull–White and many others are examples of models where the volatility is itself a random process. Along the chapter we deal with this class of models and we present the techniques of pricing European options. Comparing single factor stochastic volatility models to constant factor volatility models it seems evident that the stochastic volatility models represent nicely the movement of the asset price and its relations with changes in the risk. However, these models fail to explain the large independent fluctuations in the volatility levels and slope. Christoffersen et al. (Manag Sci 22(12):1914–1932, 2009, [4]) proposed a model with two-factor stochastic volatilities where the correlation between the underlying asset price and the volatilities varies randomly. In the last section of this chapter we introduce a variation of Chiarella and

B. Canhanga (✉)
Faculty of Sciences, Department of Mathematics and Computer Sciences,
Eduardo Mondlane University, Box 257, Maputo, Mozambique
e-mail: betuel.canhanga@mdh.se

B. Canhanga · A. Malyarenko · J.-P. Murara · S. Silvestrov
Division of Applied Mathematics, The School of Education, Culture and Communication,
Mälardalen University, Box 883, 721 23 Västerås, Sweden
e-mail: anatoliy.malyarenko@mdh.se

J.-P. Murara
e-mail: jean-paul.murara@mdh.se

S. Silvestrov
e-mail: sergei.silvestrov@mdh.se

J.-P. Murara
Department of Applied Mathematics, School of Sciences, College of Science
and Technology, University of Rwanda, P.O. Box 3900, Kigali, Rwanda

S. Silvestrov and M. Rančić (eds.), *Engineering Mathematics I*,
Springer Proceedings in Mathematics & Statistics 178,
DOI 10.1007/978-3-319-42082-0_18

Ziveyi model, which is a subclass of the model presented in [4] and we use the first order asymptotic expansion methods to determine the price of European options.

Keywords Financial markets · Option pricing · Stochastic volatilities · Asymptotic expansion

18.1 Introduction

Let $(\Omega, \mathfrak{F}, \mathbb{P})$ be a probability space with risk-neutral probability measure $\mathbb{P}$. Let $\{\mathfrak{F}_t \colon 0 \leq t \leq T\}$ be the filtration generated by a standard d-dimensional Brownian motion $\mathbf{W}_t$.

Let $\mathbf{X} = (X_1, \dots, X_m)^\top$ be the vector of stochastic variables. Assume, that under $\mathbb{P}$ the stochastic variables satisfy the following stochastic differential equation:

$$\mathrm{d}\mathbf{X}_t = \mu(t, \mathbf{X}_t)\,\mathrm{d}t + \Sigma(t, \mathbf{X}_t)\,\mathrm{d}\mathbf{W}_t, \tag{18.1}$$

where $\mu \colon [0, T] \times \mathbb{R}^m \to \mathbb{R}^m$ is the drift, and where $\Sigma \colon [0, T] \times \mathbb{R}^m \to \mathbb{R}^{m \times d}$ is the diffusion. Let $r(t, \mathbf{X}_t)$ be the instantaneous risk-free interest rate, and let $g(\mathbf{x})$ be the payoff of a financial instrument with maturity T.

By risk-neutral valuation, the price $V(t, \mathbf{x})$ of the instrument is

$$V(t, \mathbf{x}) = \mathsf{E}\left[\exp\left(-\int_0^T r(u, \mathbf{x})\,\mathrm{d}u\right) g(\mathbf{X}_T) | \mathfrak{F}_t, \mathbf{X}_t = \mathbf{x}\right].$$

In [1] it is proved that the price $V(t, \mathbf{x})$ satisfies the partial differential equation

$$\frac{\partial V}{\partial t} + \sum_{i=1}^m \mu_i(t, \mathbf{x}) \frac{\partial V}{\partial x_i} + \frac{1}{2} \sum_{i=1}^m \sum_{j=1}^m \sum_{k=1}^d \Sigma_{ik}(t, \mathbf{x}) \Sigma_{kj}(t, \mathbf{x}) \frac{\partial^2 V}{\partial x_i \partial x_j} - r(t, \mathbf{x}) V = 0$$

subject to the terminal value condition $V(T, \mathbf{x}) = g(\mathbf{x})$. The seminal Black–Scholes European option pricing model has the assumption that underlying stock price returns follow a lognormal diffusion process.

Different from the Black–Scholes, for a given stochastic process like the stock price S_t, if its variance σ_t is itself randomly distributed, then (18.1) can be written for $m = d = 2$ as

$$dS_t = \mu(S_t, t)dt + \sigma_t S_t dW_t^1 , \tag{18.2}$$

where σ_t satisfies

$$d\sigma_t = a(\sigma_t, t)dt + b(\sigma_t, t)dW_t^2 ,$$

and where W_t^1 and W_t^2 are standard one-dimensional Brownian motions defined on $(\Omega, \mathfrak{F}, \mathbb{P})$ with the covariance satisfying $d(W_t^i, W_t^j) = \rho_{ij} dt$ for some constant

$\rho_{ij} \in [-1, 1]$ and σ_t is known as the stochastic volatility or the instantaneous volatility or the spot volatility. $a(\sigma_t, t)$ and $b(\sigma_t, t)$ are smooth functions that correspond respectively to the drift and diffusion of the spot volatility. To model derivatives like European options more accurately, it is better to assume that the volatility of the underlying price is a stochastic process rather than a constant as it has been assumed for models based on Black–Scholes formula. The reason is that the latter cannot explain long-observed features of the implied volatility surface, volatility smile and skew, which indicate that the implied volatility does not tend to vary with respect to strike price K and horizon date T.

Definition 18.1 Under any martingale measure $\mathbb{P}$ and the interest rate at time t given by r_t; a model with the form

$$\begin{aligned} dS_t &= r_t S_t dt + \sigma_t S_t dW_t^1 \\ d\sigma_t &= a(\sigma_t, t)dt + b(\sigma_t, t)dW_t^2 \end{aligned}$$

is said to be a stochastic volatility model.

The sections of this chapter present different procedures to price European options with underlying asset prices governed by Constant Elasticity of Variance, Stochastic $\alpha\beta\rho$, Detemple–Tian, Grzelak–Oosterlee–Van Veeren, Jourdain–Sbai, Ilhan–Sircar and Chiarella-Ziveyimodels.

18.2 The Constant Elasticity of Variance (CEV) Model

The lognormality assumption from the Black–Scholes formula does not hold accurately. The pricing of European options has been studied recently for alternative diffusion models.

In 1976 Cox and Ross [5] focused their attention on the constant elasticity of variance diffusion class, and gave the following Constant Elasticity of Variance (CEV) Model

$$dS_t = \mu S_t dt + \sigma S_t^{\beta} dW_t. \tag{18.3}$$

They considered the drift μ to be constant and the real constant parameters are $\sigma \geq 0$ and $\beta \geq 0$. The parameter β is the main feature of this model and it is known as the elasticity factor. The relationship between volatility and price described by the CEV model is controlled by β. The payoff function is defined by $g(s) = \alpha s^{\beta}$ for positive constant α and real positive s.

Remark 18.1 Equation (18.3) becomes the Bachelier model for $\beta = 0$, and for $\beta = 1$ it becomes the Black–Scholes model.

Remark 18.2 Some say that the CEV model is not a stochastic volatility model, but a local volatility model based on the fact that it does not incorporate its own stochastic process for volatility and thus they remove it among the other stochastic volatility models.

The CEV model is used for modelling equities and commodities when attempting to capture the stochastic volatility and the leverage effect. In commodities markets, volatility rises when prices rise. This is known as the *inverse leverage effect* and for this case $\beta > 1$. Whereas in equity markets the volatility of a given stock increases when its price falls which is known as the *leverage effect* with $\beta < 1$.

Now, for cases where $0 < \beta < 1$, the infinitesimal conditional variance of the logarithmic rate of return of the stock equals $\sigma_t^2 = \alpha^2 S_t^{2(\beta-1)}$, and thus it changes inversely with the price. Under this condition the following equations hold:

$$\frac{dv_t}{dS_t}\frac{S_t}{v_t} = \frac{g'(S_t)S_t}{g(S_t)} = \frac{\alpha\beta S_t^{\beta-1} S_t}{\sigma S_t^{\beta}} = \beta, \quad v_t = g(S_t),$$

$$\frac{d\sigma_t}{dS_t}\frac{S_t}{\sigma_t} = \frac{f'(S_t)S_t}{f(S_t)} = \frac{\alpha(\beta-1)S_t^{\beta-2} S_t}{\alpha S_t^{\beta-1}} = \beta - 1.$$

Equation (18.3) corresponds to the classical Girsanov example in the theory of stochastic differential equations which is presented in [15, 16]; assuming that $\mu = 0$ then it has a unique solution for any $\beta \geq \frac{1}{2}$ and this uniqueness fails to hold for values in the interval $(0, \frac{1}{2})$.

The CEV model is complete when assuming that the filtration $\mathscr{F}$ is generated by the driving Brownian motion W_t^1. From this completeness, any European contingent claim that is $\mathscr{F}_T$-measurable and $\mathbb{P}$-integrable, with time t discounting factor B_t, possesses a unique arbitrage price given by the risk-neutral valuation formula

$$v(S_t, t) = B_t \mathbb{E}_{\mathbb{P}}(B_T^{-1} h(S_T)|\mathscr{F}_t).$$

By the Feynman–Kac theorem, the option price $v(S_t, t)$, with $v(S_T, T) = h(s)$ and $h(S_t)$ the inverse of $g(S_t)$, can be given as the solution of the following partial differential equation

$$\frac{\partial v(S_t, t)}{\partial t} + \frac{1}{2}\left(\alpha S_t^{\beta}\right)^2 \frac{\partial^2 v(S_t, t)}{\partial S_t^2} + r S_t \frac{\partial v(S_t, t)}{\partial S_t} - r v(S_t, t) = 0. \tag{18.4}$$

18.2.1 European Option Pricing Formulae Under the CEV Model

Many authors have examined option pricing equations related to the CEV model among others and mentioned that the transition probability density function for the stock price governed by the CEV model can be explicitly expressed in term of the modified Bessel functions. From this, the integration of the payoff function with respect to the transition density can be used to find the arbitrage price of any European contingent claim.

The European option pricing formula can be derived and let us have a look on a computational convenient representation of Schroder presented in [18] for the call price in the CEV model:

$$C_t(S_t, T-t) = S_t\left(1 - \sum_{n=1}^{\infty} g(n+1+\gamma, \tilde{K}_t) \sum_{m=1}^{n} g(m, \tilde{F}_t)\right) \tag{18.5}$$
$$-Ke^{-r(T-t)} \sum_{n=1}^{\infty} g(n+\gamma, \tilde{F}_t) \sum_{m=1}^{n} g(m, \tilde{K}_t)),$$

where $\gamma = \frac{1}{2(1-\beta)}$ and $g(p, x) = \frac{x^{p-1}e^{-x}}{\Gamma(p)}$ is the density function of the Gamma distribution. For the forward price of a stock $F_t = \frac{S_t}{B(t,T)}$ we have:

$$\tilde{F}_t = \frac{F_t^{2(1-\beta)}}{2\chi(t)(1-\beta)^2}, \quad \tilde{K}_t = \frac{K^{2(1-\beta)}}{2\chi(t)(1-\beta)^2}, \quad \chi(t) = \sigma^2 \int_t^T e^{2r(1-\beta)u} du$$

is the scaled expiry of an option.

18.2.2 *Implied Volatility Smile in the CEV Model*

The presence of parameter β in the CEV model is a big advantage over the classical Black-Scholes model because it is possible to make a better fit to observed market prices options with an appropriate choice of α and β. Making $\beta \neq 1$ and $\alpha \neq 0$, the CEV model yields prices of European options corresponding to smiles in the Black-Scholes implied volatility surface. Which means that, for a fixed maturity T, the implied volatility of a call option is a decreasing function of the strike K.

Considering the case when a stock price is governed by (18.3), the forward price of a stock

$$F_t = F_S(t, T) = \frac{S_t}{B(t, T)} = e^{\mu(T-t)} S_t$$

under the martingale measure $\mathbb{P}$, satisfies

$$dF_t = \alpha(t) F_t^{\beta} dW_t. \tag{18.6}$$

As presented in [16] the implied volatility $\hat{\sigma}_0(T, K)$ predicted by (18.6) is

$$\hat{\sigma}_0(T, K) = \frac{\alpha_a}{F_a^{1-\beta}}\left(1 + \frac{(1-\beta)(2+\beta)(F_0-K)^2}{24F_a^2} + \frac{(1-\beta)^2\alpha_a^2 T}{24F_a^{2(1-\beta)}} + \cdots\right).$$

$\hat{\sigma}_0(T, K)$ is the Black implied volatility, $F_a = \frac{(F_0+K)}{2}$ and $\alpha_a = (\frac{1}{T}\int_0^T \alpha^2(u)du)^{1/2}$.

18.3 The Stochastic $\alpha\beta\rho$ (SABR) Model

The SABR model can be seen as a natural extension of the CEV model. When in [11] Hagan et al. examined the issue of dynamics of the implied volatility smile, they argued that any model based on the local volatility function incorrectly predicts the future behaviour of the smile, i.e. when the price of the underlying decreases, local volatility models predict that the smile shifts to higher prices. Similarly, an increase of the price results in a shift of the smile to lower prices. It was observed that the market behaviour of the smile is precisely the opposite. Thus, the local volatility model has an inherent flaw of predicting the wrong dynamics of the Black–Scholes implied volatility. Consequently, hedging strategies based on such a model may be worse than the hedging strategies evaluated for the naive model with constant volatility that is, the Black–Scholes models.

A challenging issue is to identify a class of models that has the following essential features: a model should be easily and effectively calibrated and it should correctly capture the dynamics of the implied volatility smile.

A particular model proposed and analyzed in [11] is specified as follows: under the martingale measure $\mathbb{P}$, the forward asset price S_t is assumed to obey the equation

$$dS_t = \alpha_t S_t^{\beta} dW_t^1, \tag{18.7}$$

$$d\alpha_t = \nu_t \alpha_t dW_t^2, \tag{18.8}$$

which is the SABR model, where $\alpha_0 = 0$, $\frac{1}{2} \leq \beta \leq 1$, $\alpha_{t\neq 0} > 0$ and ν_t is the instantaneous variance of the variance process. W_t^1 and W_t^2 are two correlated Brownian motions with respect to a filtration $\mathfrak{F}$ with constant correlation $-1 < \rho < 1$. Thus, (18.8) can be written as

$$d\alpha_t = \nu_t \alpha_t (\rho dW_t^1 + \sqrt{1-\rho^2} dW_t^2),$$
$$dW_t^1 dW_t^2 = \rho dt, \quad W_t^2 = \rho W_t^1 + \sqrt{1-\rho^2} W_t^2.$$

18.3.1 European Option Pricing Formulae Under the SABR Model

Let us now assume that the overall volatility α_t and the volatility of volatility ν_t are very small. At date t, $S(t) = s$, $\alpha(0) = \alpha$ we can write the value of an European call option by

$$V(t, s, \alpha) = E\{[S(t_{ex.}) - K]^+ | S(t) = s, \alpha(t) = \alpha\}, \tag{18.9}$$

where $t_{ex.}$ is the exercise time. As presented in [19] the option price becomes

$$V(t,s,\alpha) = [s-K]^{+} + \frac{|s-K|}{2\sqrt{\pi}} \int_{\frac{x^2}{2\tau_{ex}}}^{\infty} \frac{e^{-q}}{(q)^{3/2}} dq, \tag{18.10}$$

where $q = \frac{x^2}{2\tau}$.

18.3.2 *Implied Volatility Smile in the SABR Model*

After deriving the European call option pricing formula under the SABR model, we can derive the approximate implied normal volatility and the implied Black volatility in order to utilize the pricing formula more conveniently.

At-the-money option, it is proven in [16] that the Black implied volatility formula under the SABR is as follows:

$$\hat{\sigma}_0(S_0, T) \cong \frac{\alpha}{S_0^{\hat{\beta}}} \left\{ 1 + \left[\frac{\hat{\beta}^2 \alpha^2}{24 S_0^{2\hat{\beta}}} + \frac{\rho \beta \alpha v_t}{4 S_0^{\hat{\beta}}} + \frac{(2-3\rho^2) v_t^2}{24} \right] T \right\}, \tag{18.11}$$

where $\hat{\beta} = 1 - \beta$.

18.4 The Detemple–Tian Model (DTM)

Different from most of the models we present in this chapter, the DTM is considering volatility to be constant but it assumes that the interest rate changes randomly. The underlying asset price S_t and the interest rate r_t follow the system of stochastic differential equations bellow:

$$\frac{dS_t}{S_t} = (r_t - \delta)dt + \sigma_1 dW_1(t), \tag{18.12}$$

$$dr_t = a(r - r_t)dt + \sigma_2 dW_2(t) = [\theta(t) - a r_t]dt + \sigma_2 dW_2(t), \tag{18.13}$$

where $\delta, a, \sigma_1, \sigma_2$ are constants, δ is the dividend rate, σ_1 is the asset price volatility, the speed of mean reversion of the interest rate is a and σ_2 its volatility. $\theta(t)$ is deterministic function of time and W_1, W_2 are correlated Brownian motions with correlation coefficient ρ.

Detemple and Tian in [6] use the model to compute the American option price and show that the exercise region is depending on the interest rate and dividend yield. Also the results were used to derive recursively an integral equation for the exercise region.

If we define

$$J(t,T)=e^{\left(-\int_t^T\int_t^v e^{-a(v-s)}\theta(s)dsdv+\frac{1}{2}\sigma_2^2\int_t^T(1-e^{-a(T-s)})^2ds\right)}$$

and

$$G(t,T)=\frac{1}{a}\left[1-e^{-a(T-t)}\right]$$

for the European call, the options price is given by the following formula

$$V(S_t,r_t,t)=e^{-\delta(T-t)}S_tN(h(S_t,K;t,T))-KP(t,T)\times N(h(S_t,K;t,T)-\sqrt{\omega(t,T)}), \tag{18.14}$$

where

$$h(S_t,K;t,T)=\frac{\ln(S/KP(t,T))-\delta(T-t)}{\sqrt{\omega(t,T)}}+\frac{1}{2}\sqrt{\omega(t,T)};$$

$P(t,T)$ the pure discount bond price is given by

$$P(t,T)=J(t,T)e^{-r_tG(t,T)}$$

and

$$\omega(t,T)=\int_t^T(\sigma_1^2+\sigma_2^2G(u,T)^2+2\rho\sigma_1\sigma_2G(u,T))du. \tag{18.15}$$

18.5 Grzelak–Oosterlee–Van Veeren (GOVV) Model

The particular case of (18.1) when the drift and the diffusion are defined for $m=d=3$ is known as GOVV model presented by Grzelak et al. in [10]. They considered that the price of an asset at time t is S_t and is governed by an stochastic differential equation with stochastic interest rate r_t and stochastic volatility σ_t of mean reversion type. The model evolves according to the following system:

$$\begin{aligned} dS_t &= r_tS_tdt+\sigma_t^pS_tdZ_t^1, \\ dr_t &= \lambda(\theta_t-r_t)dt+\eta dZ_t^2, \\ d\sigma_t &= k(\overline{\sigma}-\sigma_t)dt+\gamma\sigma_t^{1-p}dZ_t^3, \end{aligned} \tag{18.16}$$

where p is constant, λ and k are the speed of mean reversion processes, η is the volatility of the interest rate, γ is the volatility of volatility. θ_t is the long run mean of the interest rate and $\overline{\sigma}$ is the long run mean of the volatility. $Z_t^1,\ Z_t^2,Z_t^3$ are independent Brownian motions with correlation factors given by

$$dZ_t^i dZ_t^j = \rho_{ij} \quad \text{for} \quad i, j = 1,\ 2,\ 3.$$

Considering

$$Z_t^1 = W_t^1,$$
$$Z_t^2 = \rho_{12} W_t^1 + \sqrt{1-\rho_{12}^2}\, W_t^2,$$
$$Z_t^3 = \rho_{13} W_t^1 + \frac{\rho_{23}-\rho_{12}\rho_{13}}{\sqrt{1-\rho_{12}^2}} W_t^2 + \sqrt{1-\rho_{13}^2 - \frac{(\rho_{23}-\rho_{12}\rho_{13})}{1-\rho_{12}^2}}\, W_t^3,$$

and using the notation

$$a = \frac{\rho_{23}-\rho_{12}\rho_{13}}{\sqrt{1-\rho_{12}^2}}, \quad b = \sqrt{1-\rho_{13}^2 - \frac{(\rho_{23}-\rho_{12}\rho_{13})}{1-\rho_{12}^2}},$$

the GOVV model (18.16) can be written as

$$dS_t = r_t S_t dt + \sigma_t^p S_t dW_t^1, \tag{18.17}$$
$$dr_t = \lambda(\theta_t - r_t)dt + \eta\left(\rho_{12} dW_t^1 + \sqrt{1-\rho_{12}^2}\, dW_t^2\right),$$
$$d\sigma_t = k(\overline{\sigma} - \sigma_t)dt + \gamma \sigma_t^{1-p}\left(\rho_{13} dW_t^1 + a dW_t^2 + b dW_t^3\right).$$

If on the above model we consider that the interest rate is constant, the correlation factors ρ_{2j} and ρ_{i2} are equal to zero we generate:

- Heston Model, if $p = \frac{1}{2}$. The underlying asset price and volatilities are governed by the following system

$$dS_t = r + S_t dt + \sqrt{\sigma_t} S_t dW_t^1, \tag{18.18}$$
$$d\sigma_t = k^H(\overline{\sigma}^H - \sigma_t)dt + \gamma^H \sqrt{\sigma_t}\left(\rho_{13} dW_t^1 + \sqrt{1-\rho_{13}^2}\, dW_t^3\right),$$

  where the superscript H stands for Heston, to indicate long run volatility mean, speed of mean return and volatility of volatility.
- Schöbel–Zhu–Heston model, if $p = 1$. The underlying asset price and volatility are governed by the following system

$$dS_t = r_t S_t dt + \sigma_t S_t dW_t^1, \tag{18.19}$$
$$d\sigma_t = k^H(\overline{\sigma}^H - \sigma_t)dt + \gamma^H \left(\rho_{13} dW_t^1 + \sqrt{1-\rho_{13}^2}\, dW_t^3\right).$$

- Schöbel–Zhu model; which is a transformation of Schöbel–Zhu–Heston model that is obtained considering the variance of instantaneous stock $\sigma_t = \sqrt{v_t}$, when the speed of mean reversion of the volatility process is given by $2k$ and the long run mean is represented by $-\left(\sigma_t\overline{\sigma} + \frac{\gamma^2}{2k}\right)$ i.e.

$$dv_t = 2\sqrt{v_t}\left(k^H(\overline{\sigma}^H - \sigma_t)dt + \gamma^H\left(\rho_{13}dW_t^1 + \sqrt{1-\rho_{13}^2}dW_t^3\right)\right);$$

therefore the governing equations of the asset price and its volatility will be

$$dS_t = r_t S_t dt + v_t S_t dW_t^1, \qquad (18.20)$$
$$dv_t = 2k(v_t + \sigma_t\overline{\sigma} + \frac{\gamma^2}{2k})dt + 2\gamma\sqrt{v_t}\left(\rho_{13}dW_t^1 + \sqrt{1-\rho_{13}^2}dW_t^3\right).$$

- Black–Scholes model, if $p = 0$.

18.5.1 Pricing European Options for the GOVV Type Models

Assuming that the characteristic function of the logarithm of the underlying asset price is known; to price an European option one can choose to apply the fast Fourier transforms in a Carr–Madan technique presented in [2] or use the Fourier–Cosine explained in [8]. If from one hand Carr–Madan is a forward method and with easy computations; it requires to use a damping parameter which is only experimentally determined for some very specific classes of models. The fact that there is no any scientifically method to determine the damping parameters brings a huge limitation for the cases when dealing with models with unknown damping parameter. In the next section the pricing methodology is developed using the Fourier–Cosine method.

18.5.1.1 Pricing Method

Let us present first a theorem that will give us the approximation of the probability density function in a bounded domain.

Theorem 18.1 *For a given bounded domain $D = [a_1, a_2]$ and a Fourier expansion with N terms, the probability density function $p_Y(y|S_t)$ can be approximated by*

$$p_Y(y|S_t) = \sum_{n=0}^{N} \frac{2w_n}{|\mathbb{D}|}\Re\left[\widetilde{\phi}\left(\frac{n\pi}{|\mathbb{D}|}\right)e^{\left(-n\pi\frac{ia_1}{|\mathbb{D}|}\right)}\cos\left(n\pi\frac{y-a_1}{|\mathbb{D}|}\right)\right],$$

for $w_0 = \frac{1}{2}$, $w_n = 1$, $\forall n \in \mathbb{R}$ and $\Re$ denoting the real part.

The proof of the theorem is presented in [8].

For the European options, the general risk neutral pricing formula shows that the contingent claim $C(t,\ S_t)$ written at time t on an asset that value is S_t can be obtained by calculating the expected value under risk neutral measure $\mathbb{P}$ of the discounted payoff function $H(t, S_t)$ at maturity T, given that the information $\mathfrak{F}_t$ is known, i.e.

$$C(t, S_t) = E^{\mathbb{P}}\left(e^{-\int_t^T r_s ds} H(T, S_T)|\mathfrak{F}_t\right).$$

If the probability density function $p_Y(y|S_t)$ is known, the above expectation is given by

$$E^{\mathbb{P}}\left(e^{-\int_t^T r_s ds} H(T, S_T)|\mathfrak{F}_t\right) = \int_{\mathbb{R}} H(T, y) p_Y(y|S_t) dy,$$

where

$$p_Y(y|S_t) = \int_{\mathbb{R}} p_{YZ}(y, z|S_t) dz,$$

and

$$z = -\int_t^T r_s ds \quad \text{is the discounting exponent.}$$

Assuming that $p_Y(y|S_t)$ decays fast, it is possible to restrict the integrations to a closed and bounded domain. Therefore, the contingent claim will be approximated to

$$C(t, S_t) = \int_{\mathbb{D}} H(T, y) p_Y(y|S_t) dy, \tag{18.21}$$

where $\mathbb{D} = [a_1, a_2]$ and $|\mathbb{D}| = a_2 - a_1 > 0$.

If we set

$$\mathbf{u} = [u, 0, \ldots, 0]' \quad \text{and} \quad [\mathbf{S}_T = S_t, r_t, \ldots]$$

in order to obtain obvious boundary conditions at maturity, the discounted characteristic function is given by

$$\begin{aligned}\phi(u, S_t, t, T) &= \int\int_{\mathbb{R}} e^{z+iuy} p_{Y,Z}(y, z|S_t) dz dy \\ &= \int_{\mathbb{R}} e^{iuy} p_Y(y|S_t) dy,\end{aligned}$$

which is the transformation of the probability density function $p_Y(y|S_t)$ according to Fourier. Moreover, when considering the domain $\mathbb{D}$ instead of $\mathbb{R}$ the characteristic function is approximated to

$$\widetilde{\phi}(u,\ S_t,\ t,\ T) = \int_{\mathbb{D}} e^{iuy} p_Y(y|S_t) dy,$$

where the probability density function is determined with the use of the Theorem 18.1. The contingent claim can then be obtained by

$$C(t,\ S_t) = \int_{\mathbb{D}} H(T, y) \sum_{n=0}^{N} \frac{2w_n}{|\mathbb{D}|} \Re \left[\widetilde{\phi}\left(\frac{n\pi}{|\mathbb{D}|}\right) e^{\left(-n\pi \frac{ia_1}{|\mathbb{D}|}\right)} \cos\left(n\pi \frac{y - a_1}{|\mathbb{D}|}\right)\right] dy$$

$$= \frac{|\mathbb{D}|}{2} \sum_{n=0}^{N} \Phi_n \frac{\zeta_n^{\mathbb{D}}}{w_n},$$

where

$$\Phi_n = \sum_{n=0}^{N} \frac{2w_n}{|\mathbb{D}|} \Re \left[\widetilde{\phi}\left(\frac{n\pi}{|\mathbb{D}|}\right) e^{\left(-n\pi \frac{ia_1}{|\mathbb{D}|}\right)} \cos\left(n\pi \frac{y - a_1}{|\mathbb{D}|}\right)\right],$$

$$H(T, y) = \max(Ke^y - K; 0) \quad \text{for} \quad y = \log\left(\frac{S}{K}\right),$$

and $\zeta_n^{\mathbb{D}} = \frac{2K}{|\mathbb{D}|}(\alpha_n - \beta_n)$. α_n and β_n are defined by

$$\beta_n = \frac{|\mathbb{D}|^2}{|\mathbb{D}|^2 + (n\pi)^2} \left[\cos(a_1, a_2) + \frac{n\pi}{|\mathbb{D}|} \sin(a_1, a_2) \right]$$

for

$$\cos(a_1, a_2) = \cos(n\pi) e^{a_2} - \cos\left(\frac{-a_1 n\pi}{|\mathbb{D}|}\right),$$

$$\sin(a_1, a_2) = \sin(n\pi) e^{a_2} - \sin\left(\frac{-a_1 n\pi}{|\mathbb{D}|}\right),$$

and

$$\alpha_0 = a_2, \quad \alpha_{n\neq 0} = \frac{|\mathbb{D}|}{n\pi} \left[\sin(n\pi) - \sin\left(\frac{-a_1 n\pi}{|\mathbb{D}|}\right)\right].$$

18.5.1.2 Schöbel–Zhu–Hull–White (SZHW) Model

On a probability space $(\Omega, \mathfrak{F}_t, \mathbb{P})$, when the vector state $\mathbf{X}_t = [S_t, r_t, \sigma_t]$ is Markovian relative to filtration $\mathfrak{F}_t$ with asset price and volatility defined as in (18.16), when $p = 1$ we obtain the so called SZHW model, if interest rate process is given by

$$r_t = r_0 e^{-\lambda t} + \lambda \int_0^t e^{-\lambda(t-s)}\theta_s ds + \eta \int_0^t e^{-\lambda(t-s)} dW_s^{\mathbb{P}}.$$

From the Hull–White decomposition explained in [10], the interest rate process can be expressed by

$$r_t = \widetilde{r}_t + m_t,$$

where

$$m_t = e^{-\lambda t} r_0 + \lambda \int_0^t e^{-\lambda(t-s)}\theta_s ds,$$

and

$$d\widetilde{r}_t = -\lambda \widetilde{r}_t dt + \eta dW_s^{\mathbb{P}} \quad \text{with} \quad \widetilde{r}_0 = 0.$$

Introducing the notation $\sigma_t = \sqrt{v_t}$, $\log S_t = x_t = \widetilde{x}_t + \varphi_t$ for $\phi_t = \int_0^t m_s ds$; the SZHW model is described in an expanded vector space with the new stochastic process v_t, i.e.

$$\begin{aligned}
dx_t &= (\widetilde{r}_t + m_t - \frac{1}{2} v_t)dt + \sigma_t dW_t^1, \\
d\widetilde{r}_t &= -\lambda \widetilde{r}_t dt + \eta \left(\rho_{12} dW_t^1 + \sqrt{1-\rho_{12}^2} dW_t^2\right), \\
dv_t &= (-2 v_t k + 2k\sigma_t \overline{\sigma} + \gamma^2)dt + 2\sigma_t \gamma \left(\rho_{13} dW_t^1 + a dW_t^2 + b dW_t^3\right), \\
d\sigma_t &= k(\overline{\sigma} - \sigma_t)dt + \gamma \left(\rho_{13} dW_t^1 + a dW_t^2 + b dW_t^3\right).
\end{aligned} \tag{18.22}$$

It is shown in [7] that the characteristic function has the following form:

$$\phi^{SZHW}(\mathbf{u}, \mathbf{X}_t, t, T) = e^{-\int_t^T m_s ds + i\mathbf{u}'[\phi_T,\ m_T,\ 0,\ 0]'} e^{A(\mathbf{u},\tau) + \mathbf{B}'(\mathbf{u},\tau)[\widetilde{x}_t, \widetilde{r}_t, v_t, \sigma_t]},$$

where

$$\mathbf{B}(u, \tau) = [B_x(u, \tau), B_r(u, \tau), B_v(u, \tau), B_\sigma(u, \tau)],$$

for

$$
\begin{aligned}
B_x(u,\tau) &= iu,\\
B_r(u,\tau) &= \frac{(iu-1)}{\lambda}(1-\nu(-2\lambda)),\\
B_v(u,\tau) &= \frac{\beta-D}{2\theta}\left(\frac{1-\nu(-2D)}{1-\nu(-2D)G}\right),\\
B_\sigma(u,\tau) &= \left(\frac{\nu(D)}{\nu 2D-G}\right)\left[16k\overline{\sigma}b\sinh^2\left(\frac{\tau D}{4}\right)D^{-1}+\frac{iu-1}{\lambda}\mathrm{F}(u,\tau)\right],\\
A(u,\tau) &= \frac{(\beta-D)\tau-2\log\left(\frac{G\nu(-2D)-1}{G-1}\right)}{4\gamma^2}-\\
&\quad -\frac{(iu-1)^2(3+\nu(-4\lambda)-4\nu(-2\lambda)-2\tau\lambda)}{2\lambda^3},
\end{aligned}
$$

where

$$
\begin{aligned}
\mathrm{F}(u,\tau) &= \eta\rho_{12}iuF_1(u,\tau)+2\eta\gamma\rho_{23}bF_2(u,\tau),\\
F_1(u,\tau) &= \frac{2}{D}(\nu(D)-1)+\frac{2G}{D}(\nu(-D)-1)-\frac{2(\nu(D-2\lambda)-1)}{D-2\lambda}+\\
&\quad +\frac{2G(1-\nu(2\lambda-D))}{D+2\lambda},\\
F_2(u,\tau) &= \frac{2}{D-2\tau}-\frac{4}{D}+\frac{2}{D+2\lambda}+\\
&\quad +\nu(2\lambda-D)\left(\frac{2\nu(2\lambda)(1+\nu(2D))}{D}-\frac{2\nu(2D)}{D-2\lambda}-\frac{2}{D+2\lambda}\right),\\
F_3(u,\tau) &= \int_0^\tau B_\sigma(u,s)\left(k\overline{\sigma}+\frac{1}{2}\gamma^2B_\sigma(u,s)+\eta\rho_{23}\gamma B_r(u,s)\right)ds,
\end{aligned}
$$

and

$$
\beta=(k-\rho_{13}\gamma ui),\quad D=\sqrt{\beta^2-4\alpha\gamma},\quad \theta=2\gamma^2,\quad \alpha=\frac{1}{2}u(i+u),
$$
$$
G=\frac{\beta-D}{\beta+D},\quad \nu(x)=e^{\frac{x\tau}{2}},\quad b=\frac{\beta-D}{2\theta}.
$$

Making $\mathbf{U}=[u,\ 0,\ 0,\ 0]$ the boundary conditions at maturity will be

$$
\phi^{SZHW}(\mathbf{u},\ [\widetilde{x}_t,\widetilde{r}_t,v_t,\sigma_t],\ T,\ T)=e^{iu\widetilde{x}_T},\quad B_x(u,0)=iu,
$$
$$
A(u,0)=B_r(u,0)=B_\sigma(u,0)=B_v(u,0)=0.
$$

This implies that, for the log S_T, the discounted characteristic function is

$$\phi^{SZHW}(u, \mathbf{X}_t, t, T) = e^{\widetilde{A}(u,\tau)+B_x(u,\tau)x_t+B_r(u,\tau)r_t+B_v(u,\tau)v_t+B_\sigma(u,\tau)\sigma_t},$$

for

$$\widetilde{A}(u,\tau) = -\int_t^T m_s ds + iu\int_t^T m_s ds + A(u,\tau) = \Theta(u,\tau) + A(u,\tau),$$

$$\Theta(u,\tau) = (1-iu)\left\{\log\left(\frac{P(0,T)}{P(0,t)}\right) + \frac{\eta^2}{2\lambda^2}\left(\tau + \frac{2}{\lambda}\left(e^{-2\lambda T - e^{-2\lambda t}}\right)\right)\right\},$$

and

$$P(0,t) = e^{-\int_0^t m_s ds} e^{A(0,\tau)+B_x(0,\tau)x_0+B_v(0,\tau)v_0+B_\sigma(0,\tau)\sigma_0}.$$

18.6 Jourdain–Sbai Model (JSM)

Another particular case of (18.16) can be obtained by considering constant interest rate. In this particular model, let us denote volatility by Y

$$Y_t^p = f(Y_t), \quad k(\overline{Y} - Y_t) = b(Y_t), \quad \gamma Y_t^{1-p} = c(Y_t),$$

with

$$Z_t^1 = \rho W_t^2 + \sqrt{1-\rho^2}W_t^1, \quad Z_t^2 = W_t^2$$

for independent correlated Brownian motions W_t^1 and W_t^2. Under these conditions, the underlying asset price is governed the following model:

$$dS_t = rS_t dt + f(Y_t)S_t\left(\rho dW_t^2 + \sqrt{1-\rho^2}dW_t^1\right), \tag{18.23}$$
$$dY_t = b(Y_t)dt + c(Y_t)dW_t^2, \quad Y_0 = y_0.$$

In [14] the above model was treated considering a particular case of Ornstein–Uhlenbeck process and introducing higher order discretization schemes. JSM considers function f to be positive and strictly monotonic allowing that the effective correlation between the asset price and the volatility remain with the same signal (positive). It also considers that function b and c are also smooth functions. This generalizes a group of model, for example Quadratic Gaussian, Stein & Stein, Scotts, Hull and White, Cox and Ross and Detemple–Tian model. When considering the log-price of the asset return, model (18.23) is transformed to

$$dX_t = \left(r - \frac{1}{2}f^2(Y_t)\right)dt + f(Y_t)\left(\rho dW_t^2 + \sqrt{1-\rho^2}dW_t^1\right), \qquad (18.24)$$
$$dY_t = b(Y_t)dt + c(Y_t)dW_t^2, \quad Y_0 = y_0.$$

The goal is to use the second equation of (18.24) into the first equation and make it free of the stochastic integral involving the common Brownian motion W_t^2. Assuming that the volatility of volatility is positive, the drift function of the volatility and the underlying asset volatility are first order differentiable functions with continuous derivatives, then one can define a primitive

$$F(y) = \int_0^y \frac{f}{c}(z)dz,$$

and using Ito's formula, the differential of the primitive is

$$dF(Y_t) = \frac{f}{c}(Y_t)dY_t + 0.5\left(c\frac{\partial f}{\partial y} - f\frac{\partial c}{\partial y}\right)(Y_t)dt,$$

which transforms (18.24) into

$$dX_t = \rho dF(Y_t) + h(Y_t)dt + \sqrt{1-\rho^2}f(Y_t)dW_t^1, \qquad (18.25)$$
$$dY_t = b(Y_t)dt + c(Y_t)dW_t^2,$$

where

$$h(y) = r - 0.5f^2(y) - \rho\left(\frac{b}{c}f + 0.5\left(c\frac{\partial f}{\partial y} - f\frac{\partial c}{\partial y}\right)\right)y.$$

For simplicity, functions $c(Y_t)$, $b(Y_y)$ are denoted by c and b respectively. Bellow we present the discretization of the SDE satisfied by Y_t constructing a scheme which converges to order 2. The details can be found in [14].

18.6.1 The Weak Scheme of Second Order

In the system (18.25), the integration of both sides of the first integral when time goes from 0 to t, gives

$$X_t = \log(S_0) + \rho\left[F(Y_t) - F(y_0)\right] + \int_0^t h(Y_s)ds + \sqrt{1-\rho^2}\int_0^t f(Y_s)dW_t^1,$$

which is not dependent on the Brownian motion W_t^2. The challenge now is to solve one integral with respect to time and another integral with respect to a Brownian motion W_t^1. This is done using numerical techniques (i.e. numerical integration). The weak scheme is defined as:

$$\overline{X}_T^N = \log(S_0) + \rho\left[F(\overline{Y}_T^N) - F(y_0)\right] + \overline{a}_T^N + \sqrt{1-\rho^2\overline{u}_T^N}\,dW_t^1,$$

where

$$\overline{a}_T^N = \delta_N \sum_{k=0}^{N-1} \frac{h\left(\overline{Y}_{tk}^N\right) + h\left(\overline{Y}_{tk+1}^N\right)}{2}, \quad \delta_N = \frac{T}{N},$$

$$\overline{u}_T^N = \delta_N \sum_{k=0}^{N-1} \frac{f^2\left(\overline{Y}_{t_k}^N\right) + f^2\left(\overline{Y}_{t_{k+1}}^N\right)}{2},$$

$$\overline{Y}_0^N = y_0,$$

$$\overline{Y}_{t_{k+1}}^N = e^{\frac{T}{2N}V_0} e^{\left[(W_{t_{k+1}} - W_{t_k})V\right]} e^{\frac{T}{2N}V_0}\,\overline{Y}_{t_k}^N,$$

for

$$V_0 = b(x) - \frac{1}{2}c \times c'(x) \quad \text{and} \quad v = c(x).$$

The notation $e^{tV(x)}$ means the solution of an ordinary differential equation of order one in the form $\zeta'(t) = V(\zeta(t))$ at time t and starting from x.

On the other hand if $Z_t = X_t - \rho F(Y_t)$ the system on our scheme will be

$$\begin{aligned} dZ_t &= h(Y_t)dt + \sqrt{1-\rho^2} f(Y_t)dW_t^1, \\ dY_t &= b(Y_t)dt + c(Y_t)dW_t^2. \end{aligned} \tag{18.26}$$

Applying Feynman–Kac theorem the differential operator associated with (18.26) will be

$$\begin{aligned} \mathscr{L}v(z,y) &= h(y)\frac{\partial v}{\partial z} + b(y)\frac{\partial v}{\partial y} + \frac{c^2(y)}{2}\frac{\partial^2 v}{\partial y^2} + \frac{1-\rho^2}{2} f^2(y)\frac{\partial^2 v}{\partial z^2} \\ &= \mathscr{L}_1 v(z,y) + \mathscr{L}_2 v(z,y), \end{aligned} \tag{18.27}$$

with

$$\mathscr{L}_1 v(z,y) = b(y)\frac{\partial v}{\partial y} + \frac{c^2(y)}{2}\frac{\partial^2 v}{\partial y^2},$$

$$\mathscr{L}_2 v(z,y) = h(y)\frac{\partial v}{\partial z} + \frac{1-\rho^2}{2} f^2(y)\frac{\partial^2 v}{\partial z^2}.$$

In the case of plain vanilla, the option price is given in [17] by

$$BS_{\alpha,T}\left(s_0 e^{\rho(F(Y_T)-F(y_0))+a_T+\left(\frac{(1-\rho^2)v_T}{2T}-r\right)T}, \frac{(1-\rho^2)v_T}{T}\right),$$

where α is the payoff function depending on the underlying asset and the strike price. $BS_{\alpha,T}(s,v)$ is the price of a European option with payoff function α which matures at T, initial stock price s, volatility $\sqrt{v}$, constant interest rate r, given by Black - Scholes formula. For the case of call or put option, $BS_{\alpha,T}$ is given in a closed formula and the option price can be approximated by

$$P(s,T,r,v,K) \cong \frac{1}{M}\sum_{i=1}^{M} BS_{\alpha,T}\left(s_0 e^{\rho(F(\overline{Y}_T^{N,i})-F(y_0))+a_T^{N,i}+\left(\frac{(1-\rho^2)v_T^{N,i}}{2T}-r\right)T}, \frac{(1-\rho^2)v_T^{N,i}}{T}\right),$$

where M is the total number of Monte Carlo samples and the index i refers to independent draws.

18.7 Ilhan–Sircar Model (ISM)

Barrier options are contingent claims that are activated or deactivated if the underlying asset price hits the barrier during the life time of the option. These options are qualified as:

- *up in* - the underlying asset price in the beginning is lower than the barrier level and the option will be activated only if before the maturity the asset price hits the barrier;
- *up out* - the underlying asset price in the beginning is lower than the barrier level and the option starts activated. If the asset price hits the barrier before the maturity the option is deactivated;
- *down in* - the underlying asset price in the beginning is greater than the barrier level and the option will be activated only if before the maturity the asset price hits the barrier;
- *down out* - the underlying asset price in the beginning is greater than the barrier level and the option starts activated. If the asset price hits the barrier before the maturity the option is deactivated.

The activation or deactivation of an barrier option is for its life, meaning that if the option hits the barrier and is activated or deactivate doesn't matter if afterwards it returns to the barrier. For the execution or not is only considered the position the option took at the first time it hits the barrier level.

In a model presented by Ilhan and Sircar in [13] the stock price process and the volatility driving process are solutions of the following stochastic differential equations:

$$dS_t = \mu S_t \, dt + \sigma(t, Y_t) \, dW_t^1, \quad S_0 = xe^{-rT},$$
$$dY_t = b(t, Y_t) \, dt + a(t, Y_t)(\rho \, dW_t^1 + \rho' \, dW_t^2), \quad Y_0 = y,$$

where ρ is the instantaneous correlation between shocks to S and Y and the symbol ρ' denote $\sqrt{1-\rho^2}$. Assuming that $a(t, Y_t)$ and $\sigma(t, Y_t)$ are bounded above and bellow away from zero and smooth with bounded derivatives, and also that $b(t, Y_t)$ is smooth with bounded derivatives. The utility indifference price of the contingent claim D at time $t = 0$ of an investor who has initial wealth z, is the solution $\tilde{h}(z, D)$ to the following equation:

$$u(z, D) = u(z - e^{rT} \tilde{h}(z, D), 0).$$

Let $h(z, D) = e^{rT} \tilde{h}(z, D)$ be the T-forward value of indifference price. According to

$$h(z, D) = \frac{1}{\gamma} \log \left(\frac{u(0, D)}{u(0, 0)} \right),$$

the indifference price does not depend on the initial wealth. Therefore, we omit the dependence on z in the notation.

According to [13] under some regularity conditions, the optimal static hedging position exists, is unique, and satisfies the following equation:

$$\tilde{h}'(B^{\alpha^*}) = \tilde{p}.$$

It remains to find $\tilde{h}(B^{\alpha})$.

Let $\mathscr{L}_y^0$ be the following differential operator:

$$\mathscr{L}_y^0 = \frac{1}{2} a^2(t, y) \frac{\partial^2}{\partial y^2} + \left(b(t, y) - \rho a(t, y) \frac{\mu - r}{\sigma(t, y)} \right) \frac{\partial}{\partial y},$$

and $f(t, y)$ be the solution to the following problem:

$$\frac{\partial f}{\partial t} + \mathscr{L}_y^0 f = (1 - \rho^2) \frac{(\mu - r)^2}{2\sigma^2(t, y)} f, \quad t < T,$$
$$f(T, y) = 1.$$

Denoting

$$\psi(t, y) = \frac{1}{1 - \rho^2} \log f(t, y),$$

for the differential operator $\mathscr{L}^E_{x,y}$ defined as:

$$\mathscr{L}^E_{x,y} = \mathscr{L}^0_y + \rho'^2 a^2(t,y)\frac{\partial \psi}{\partial y}(t,y)\frac{\partial}{\partial y} + \frac{1}{2}\sigma^2(t,y)x^2\frac{\partial^2}{\partial x^2} + \rho\sigma(t,y)a(t,y)x\frac{\partial^2}{\partial x \partial y},$$

if $\Phi(t,x,y)$ is the solution to the following problem:

$$\frac{\partial \Phi}{\partial t} + \mathscr{L}^E_{x,y}\Phi + \frac{1}{2}\gamma\rho'^2 a^2(t,y)\left(\frac{\partial \Phi}{\partial y}\right)^2 = 0, \quad t < T, \quad x > 0,$$
$$\Phi(T,x,y) = \alpha(K'-x)^+ - (x-K)^-,$$

and $\varphi(t,x,y)$ the solution to the following problem:

$$\frac{\partial \varphi}{\partial t} + \mathscr{L}^E_{x,y}\varphi + \frac{1}{2}\gamma\rho'^2 a^2(t,y)\left(\frac{\partial \varphi}{\partial y}\right)^2 = 0, \quad t < T, \quad x > Be^{r(T-t)},$$
$$\varphi(T,x,y) = \alpha(K'-x)^+,$$
$$\varphi(t, Be^{r(T-t)}, y) = \Phi(t, Be^{r(T-t)}, y),$$

then, the indifference price at time $t = 0$ is

$$\tilde{h}(B^\alpha) = e^{-rT}\varphi(0,x,y).$$

18.8 Two Stochastic Volatilities Model

The previous models considered an underlying asset governed by one stochastic variance. Some models considered stochastic interest rate and others assume interest rate as constant. We consider here the price evolution of an asset (for example an equity stock) that is governed by the following stochastic differential equation

$$\mathrm{d}S_t = \mu S_t\,\mathrm{d}t + \sqrt{V_{1,t}}S_t\,\mathrm{d}W_1 + \sqrt{V_{2,t}}S_t\,\mathrm{d}W_2, \tag{18.28}$$

where μ is the mean return of the asset, $V_{1,t}$ and $V_{2,t}$ are two uncorrelated and finite variance processes described by Heston [12] that also change stochastically according to the following equations

$$\mathrm{d}V_{1,t} = \frac{1}{\varepsilon}(\theta_1 - V_{1,t})\,\mathrm{d}t + \rho_{13}\sqrt{\frac{1}{\varepsilon}V_{1,t}}\,\mathrm{d}W_1 + \sqrt{\frac{1}{\varepsilon}(1-\rho_{13}^2)V_{1,t}}\,\mathrm{d}W_3, \tag{18.29}$$
$$\mathrm{d}V_{2,t} = \delta(\theta_2 - V_{2,t})dt + \rho_{24}\sqrt{\delta V_{2,t}}\,\mathrm{d}W_2 + \sqrt{\delta(1-\rho_{24}^2)V_{2,t}}\,\mathrm{d}W_4.$$

Here $\frac{1}{\varepsilon}$ and δ are the speeds of mean reversion; θ_1 and θ_2 are the long run means; $\sqrt{\frac{1}{\varepsilon}}$ and $\sqrt{\delta}$ the instantaneous volatilities of $V_{1,t}$ and $V_{2,t}$ respectively and W_i, for $i = \{1, 2, 3, 4\}$ are Wiener processes. The correlations between the asset price S_t and the variance processes $V_{1,t}$ and $V_{2,t}$ are given respectively by $\rho_{13}\sqrt{\frac{V_{1,t}}{\varepsilon}}$ and $\rho_{24}\sqrt{V_{2,t}\delta}$ which are chosen as in Chiarella and Ziveyi [3] to avoid the product term $\sqrt{V_{1,t}V_{2,t}}$.

In Eq. (18.29) choosing ε and δ to be small and to follow Feller [9] conditions, we have a fast mean reversion speed for $V_{1,t}$ and a slow mean reversion speed for $V_{2,t}$. Therefore in our model the underlying asset price S_t is influenced by two volatility terms that behave completely differently. For example, one may change each month whereas the other one may change twice a day.

The finiteness of the two variances gives guarantee that (18.28) has a solution under the real-world probability measure. In addition it ensures that there exists an equivalent risk neutral measure under which the same equation has a solution and the discounted stock price process under this measure is a martingale. Girsanov theorem presented in [15] allow to transform the presented environment into risk neutral probability world. Feynman–Kac theorem also presented in [15], proves that the option price of the underlying asset described above can be given as the solution of the following partial differential equation

$$
\begin{aligned}
&(r-q)S_t\frac{\partial U}{\partial S_t} + \left[\frac{1}{\varepsilon}(\theta_1 - V_{1,t}) - \lambda_1 V_{1,t}\right]\frac{\partial U}{\partial V_{1,t}} + [\delta(\theta_2 - V_{2,t}) - \lambda_2 V_{2,t}]\frac{\partial U}{\partial V_{2,t}} \\
&+\frac{1}{2}\left[(V_{1,t}+V_{2,t})S_t^2\frac{\partial^2 U}{\partial S_t^2} + \frac{1}{\varepsilon}V_{1,t}\frac{\partial^2 U}{\partial V_{1,t}^2} + \delta V_{2,t}\frac{\partial^2 U}{\partial V_{2,t}^2}\right] + \frac{1}{\sqrt{\varepsilon}}\rho_{13}S_t V_{1,t}\frac{\partial^2 U}{\partial S_t \partial V_{1,t}} \\
&\qquad + \sqrt{\delta}\rho_{24}S_t V_{2,t}\frac{\partial^2 U}{\partial S_t \partial V_{2,t}} = rU - \frac{\partial U}{\partial t},
\end{aligned}
$$

subject to the terminal value condition $U(T, S_t, V_{1,t}, V_{2,t}) = h(S_t)$. λ_1 and λ_2 are the market prices of risk; r and q are constant interest rate and dividend factor respectively. Consider that the solution of the partial differential equation U depends on the values of ε and δ, i.e. $U = U^{\varepsilon,\delta}$; collecting terms with the same power of $\frac{1}{\sqrt{\varepsilon}}$ and $\sqrt{\delta}$ will transform the above partial differential equation into

$$
\left(\frac{1}{\varepsilon}\mathscr{L}_0 + \frac{1}{\sqrt{\varepsilon}}\mathscr{L}_1 + \mathscr{L}_2 + \sqrt{\delta}\mathscr{M}_1 + \delta\mathscr{M}_2\right)U^{\varepsilon,\delta} = 0 \qquad (18.30)
$$

for

$$\mathcal{L}_0 = (\theta_1 - V_{1,t})\frac{\partial}{\partial V_{1,t}} + \frac{1}{2}V_{1,t}\frac{\partial^2}{\partial V_{1,t}^2}, \tag{18.31}$$

$$\mathcal{L}_1 = \rho_{13} S_t V_{1,t}\frac{\partial^2}{\partial S_t \partial V_{1,t}},$$

$$\mathcal{L}_2 = \frac{\partial}{\partial t} + (r-q)S_t\frac{\partial}{\partial S_t} + \frac{1}{2}(V_{1,t} + V_{2,t})S_t^2\frac{\partial^2}{\partial S_t^2} - r - $$

$$-\lambda_1 V_{1,t}\frac{\partial}{\partial V_{1,t}} - \lambda_2 V_{2,t}\frac{\partial}{\partial V_{2,t}},$$

$$\mathcal{M}_1 = \rho_{24} S_t V_{2,t}\frac{\partial^2}{\partial S_t \partial V_{2,t}},$$

$$\mathcal{M}_2 = (\theta_2 - V_{2,t})\frac{\partial}{\partial V_{2,t}} + \frac{1}{2}V_{2,t}\frac{\partial^2}{\partial V_{2,t}^2}.$$

Our aim is to find the price of a European option with payoff function $h(S_t)$ at maturity T. Taking into account the Markov property and the fact that our system is considered under the risk neutral probability measure, we can apply Feynman–Kac theorem to obtain the option price as

$$U(t, S_t, V_{1,t}, V_{2,t}) = e^{-(T-t)}\mathsf{E}\left[h(S_t) \mid S_t = s, V_{1,t} = v_1, V_{2,t} = v_2\right].$$

Calculation of this expectation is very complicated because it involves many parameters that have to be clearly measured and applied. To avoid this complication, we present a perturbation method that approximates the option price by a quantity that depends on much less parameters than those imposed by Feynman–Kac theorem. From our system and also our partial differential equation, it is clear that $U(t, s, v_1, v_2)$ depends on ε and δ. From now on, to make this dependence clear, we write $U^{\varepsilon;\delta}(t, s, v_1, v_2)$ instead of $U(t, s, v_1, v_2)$. Our assumption is that if ε and δ are small, the associated operators will diverge and be small respectively. Therefore we use the approach of singular and regular perturbations. Assume that our solution can be expressed in the following form

$$U^{\varepsilon,\delta} = \sum_{i\geq 0}\sum_{j\geq 0}(\sqrt{\delta})^i(\sqrt{\varepsilon})^j U_{j,i}. \tag{18.32}$$

Applying this expansion in (18.30) we generate systems of partial differential equations that can be solved to obtain the prices of European option in the following form

$$U^{\varepsilon,\delta} = U_{BS} + (T-t)\left(\mathscr{A}^{\delta} + \mathscr{B}^{\varepsilon}\right)U_{BS},$$

where the notation U_{BS} stands for the solution to the corresponding two-dimensional Black–Scholes model.

$$\mathscr{B}^{\varepsilon} = -\Upsilon_2^{\varepsilon}(v_2)D_1D_2, \quad D_k = x_i^k \frac{\partial^k}{\partial x_i^k}, \quad i = 1, 2, \quad \Upsilon_2^{\varepsilon}(v_2) = -\frac{\sqrt{\varepsilon}\rho_{13}}{2}\left\langle v_1 \frac{\partial \phi(v_1, v_2)}{\partial v_1} \right\rangle,$$

and $\phi(v_1, v_2)$ is the solution of

$$\mathscr{L}_0\phi(v_1, v_2) = f^2(v_1, v_2) - \sigma^2(v_2),$$
$$\mathscr{A}^{\delta} = \frac{1}{2}\sqrt{\delta}\rho_{24}\langle v_2 \rangle \frac{\partial \sigma(v_2)}{\partial v_2}, \text{ and}$$
$$\sigma^2(v_2) = \int (v_1 + v_2)\Pi(dv_1),$$

where $\langle \cdot \rangle = \int \cdot\, \pi(s)ds$ denotes the averaging over the invariant distribution Π of the variance process $V_{1,t}$.

Acknowledgements This work was partially supported by Swedish SIDA Foundation International Science Program. Betuel Canhanga and Jean-Paul Murara thanks Division of Applied Mathematics, School of Education, Culture and Communication, Mälardalen University for creating excellent research and educational environment.

References

1. Andersen, L., Piterbarg, V.: Interest Rate Modeling. Volume 1: Foundations and Vanilla Models. Atlantic Financial Press, Boston (2010)
2. Carr, P.P., Madan, D.B.: Option valuation using fast fourier transform. J. Comput. Financ. **2**, 61–73 (1999)
3. Chiarella, C., Ziveyi, J.: American option pricing under two stochastic volatility processes. J. Appl. Math. Comput. **224**, 283–310 (2013)
4. Christoffersen, P., Heston, S., Jacobs, K.: The shape and term structure of index option smirk: why multifactor stochastic volatilities models work so well. Manag. Sci. **22**(12), 1914–1932 (2009)
5. Cox, J., Ross, S.: The valuation of options alternative stochastic processes. J. Financ. Econ. **3**, 145–166 (1976)
6. Detemple, J., Tian, W.: The valuation of American options for a class of diffusion processes. Manag. Sci. **48**(7), 917–937 (2002)
7. Duffie, D., Pan, J., Singleton, K.: Transform analysis and asset pricing for affine jump-diffusions. Econometrica **68**, 1343–1376 (2000)
8. Fang, F., Oosterlee, C.W.: A Novel Option Pricing Method based on Fourier-Cosine Series Expansions. Technical Report 08-02, Delft University Technical, The Netherlands. http://ta.twi.tudelft.nl/mf/users/oosterlee/oosterlee/COS.pdf (2008)
9. Feller, W.: Two singular diffusion problems. Ann. Math. **54**, 173–182 (1951)
10. Grzelak, L., Oosterlee, C.W, Van, V.: Extension of stochastic volatility models with Hull–White interest rate process. Report 08-04, Delft University of technology, The Netherlands (2008)
11. Managing smile risk: Hegan, P.S., Kumar, D., Lesniewski, S.S., Woodward. D.E. Wilmott Mag. **1**, 84–108 (2003)
12. Heston, S.L.: A closed-form solution for options with stochastic volatility with applications to bonds and currency options. Rev. Financ. Stud. **6**, 327–343 (1993)

13. Ilham, A., Sircar, R.: Optimal static-dynamic hedges for barrier options. Math. Financ. **16**, 359–385 (2006)
14. Jourdain, B., Sbai, M.: Higher order discretization schemes for stochastic volatility models. AeXiv e-prrint. http://arxiv.org/abs/0908.1926v3 (2011)
15. Kijima, M.: Stochastic Processes with Applications to Finance, 2nd edn. Capman and Hall/CRC, New York (2013)
16. Musiela, M., Rutkowski, M.: Martingale Methods in Financial Modelling, 2nd edn. Springer, Berlin (2005)
17. Romano, M., Touzi, N.: Contingent claim and market completeness in a stochastic volatility model. Math. Financ. **7**(4), 399–412 (1997)
18. Schroder, M.: Computing the constant elasticity of variance option princing formula. J. Financ. **44**, 211–219 (1989)
19. Zhang, N.: Properties of the SABR model. Departement of Mathematics. Uppsala University. http://uu.diva-portal.org/smash/get/diva2:430537/FULLTEXT01.pdf (2011)

Index

S. Silvestrov and M. Rančić (eds.), *Engineering Mathematics I*, Springer Proceedings in Mathematics & Statistics 178, DOI 10.1007/978-3-319-42082-0

Printed in the United States
By Bookmasters